Reliability-based Structural Design

Seung-Kyum Choi, Ramana V. Grandhi and
Robert A. Canfield

Reliability-based Structural Design

With 88 Figures

Seung-Kyum Choi, PhD
George W. Woodruff School of
  Mechanical Engineering
Georgia Tech Savannah
Savannah, Georgia 31407
USA

Ramana V. Grandhi, PhD
Department of Mechanical and
  Materials Engineering
Wright State University
Dayton, Ohio 45435
USA

Robert A. Canfield, PhD
Department of Aeronautics and
  Astronautics, Air Force Institute
  of Technology
WPAFB, Ohio 45433
USA

British Library Cataloguing in Publication Data
Choi, Seung-Kyum
  Reliability-based structural design
  1. Structural optimization 2. Reliability (Engineering)
  3. Structural analysis (Engineering)
  I. Title II. Grandhi, R. V. III. Canfield, Robert A.
  624.1'7713
ISBN-13: 9781846284441
ISBN-10: 1846284449

Library of Congress Control Number: 2006933376

ISBN-10: 1-84628-444-9   e-ISBN 1-84628-445-7   Printed on acid-free paper
ISBN-13: 978-1-84628-444-1

© Springer-Verlag London Limited 2007

MATLAB® and Simulink® are registered trademarks of The MathWorks, Inc., 3 Apple Hill Drive, Natick, MA 01760-2098, U.S.A. http://www.mathworks.com

Mathematica® is a registered trademark of Wolfram Research, Inc., 100 Trade Center Drive Champaign, IL 61820-7237 USA. http:www.wolfram.com

Apart from any fair dealing for the purposes of research or private study, or criticism or review, as permitted under the Copyright, Designs and Patents Act 1988, this publication may only be reproduced, stored or transmitted, in any form or by any means, with the prior permission in writing of the publishers, or in the case of reprographic reproduction in accordance with the terms of licences issued by the Copyright Licensing Agency. Enquiries concerning reproduction outside those terms should be sent to the publishers.

The use of registered names, trademarks, etc. in this publication does not imply, even in the absence of a specific statement, that such names are exempt from the relevant laws and regulations and therefore free for general use.

The publisher makes no representation, express or implied, with regard to the accuracy of the information contained in this book and cannot accept any legal responsibility or liability for any errors or omissions that may be made.

9 8 7 6 5 4 3 2

Springer Science+Business Media
springer.com

Preface

As modern structures require more critical and complex designs, the need for accurate and efficient approaches to assess uncertainties in loads, geometry, material properties, manufacturing processes and operational environments has increased significantly. Reliability assessment techniques help to develop initial guidance for robust designs. They also can be used to identify where significant contributors of uncertainty occur in structural systems or where further research, testing and quality control could increase the safety and efficiency of the structure. This book provides engineers intuitive appreciation for probability theory, statistic methods, and reliability analysis methods, including Monte Carlo Sampling, Latin Hypercube Sampling, First and Second-order Reliability Methods, Stochastic Finite Element Method, and Stochastic Optimization. In addition, this book explains how to use stochastic expansions, including Polynomial Chaos Expansion and Karhunen-Loeve Expansion, for the optimization and the reliability analysis of practical engineering problems. Example problems are presented for demonstrating the application of theoretical formulations using truss, beam and plate structures. Several practical engineering applications, e.g., an uninhabited joined-wing aircraft and a supercavitating torpedo, are also presented to demonstrate the effectiveness of these methods on large-scale physical systems.

The authors would like to acknowledge the anonymous reviewers whose comments on the preliminary draft of the book led to a much better presentation of the material. During the growth of the final version, many colleagues reviewed and commented on various chapters, including Dr. Mark Cesaer of Applied Research Associates, Inc., Prof. George Karniadakis of Brown University, Prof. Efstratios Nikolaidis of the University of Toledo, Prof. Chris Pettit of the United States Naval Academy, Dr. Jon Wallace of Exxon Mobil Corp., and Professors Richard Bethke and Ravi Penmetsa of Wright State University. In addition, Dr. V.B.Venkayya, U.S. Air Force (retired), presented challenging ideas for developing uncertainty quantification techniques for computer-intensive, large-scale finite element analysis and for multi-physics problems, which were very useful and greatly appreciated.

Many current and prior graduate students and research scientists assisted in the development of new reliability analysis methods and in validating the methods on engineering problems. These include Dr. Liping Wang of General Electric Corp., Dr. Ed Alyanak of CFDRC Corp., Dr. Ha-rok Bae of Caterpillar Inc., and Dr. Brian Beachkofski, Jeff Brown, and Mark Haney, all of the Air Force Research Laboratory. In addition, students from Wright State University's Computational Design Optimization Center (CDOC), including Hemanth Amarchinta, Todd Benanzer, Arif Malik, Justin Maurer, Sang-ki Park, Jalaja Repalle, Gulshan Singh, and Randy Tobe, contributed to the work. We would also like to thank graduate students from AFIT, including Capt. Ronald Roberts and Capt. Ben Smallwood, and intern Jeremiah Allen. The detailed editing of this book was smoothly accomplished by Brandy Foster, Chris Massey, and Alysoun Taylor.

The research developments presented in this book were partially sponsored by multiple organizations over the last fifteen years, including the NASA Glenn Research Center, Cleveland, OH, the Air Force Office of Scientific Research, the Office of Naval Research, the National Institute of Standards, Wright Patterson Air Force Base, and the Dayton Area Graduate Studies Institute (DAGSI).

March 2006

Seung-Kyum Choi
Ramana V. Grandhi
Robert A. Canfield

Contents

1. Introduction 1
  1.1 Motivations 1
  1.2 Uncertainty and Its Analysis 2
  1.3 Reliability and Its Importance 4
  1.4 Outline of Chapters 6
  1.5 References 7

2. Preliminaries 9
  2.1 Basic Probabilistic Description 9
    2.1.1 Characteristics of Probability Distribution 9
      Random Variable 9
      Probability Density and Cumulative Distribution Function 10
      Joint Density and Distribution Functions 12
      Central Measures 13
      Dispersion Measures 14
      Measures of Correlation 15
      Other Measures 17
    2.1.2 Common Probability Distributions 20
      Gaussian Distribution 20
      Lognormal Distribution 25
      Gamma Distribution 28
      Extreme Value Distribution 29
      Weibull Distribution 31
      Exponential Distribution 34
  2.2 Random Field 36
    2.2.1 Random Field and Its Discretization 36
    2.2.2 Covariance Function 41
      Exponential Model 42
      Gaussian Model 42
      Nugget-effect Model 42
  2.3 Fitting Regression Models 43

    2.3.1 Linear Regression Procedure 44
    2.3.2 Linear Regression with Polynomial Fit.................................... 45
    2.3.3 ANOVA and Other Statistical Tests 46
2.4 References.. 50

3. Probabilistic Analysis 51
3.1 Solution Techniques for Structural Reliability 51
    3.1.1 Structural Reliability Assessment 51
    3.1.2 Historical Developments of Probabilistic Analysis.................. 56
        First- and Second-order Reliability Method 56
        Stochastic Expansions 58
3.2 Sampling Methods 60
    3.2.1 Monte Carlo Simulation (MCS).. 60
        Generation of Random Variables 62
        Calculation of the Probability of Failure 65
    3.2.2 Importance Sampling 68
    3.2.3 Latin Hypercube Sampling (LHS) 70
3.3 Stochastic Finite Element Method (SFEM) 72
    3.3.1 Background 73
    3.3.2 Perturbation Method 73
        Basic Formulations 74
    3.3.3 Neumann Expansion Method 75
        Basic Procedure 75
    3.3.4 Weighted Integral Method 77
        Formulation of Weighted Integral Method 77
    3.3.5 Spectral Stochastic Finite Element Method 79
3.4 References.. 79

4. Methods of Structural Reliability 81
4.1 First-order Reliability Method (FORM) 81
    4.1.1 First-order Second Moment (FOSM) Method 81
    4.1.2 Hasofer and Lind (HL) Safety-index 86
    4.1.3 Hasofer and Lind Iteration Method.. 88
    4.1.4 Sensitivity Factors... 97
    4.1.5 Hasofer Lind - Rackwitz Fiessler (HL-RF) Method 99
    4.1.6 FORM with Adaptive Approximations.................................. 110
        TANA 111
        TANA2 111
4.2 Second-order Reliability Method (SORM)... 124
    4.2.1 First- and Second-order Approximation of Limit-state Function....... 125
        Orthogonal Transformations.. 125
        First-order Approximation... 126
        Second-order Approximation 128
    4.2.2 Breitung's Formulation 130
    4.2.3 Tvedt's Formulation 133
    4.2.4 SORM with Adaptive Approximations.................................. 136
4.3 Engineering Applications 138

    4.3.1 Ten-bar Truss 138
    4.3.2 Fatigue Crack Growth 142
    4.3.3 Disk Burst Margin 144
    4.3.4 Two-member Frame 146
4.4 References 150

5. Reliability-based Structural Optimization 153
5.1 Multidisciplinary Optimization 153
5.2 Mathematical Problem Statement and Algorithms 155
5.3 Mathematical Optimization Process 157
    5.3.1 Feasible Directions Algorithm 157
    5.3.2 Penalty Function Methods 160
        Interior Penalty Function Method 160
        Exterior and Quadratic Extended Interior Penalty Functions 162
        Quadratic Extended Interior Penalty Functions Method 163
5.4 Sensitivity Analysis 178
    5.4.1 Sensitivity with Respect to Means 181
    5.4.2 Sensitivity with Respect to Standard Deviations 182
    5.4.3 Failure Probability Sensitivity in Terms of β 183
5.5 Practical Aspects of Structural Optimization 197
    5.5.1 Design Variable Linking 197
    5.5.2 Reduction of Number of Constraints 198
    5.5.3 Approximation Concepts 198
    5.5.4 Move Limits 198
5.6 Convergence to Local Optimum 200
5.7 Reliability-based Design Optimization 200
5.8 References 201

6. Stochastic Expansion for Probabilistic Analysis 203
6.1 Polynomial Chaos Expansion (PCE) 203
    6.1.1 Fundamentals of PCE 203
    6.1.2 Stochastic Approximation 209
    6.1.3 Non-Gaussian Random Variate Generation 211
        Generalized Polynomial Chaos Expansion 212
        Transformation Technique 212
    6.1.4 Hermite Polynomials and Gram-Charlier Series 213
6.2 Karhunen-Loeve (KL) Transform 218
    6.2.1 Historical Developments of KL Transform 219
    6.2.2 KL Transform for Random Fields 220
    6.2.3 KL Expansion to Solve Eigenvalue Problems 226
6.3 Spectral Stochastic Finite Element Method (SSFEM) 229
    6.3.1 Role of KL Expansion in SSFEM 230
    6.3.2 Role of PCE in SSFEM 231
6.4 References 233

7. Probabilistic Analysis Examples via Stochastic Expansion 237
7.1 Gaussian and Non-Gaussian Distributions 237

x   Contents

    7.1.1 Stochastic Analysis Procedure 237
    7.1.2 Gaussian Distribution Examples 239
        Demonstration Examples 239
        Joined-wing Example 244
    7.1.3 Non-Gaussian Distribution Examples 248
        Pin-connected Three-bar Truss Structure 248
        Joined-wing Example 251
7.2 Random Field 252
    7.2.1 Simulation Procedure of Random Field 253
    7.2.2 Cantilever Plate Example 253
    7.2.3 Supercavitating Torpedo Example 256
7.3 Stochastic Optimization 260
    7.3.1 Overview of Stochastic Optimization 261
    7.3.2 Implementation of Stochastic Optimization 261
    7.3.3 Three-bar Truss Structure 264
    7.3.4 Joined-wing SensorCraft Structure 267
7.4 References 270

8. Summary 273

Appendices 275
  A. Function Approximation Tools 275
    A.1 Use of Approximations and Advantages 276
    A.2 One-point Approximations 277
        A.2.1 Linear Approximation 278
        A.2.2 Reciprocal Approximation 278
        A.2.3 Conservative Approximation 279
    A.3 Two-point Adaptive Nonlinear Approximations 280
        A.3.1 Two-point Adaptive Nonlinear Approximation 280
        A.3.2 TANA1 281
        A.3.3 TANA2 283
    A.4 References 289
  B. Asymptotic of Multinormal Integrals 291
    B.1 References 293
  C. Cumulative Standard Normal Distribution Table 295
  D. F Distribution Table 297

Index 301

1
Introduction

1.1 Motivations

As modern structures require more critical and complex designs, the need for accurate approaches to assess uncertainties in computer models, loads, geometry, material properties, manufacturing processes, and operational environments has increased significantly. For problems in which randomness is relatively small, a deterministic model is usually used rather than a stochastic model. However, when the level of uncertainty is high, stochastic approaches are necessary for system analysis and design.

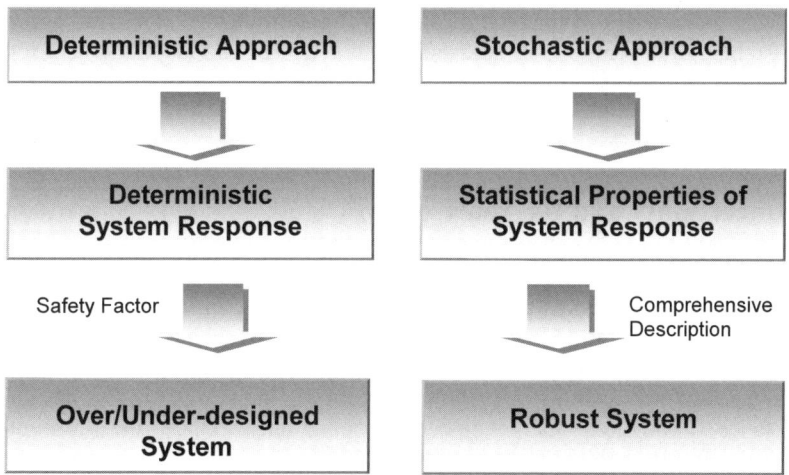

Figure 1.1. Tools for Design under Uncertainty Analysis

A number of probabilistic analysis tools have been developed to quantify uncertainties, but the most complex systems are still designed with simplified rules and schemes, such as safety factor design (Figure 1.1). However, these traditional design processes do not directly account for the random nature of most input parameters. Factor of safety is used to maintain a some degree of safety in structural design. Generally, the factor of safety is understood to be the ratio of the expected strength of response to the expected load. In practice, both the strength and load are variables, the values of which are scattered about their respective mean values. When the scatter in the variables is considered, the factor of safety could potentially be less than unity, and the traditional factor of safety based design would fail. More likely, the factor of safety is too conservative, leading to an overly expensive design.

According to Elishakoff [1],

This factor allowed continues to enable constructing safe or nearly safe structures that work. "If it works, use it," one would say. The question is: Could such a methodology be improved? Can we do it better, even though the American proverb advises, "If it ain't broke, don't fix it."?

In the modern competitive world, the engineering community's motto should be, *"If it works, make it better."* Compared to the deterministic approach based on safety factors, the stochastic approach improves design reliability. The stochastic approach provides a number of advantages to engineers. The various statistical results, which include mean value, variance, and confidence interval, can provide a broader perspective and a more complete description of the given structural system, one that takes more factors and uncertainties into account. Such an approach can accommodate a sensitivity analysis of the system, allowing engineers to find significant parameters of uncertainty models. In addition, the stochastic approach can also help to develop initial guidance for safe design and identify where further inspections and investigations could increase the safety and efficiency of the structure.

1.2 Uncertainty and Its Analysis

Two French mathematicians, Blaise Pascal and Pierre de Fermat, began to formulate probability theory in the 17th century. They explored games of chance as mathematical problems [3]. *Probability theory* treats the likelihood of a given event's occurrence and quantifies uncertain measures of random events. The appearance and applicability of probability theory in the design process has gained importance throughout the engineering community. Once the concept of probability has been incorporated, however, it is still quite difficult to explicitly define uncertainty and accurately evaluate it for large structural systems. The advent of high-powered computers makes it feasible to find numerical solutions to realistic problems of large-scale, complex systems involving uncertainties in their behavior. This feasibility has sparked an interest among researchers in combining traditional analysis methods with uncertainty quantification measures. These new methodologies, which can consider the randomness or uncertainty in the data or model, are known as uncertainty analysis or stochastic analysis. These methods facilitate robust designs that provide the designer with a guarantee of satisfaction in the presence of a given amount of uncertainty. Contemporary methods of stochastic analysis are being introduced into the whole gamut of science and engineering fields (*i.e.*, physics, meteorology, medicine, human inquiry, computer science, *etc.*).

Uncertainty has several connotations, such as the likelihood of events, degree of belief, lack of knowledge, inaccuracy, variability, *etc*. An accurate representation of uncertainties for given systems is crucial because different representations of uncertainty may yield different interpretations for the given system. The competence and limitations of these representations have been delineated by classifying uncertainties into two categories: aleatory and epistemic. *Aleatory* (*Random* or *Objective*) *uncertainty* is also called irreducible or inherent uncertainty. *Epistemic* (*Subjective*) *uncertainty* is a reducible uncertainty that stems from lack of knowledge and data. The birthday problem found in common elementary probabilistic books illustrates the difference between subjective and objective uncertainty: "What is the probability that a selected person has a birthday on July 4th?" One objective person may answer that the probability is 1/365. And the other person, who is a close friend of the selected person, may have a different answer of 1/62, because he is sure that his friend's birthday is in July or August. The second person provides higher probability (narrower bounds) compared to the first person's answer; however, the accuracy of his answer depends on his degree of belief. Since subjective uncertainty is viewed as reducible as more information is gathered–based on past experience or expert judgement–it requires more attention and careful judgement.

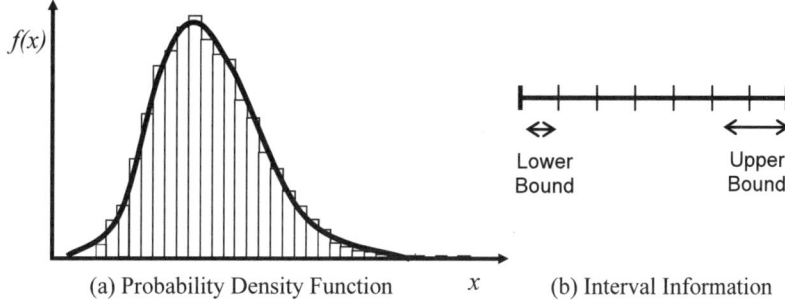

Figure 1.2. Uncertainty Representation

Two types of uncertainty characterization (probability density, or frequency; and interval information) are commonly used to represent aleatory and epistemic uncertainties, as shown in Figure 1.2. The *Probability Density Function* (PDF) represents the relative frequency of certain realizations for random variables: the center of the PDF indicates a most probable point and the tail region of the PDF indicates less probable events. If the information for a PDF is not available, the interval of upper and lower bounds of random variables may be appropriate to represent these kinds of uncertainties. The interval information better reflects incomplete, imperfect data and knowledge.

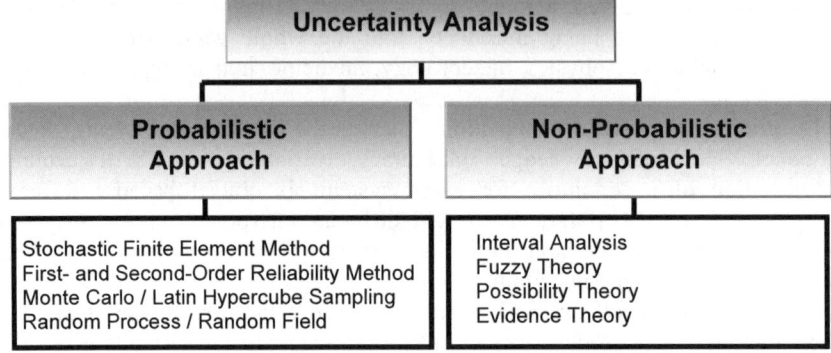

Figure 1.3. Uncertainty Analysis Categories

The *probabilistic approach* is based on the theoretical foundation of the PDF information and introduces the use of random variables, processes, and fields to represent uncertainty. The *non-probabilistic approach* manages imprecise knowledge about the true value of parameters. Figure 1.3 shows various methods of uncertainty analysis based on the representation of uncertainties. The later chapters describe details of each class of method, and further details can be found in [2],[4], and [5].

1.3 Reliability and Its Importance

Reliability is the probability that a system will perform its function over a specified period of time and under specified service conditions. Reliability theory was originally developed by maritime and life insurance companies in the 19th century to compute profitable rates to charge customers. The goal was to predict the probability of death for a given population or an individual. In many ways, the failure of structural systems, (*i.e.*, aircrafts, cars, ships, bridges, *etc.*), is similar to the life or death of biological organisms. Although there are many definitions and classifications of structural failure [1], a distinctive fact is that structural failure can cause tragic losses of life and property.

Technological defects and incongruent attitudes of risk management led to space shuttle catastrophes in 1986 and 2003. The aging problem–it is an inevitable problem of all structural systems–caused critical damage of an aircraft for Aloha Airlines Flight 243 in 1988. These failures are illustrated in Figure 1.4.

Introduction 5

(a) Space Shuttle Catastrophes, USA, 1986 and 2003: Unforseen variations of system conditions cause of two shuttle accidents (Challenger and Columbia)

(b) Risk of Aging Aircraft, Aloha Airlines Flight 243 (19-year-old aircraft), Hawaii, 1988: Undetected fatigue causes critical damage

Figure 1.4. Examples of Structural Failures

Even though these designs all satisfied structural requirements, those restrictions did not directly consider the uncertainty factors of each system. An engineering structure's response depends on many uncertain factors such as loads, boundary conditions, stiffness, and mass properties. The response (e.g., critical location stresses, resonant frequencies, *etc*.) is considered satisfactory when the design requirements imposed on the structural behavior are met within an acceptable degree of certainty. Each of these requirements is termed as a *limit-state* or *constraint*.

The study of *structural reliability* is concerned with the calculation and prediction of the probability of limit-state violations at any stage during a structure's life. The probability of the occurrence of an event such as a limit-state violation is a numerical measure of the chance of its occurring. Once the probability is determined, the next goal is to choose design alternatives that improve structural reliability and minimize the risk of failure.

Methods of reliability analysis are rapidly finding application in the multidisciplinary design environment because of the engineering system's stringent performance requirements, narrow margins of safety, liability, and market competition. In a structural design problem involving uncertainties, a structure designed using a deterministic approach may have a greater probability of failure than a structure of the same cost designed using a probabilistic approach that accounts for uncertainties. This is because the design requirements are precisely satisfied in the deterministic approach, and any variation of the parameters could potentially violate the system constraints.

When unconventional structures are designed, there is little relevant data or sufficient prior knowledge. Appropriate perceptions of uncertainty are essential for safe and efficient decisions. Probabilistic methods are convenient tools to describe or model physical phenomena that are too complex to treat with the present level of scientific knowledge. Probabilistic design procedures promise to improve the product quality of engineering systems for several reasons. Probabilistic design explicitly incorporates given statistical data into the design algorithms, whereas conventional deterministic design discards such data. In the absence of other considerations, the engineer chooses the design having the lowest failure probability. Probabilistic-based information about mechanical performance can be used to develop rational policies towards pricing, warranties, component life, spare parts requirements, *etc*. The critical aspects of several probabilistic design methodologies can be found in later chapters.

1.4 Outline of Chapters

Figure 1.5 shows the uncertainty analysis framework and the layout of chapters. It also shows how the chapters relate to each other. The first three chapters lay the foundations for the more advanced developments, which are given in Chapters 4, 5, 6, and 7. Chapter 1 summarizes the objectives, provides an overview of this book, and discusses the importance of uncertainty analysis. Chapter 2 describes preliminaries of the descriptions for probabilistic characteristics, such as first and second statistics, random fields, and regression procedures. Chapters 3 and 4 contain reviews of probabilistic analysis, including sampling methods, reliability analysis, and stochastic finite element methods. The most critical content of this book is found in Chapters 4, 5, 6, and 7, which include state-of-the-art computational methods using stochastic expansions and practical examples. Chapter 6 presents the theoretical foundation and useful properties of stochastic expansion and its developments points. Chapter 7 demonstrates the capability of the presented methods with several numerical examples and large-scale structural systems.

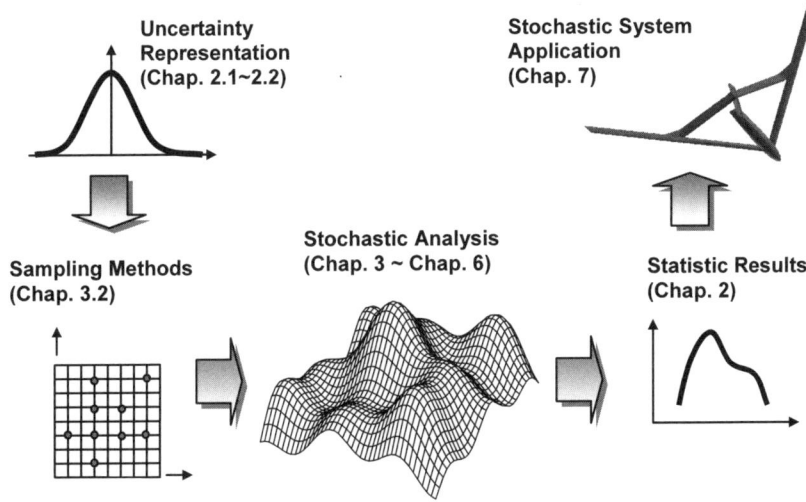

Figure 1.5. Content Layout

1.5 References

[1] Elishakoff I., *Safety Factors and Reliability: Friends or Foes?*, Kluwer Academic Publishers, Boston, 2004.
[2] Ghanem, R., and Spanos, P.D., *Stochastic Finite Elements: A Spectral Approach*, Springer-Verlag, NY, 1991.
[3] Renyi, A., *Letters on Probability*, Wayne State University Press, Detroit, 1973.
[4] Schuëller G.I. (Ed.), "A State-of-the-Art Report on Computational Stochastic Mechanics," *Journal of Probabilistic Engineering Mechanics*, Vol. 12, (4), 1997, pp. 197-313.
[5] Tatang, M.A., Direct Incorporation of Uncertainty in Chemical and Environmental Engineering Systems, Ph.D. Dissertation, Massachusetts Institute of Technology, Cambridge, MA, 1995.

2
Preliminaries

This chapter presents several probabilistic representation methods of the random nature of input parameters for structural models. The concept of the random field and its discretization are discussed with graphical interpretations. In later sections, we discuss linear regression and polynomial regression procedures which can be applied to stochastic approximation. A procedure for checking the adequacy of a regression model is also given with a representative example of the regression problem.

2.1 Basic Probabilistic Description

There are many ways to specify probabilistic characteristics of systems under uncertainty. Random variables are measurable values in the probability space associated with events of experiments. Accordingly, random vectors are sequences of measurements in the context of random experiments. Random variables are analyzed by examining underlying features of their probability distributions. A PDF indicates a relative probability of observing each random variable x and can be expressed as a formula, graph, or table. Since the computation of the PDF is not always easy, describing the data through numerical descriptive measures, such as the mean and variance, is also popular. In this section, elementary statistical formulas and several definitions of probability theory, random field, and regression analysis are briefly described in order to facilitate an introduction to the later sections.

2.1.1 Characteristics of Probability Distribution

Random Variable

A random variable X takes on various values x within the range $-\infty < x < \infty$. A random variable is denoted by an uppercase letter, and its particular value is represented by a lowercase letter. Random variables are of two types: discrete and continuous. If the random variable is allowed to take only discrete values, $x_1, x_2, x_3, ..., x_n$, it is called a *discrete random variable*. On the other hand, if the random variable is permitted to take any real value within a specified range, it is called a *continuous random variable*.

Probability Density and Cumulative Distribution Function

If a large number of observations or data records exist, then a *frequency diagram* or *histogram* can be drawn. A histogram is constructed by dividing the range of data into intervals of approximately similar size and then constructing a rectangle over each interval with an area proportional to the number of observations that fell within the interval.

The histogram is a useful tool for visualizing characteristics of the data such as the spread in the data and locations. If the rectangular areas are normalized so that the total sum of their areas is unity, then the histogram would represent the probability distribution of the sample population, and the ordinate would represent the probability density. The probability that a randomly chosen sample will fall within a certain range can be calculated by summing up the total area within that range. In this sense, it is analogous to calculating mass as density times volume where

Probability = Probability density × Interval size.

There are an infinite number of values a continuous variable can take within an interval, although there is a limit on measurement resolution. One can see that if the histogram were constructed with a very large number of observations and the intervals were to become infinitesimally small as the number of observations grew, the probability distribution would become a continuous curve. The mathematical function that describes the distribution of a random variable over the sample space of the continuous random variable, X, is called the probability density function and is designated as $f_X(x)$. The PDF is only defined for continuous random variables. The *Probability Mass Function* (PMF) describes the distribution of discrete random variables and is denoted as $p_X(x)$. Another way to describe the probability distribution for both discrete and continuous random variables is the *Cumulative Distribution Function* (CDF), $F_X(x)$. The CDF is defined for all values of random variables X from $-\infty$ to $+\infty$ and is equal to the probability that X is less than or equal to a realized value x.

For a continuous random variable, $F_X(x)$ is calculated by integrating the PDF for all values of X less than or equal to x:

$$F_X(x) = \int_{-\infty}^{x} f_X(s)ds \qquad (2.1)$$

Furthermore, if $F_X(x)$ is continuous, then the probability of X having a value between a and b can be calculated as

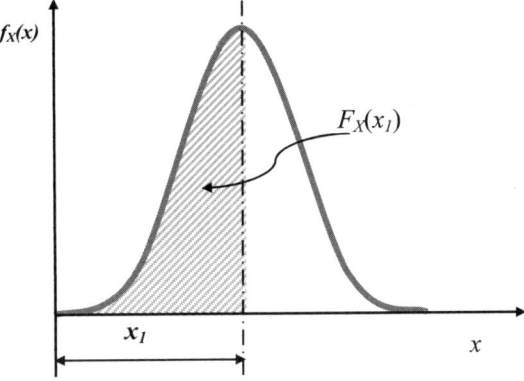

(a) Probability Density Function

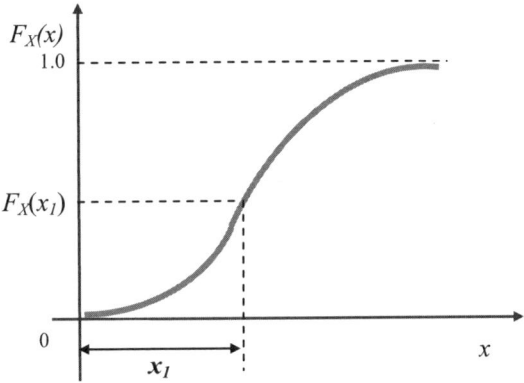

(b) Cumulative Distribution Function

Figure 2.1. PDF and Associated CDF

$$F_X(b) - F_X(a) = \int_a^b f_X(x)dx \quad \text{(for all real numbers } a \text{ and } b\text{)} \tag{2.2}$$

If the random variable X is continuous and if the first derivative of the distribution function exists, then the probability density function $f_X(x)$ is given by the first derivative of the CDF, $F_x(x)$:

$$f_X(x) = \frac{dF_X(x)}{dx} \tag{2.3}$$

If Y is a one-to-one function of the random variable X, $Y=h(X)$; then the *derived density function* of Y is given by [1]

$$f_Y(y) = \frac{dF_Y(y)}{dy} = f_X(h^{-1})\left|\frac{dh^{-1}}{dy}\right| \qquad (2.4a)$$

If Y is a many-to-one function of X, then $Y=h(X)$ may be inverted for each case,

$$f_Y(y) = \sum_{i=1}^{k} \frac{dF_Y(y)}{dy} = \sum_{i=1}^{k} f_X(h_i^{-1})\left|\frac{dh_i^{-1}}{dy}\right| \qquad (2.4b)$$

where h_i^{-1} is the inverse function of h_i, that is, $h_i^{-1}(y) = x_i$. For example, if $y = x^2 = h(X)$, then $x = \pm\sqrt{y}$ or $x_i = h_i^{-1}(y)$ where $h_1^{-1}(y) = \sqrt{y}$ and $h_2^{-1}(y) = -\sqrt{y}$.

If X is a discrete random variable, then the CDF of X is discontinuous at points x_i and is obtained as

$$F_X(x) = \sum_{x_i \leq x} p_X(x_i) \qquad (2.5)$$

The CDF is a non-decreasing function of x (its slope is always greater than or equal to zero) with lower and upper limits of 0 and 1, respectively. The CDF is also referred to at times as a distribution function, and the corresponding distribution functions are shown in Figure 2.1. Because the CDF is defined by integrating the PDF, $F_X(x_1)$ is obtained by integrating the PDF $f_X(x)$ between the limits $-\infty$ and x_1, as shown in Figure 2.1.

Joint Density and Distribution Functions

Joint probability expresses the probability that two or more random events will happen simultaneously. In general, if there are n random variables, the outcome is an n-dimensional random vector. For instance, the probability of the two-dimensional case is calculated as

$$P[a<X<b,\ c<Y<d] = \int_c^d \int_a^b f_{XY}(x,y)\,dx\,dy \qquad (2.6)$$

where $f_{XY}(x,y)$ is the joint PDF of the random variables X and Y ($f_{XY}(x,y) \geq 0$, $\int_{-\infty}^{\infty}\int_{-\infty}^{\infty} f_{XY}(x,y)\,dx\,dy = 1$).

The probability density of X for all possible values of y is the *marginal density* of x. The marginal density of x is determined by

$$f_X(x) = \int_{-\infty}^{\infty} f_{XY}(x,y)dy \tag{2.7}$$

The PDF of X for a specified y represents the *conditional probability* of X given by:

$$f_{X|Y}(x|y) = \frac{f_{XY}(x,y)}{f_Y(y)}, \ f_Y > 0 \tag{2.8}$$

If X and Y are independent, then

$$f_{X|Y}(x|y) = f_X(x) \ \text{and} \ f_{Y|X}(y|x) = f_Y(y) \tag{2.9}$$

The conditional PDF becomes the marginal PDF, and the joint PDF becomes the product of the marginals:

$$f_{XY}(x,y) = f_X(x)f_Y(y) \tag{2.10}$$

In general, the joint PDF is equal to the product of the marginals when all the variables are mutually independent:

$$f_X(X) = f_{X_1}(x_1)f_{X_1}(x_2)...f_{X_{n-1}}(x_{n-1})f_{X_n}(x_n) = \prod_{i=1}^{n} f_{X_i}(x_i) \tag{2.11}$$

Central Measures

The population *mean*, also referred to as the *expected value* or *average*, is used to describe the central tendency of a random variable. This is a weighted average of all the values that a random variable may take. If $f_X(x)$ is the probability density function of X, the mean is given by

$$\mu_X = E(X) = \int_{-\infty}^{\infty} x \ f_X(x)dx \tag{2.12}$$

Thus, μ_X is the distance from the origin to the centroid of the PDF. It is called the *first moment* since it is the first moment of area of the PDF. The mean is analogous to the centroidal distance of a cross-section.

According to the definition of a random variable, any function of a random variable is itself a random variable. Therefore, if $g(x)$ is an arbitrary function of x, the expected value of $g(x)$ is defined as

$$E[g(X)] = \int_{-\infty}^{\infty} g(x)f_X(x)dx \tag{2.13}$$

The *expectation operator*, $E[\cdot]$, possesses the following useful properties: If X and Y are independent,

$$E[XY] = E[X]E[Y] \tag{2.14}$$

And if c is a constant

$$E[c] = c \tag{2.15}$$

$$E[cX] = cE[X] \tag{2.16}$$

But, in general

$$E[g(X)] \neq g(E[X]) \tag{2.17}$$

Given $Z = X_1 + X_2 + ... + X_n$, the expected value of Z is a linear combination of individual values:

$$E(Z) = E(X_1) + E(X_2)..... + E(X_n) \tag{2.18}$$

Other useful central measures are the *median* and *mode* of the data: the median is the value of X at which the cumulative distribution function has a value of 0.5, and the mode is the value of X corresponding to the peak value of the probability density function.

Dispersion Measures

The expected value or mean value is a measure of the central tendency, which indicates the location of the distribution on the coordinate axis representing the random variable. The *variance*, $V(X)$, a second central moment of X, is a measure of spread in the data about the mean:

$$\begin{aligned} V(X) &= E[(X - \mu_X)^2] \\ &= E(X^2) - 2E(X)\mu_X + \mu_X^2 = E(X^2) - \mu_X^2 \end{aligned} \tag{2.19}$$

Geometrically, it represents the moment of inertia of the probability density function about the mean value. The variance of a random variable is analogous to the moment of inertia of a weight about its centroid. A measure of the variability of the random variable is usually given by a quantity known as the *standard deviation*. The standard deviation is a square root of the variance:

$$\sigma_X = \sqrt{V(X)} \tag{2.20}$$

The standard deviation is often preferred over the variance as a measure of dispersion because the units are consistent with the variable X and its mean value μ_X.

Nondimensionalizing the standard deviation will result in the *Coefficient of Variation* (COV), δ_X, which indicates the relative amount of uncertainty or randomness:

$$\delta_X = \frac{\sigma_X}{\mu_X} \tag{2.21}$$

Therefore, if we know any two of the mean (expected value), standard deviation, or coefficient of variation, the third term can be determined.

Measures of Correlation

If two random variables (X and Y) are correlated, the likelihood of X can be affected by the value taken by Y. In this case, the *covariance*, σ_{XY}, can be used as a measure to describe a linear association between two random variables:

$$\sigma_{XY} = Cov(X,Y) = E[(X - \mu_X)(Y - \mu_Y)] \tag{2.22}$$
$$= \int_{-\infty}^{\infty} \int_{-\infty}^{\infty} (x - \mu_X)(y - \mu_Y) f_{XY}(x,y) dx dy$$

The *correlation coefficient* is a nondimentional measure of the correlation

$$\rho_{XY} = \frac{\sigma_{XY}}{\sigma_X \sigma_Y} \tag{2.23}$$

If x and y are statistically independent, the variables are uncorrelated and the covariance is 0 (Figure 2.2a). Therefore, the correlation coefficients of ±1 indicate a perfect correlation (Figure 2.2b).

If $Y = a_1 X_1 + a_2 X_2$, where a_1 and a_2 are constants, the variance of Y can be obtained as

$$Var[Y] = E[\{a_1 X_1 + a_2 X_2 - (a_1 \mu_{X_1} + a_2 \mu_{X_2})\}^2] \tag{2.24}$$

$$= E[\{a_1(X_1 - \mu_{X_1}) + a_2(X_2 - \mu_{X_2})\}^2]$$

$$= E[a_1^2 (X_1 - \mu_{X_1})^2 + 2a_1 a_2 (X_1 - \mu_{X_1})(X_2 - \mu_{X_2}) + a_2^2 (X_2 - \mu_{X_2})^2]$$

$$= a_1^2 Var[X_1] + a_2^2 Var[X_2] + 2a_1 a_2 Cov(X_1, X_2)$$

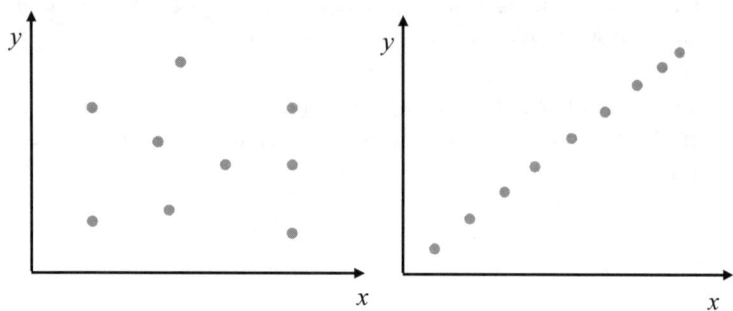

(a) Covariance near Zero (b) Positive Covariance

Figure 2.2. Examples of Paired Data Sets

Table 2.1. Properties of Central and Dispersion Measures

Central	$E[a_0] = a_0$, $E[a_1 X_1] = a_1 E[X_1]$ $E[X_1 X_2] = E[X_1] E[X_2]$ $E[a_0 + a_1 X_1 + a_2 X_2] = a_0 + a_1 E[X_1] + a_2 E[X_2]$
Dispersion	$Var[a_0] = 0$, $Var[a_1 X_1] = a_1^2 [X_1]$ $Var[a_0 + a_1 X_1 + a_2 X_2]$ $= a_1^2 Var[X_1] + a_2^2 Var[X_2] + 2 a_1 a_2 Cov(X_1, X_2)$ $= a_1^2 Var[X_1] + a_2^2 Var[X_2] + 2 a_1 a_2 \rho_{X_1 X_2} \sigma_{X_1} \sigma_{X_2}$ $Cov[a_1 X_1, X_2] = a_1 Cov[X_1, X_2]$ $Cov[X_1, X_2 + X_3] = Cov[X_1, X_2] + Cov[X_1, X_3]$ $Cov[a_1 + X_1, a_2 + X_2] = Cov[X_1, X_2]$

In general, if $Y = \sum_{i=1}^{n} a_i X_i$, then the corresponding variance is

$$Var[Y] = \sum_{i=1}^{n} a_i^2 Var[X_i] + \sum_{i=1}^{n} \sum_{j=1}^{n} a_i a_j Cov(X_i, X_j), \; i \neq j \qquad (2.25)$$

$$= \sum_{i=1}^{n} a_i^2 \sigma_{X_i}^2 + \sum_{i=1}^{n} \sum_{j=1}^{n} a_i a_j \rho_{ij} \sigma_{X_i} \sigma_{X_j}$$

Furthermore, if another linear function of X is given as $Z = \sum_{i=1}^{n} b_i X_i$, the covariance between Y and Z can be obtained as [1]

$$Cov[Y,Z] = \sum_{i=1}^{n} a_i b_i Var[X_i] + \sum_{i=1}^{n}\sum_{j=1}^{n} a_i b_j Cov(X_i, X_j), i \neq j \quad (2.26)$$

$$= \sum_{i=1}^{n} a_i b_i \sigma_{X_i}^2 + \sum_{i=1}^{n}\sum_{j=1}^{n} a_i b_j \rho_{ij} \sigma_{X_i} \sigma_{X_j}$$

Useful properties for the central and dispersion measures of the random variables X_1, X_2 and X_3 are summarized in Table 2.1 (a_0, a_1, and a_2 are constants).

Other Measures

The expected value of the cube of the deviation of the random variable from its mean value (also known as the third moment of the distribution about the mean) is taken as a measure of the skewness, or lack of symmetry, of the distribution. Therefore, the *skewness*, the third central moment of X, describes the degree of asymmetry of a distribution around its mean:

$$skewness = E[(X - \mu_x)^3] = \int_{-\infty}^{\infty} (X - \mu_x)^3 f_X(x)dx \quad (2.27)$$

The value of $E[(X - \mu_X)^3]$ can be positive or negative.

A nondimensional measure of skewness known as the *skewness coefficient* is denoted as

$$\theta_X = \frac{E[(X - \mu_X)^3]}{\sigma_X^3} \quad (2.28)$$

Any symmetric data have zero θ_x; if θ_x is positive, the dispersion is more above the mean than below the mean (Figure 2.3a); and, if it is negative, the dispersion is more below than above the mean (Figure 2.3b). Therefore, the skewness coefficient is known as a measure of the symmetry of density functions.

The *kurtosis*, the fourth central moment of X, is a measure of the flatness of a distribution:

$$kurtosis = \frac{E[(X - \mu_X)^4]}{\sigma_X^4} \quad (2.29)$$

An alternative definition of the kurtosis is given by

$$\text{kurtosis} = \frac{1}{n}\sum_{i=1}^{n}\frac{(x_i - \mu_X)^4}{\sigma_X} - 3 \qquad (2.30)$$

In this definition, the kurtosis of the normal distribution is zero, a positive value of the kurtosis describes a distribution that has a sharp peak, and a negative value of the kurtosis indicates a flat distribution compared to the normal distribution.

Recall that the first and second moments of X are defined in Equation 2.12 and Equation 2.19, respectively. The nth-order central moments are traditionally defined in terms of differences from the mean:

$$m_X^n = E[(X - \mu_x)^n] = \int_{-\infty}^{\infty}(X - \mu_x)^n f_X(x)dx \qquad (2.31)$$

where, $\mu_X = E(X) = \int_{-\infty}^{\infty} x\, f_X(x)dx$.

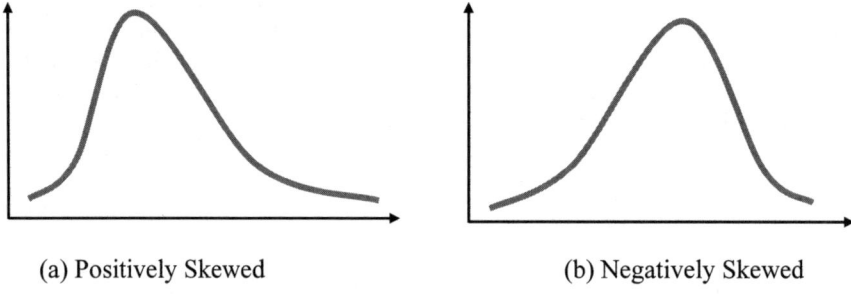

(a) Positively Skewed (b) Negatively Skewed

Figure 2.3. Skewed Density Functions

Example 2.1

The probability that a given number of cars per minute will arrive at a tollbooth is given in the table below. (a) Sketch the probability distribution as a function of X and find the mean, median, and mode. (b) Determine $E(X^2)$ and $E(X^3)$, the standard deviation, and the skewness coefficient.

No. of cars arriving per minute (X)	1	2	3	4	5	6	7	8
Probability per minute	0.025	0.075	0.125	0.150	0.200	0.275	0.100	0.050

Solution:
(a)

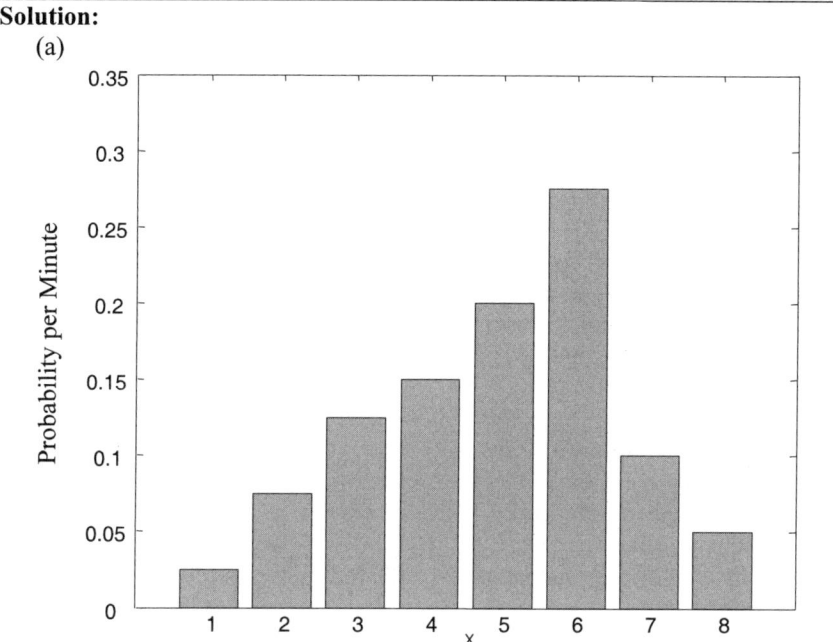

Mean:
$$\mu = \sum_{i=1}^{8} x_i P_i = 1(0.025) + 2(0.075) + 3(0.125) + 4(0.150)$$
$$+ 5(0.2) + 6(0.275) + 7(0.10) + 8(0.05) = 4.9$$

Mode: The peak in the probability density function is at $x = 6$, therefore this is the mode.

Median: Examination of the data shows that the cumulative probability of 0.5 lies between 4 and 5 cars per minute. A quadratic interpolation of CDF using 4, 5, and 6 provides a value of 4.75.

(b) $E(X^2) = \sum_{i=1}^{8} x_i^2 P_i = 1(0.025)+4(0.075)+9(0.125)+16(0.150)$
$$+25(0.2)+36(0.275)+49(0.10)+64(0.05) = 26.85$$

$E(X^3) = \sum_{i=1}^{8} x_i^3 P_i = 1(0.025)+8(0.075)+27(0.125)+64(0.150)+125(0.2)$
$$+216(0.275)+343(0.10)+512(0.05) = 157.9$$

$$\sigma_X^2 = \sum_{i=1}^{8}(x_i - \mu)^2 P_i = E(X^2) - \mu_X^2 = 26.85 - 4.9^2 = 2.84$$

$$\sigma_X = \sqrt{2.84} = 1.69$$

$$\theta_X = \frac{E[(x-\mu)^3]}{\sigma_X^3} = \frac{\sum_{i=1}^{8}(x_i-\mu)^3 P_i}{\sigma_X^3} = \frac{-1.497}{1.69^3} = -0.313$$

2.1.2 Common Probability Distributions

In evaluating structural reliability, several types of standardized probability distributions are used to model the design parameters or random variables. Selection of the distribution function is an essential part of obtaining probabilistic characteristics of structural systems. The selection of a particular type of distribution depends on

- The nature of the problem
- The underlying assumptions associated with the distribution
- The shape of the curve between $f_X(x)$ or $F_X(x)$ and x obtained after estimating data
- The convenience and simplicity afforded by the distribution in subsequent computations

The selection or determination of the distribution functions of random variables is known as *statistical tolerancing*. In general, the first few moments (mean, variance, skewness, etc.) of the distribution need to be estimated and matched through the use of several techniques, including the Taylor series approximation, the Taguchi method, and the Monte Carlo method. Detailed discussions of these methods can be found in [3] and [6]. In this section, the properties of some of the more commonly used distributions are presented.

Gaussian Distribution

The *Gaussian* (or *normal*) *distribution* is used in many engineering and science fields due to its simplicity and convenience, especially a theoritical basis of the central limit theorem. The *central limit theorem* states that the sum of many arbitrary distribution random variables asymptotically follows a normal distribution when the sample size becomes large.

This distribution is often used for small coefficients of variation cases, such as Young's modulus, Poisson's ratio, and other material properties. The Gaussian distribution is given by

$$f_X(x) = \frac{1}{\sigma_X \sqrt{2\pi}} \exp\left[-\frac{1}{2}\left(\frac{x-\mu_X}{\sigma_X}\right)^2\right], \quad -\infty < x < \infty \qquad (2.32)$$

where the parameters of the distribution μ_X and σ_X denote the mean and standard deviation of the variable X, respectively, and X is identified as $N(\mu_X, \sigma_X)$. The location (μ_X) and scale (σ_X) parameters generate a family of distributions.

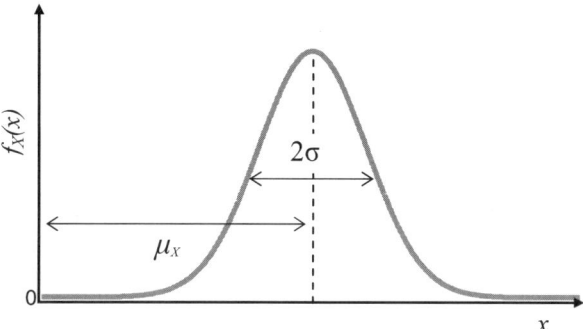

Figure 2.4. Normal Density Function

The density function and corresponding parameters are shown in Figure 2.4. The PDF of the Gaussian distribution is also known as a *bell curve* because of its shape in the graph. The Gaussian distribution is symmetric with respect to the mean and has inflection points at $x = \mu \pm \sigma$. The areas under the curve within one, two, and three standard deviations are about 68%, 95.5%, and 99.7% of the total area, respectively.

The Gaussian distribution has the following useful properties:

1) Any linear functions of normally distributed random variables are also normally distributed. For instance, let Z be the sum of n normally distributed random variables:

$$Z = a_0 + a_1 X_1 + a_2 X_2 + \ldots + a_n X_n \tag{2.33}$$

where X_i are independent random variables, and a_i's are constants.

Then, Z will also be normal with the following properties:

$$\mu_Z = a_0 + \sum_{i=1}^{n} a_i \mu_i \quad \sigma_Z = \sqrt{\sum_{i=1}^{n} (a_i \sigma_i)^2} \tag{2.34}$$

2) A nonlinear function of normally distributed random variables may or may not be normal. For Example, the function $y = \sqrt{X_1^2 + X_2^2}$ of two independent standard normally distributed random variables X_1 and X_2 with $N(0, \sigma^2)$ is a Rayleigh distribution. Its density and distribution functions are computed as

$$f_Y(y) = \frac{y}{\sigma^2} e^{-\frac{y^2}{2\sigma^2}}, \quad y \geq 0 \tag{2.35}$$

$$F_Y(y) = 1 - e^{-\frac{y^2}{2\sigma^2}}, \quad y \geq 0 \tag{2.36}$$

The Gaussian distribution can be normalized by defining $\xi = (x-\mu)/\sigma$ and yields the *standard normal distribution* $N(0,1)$. The density function of a *standard normally distributed variable* ξ is given by

$$f_\Xi(\xi) = \frac{1}{\sqrt{2\pi}} \exp\left(\frac{-\xi^2}{2}\right), \quad -\infty < \xi < \infty \tag{2.37}$$

The notation $\Phi(\cdot)$ is commonly used for the cumulative distribution function of the standard normally distributed variable ξ and is given by

$$\Phi(\xi) = F_\Xi(\xi) = \int_{-\infty}^{\xi} \frac{1}{\sqrt{2\pi}} \exp\left(\frac{-\xi^2}{2}\right) d\xi \tag{2.38}$$

If $\Phi(\xi_p) = p$ is given, the standard normal variate ξ_p corresponding to the cumulative probability (p) is denoted as

$$\xi_p = \Phi^{-1}(p) \tag{2.39}$$

The values of the standard normal cumulative distribution function, $\Phi(\cdot)$, are tabulated (Appendix C). Usually, the probabilities are given in tables only for positive values of ξ and for negative values

$$\Phi(-\xi) = 1 - \Phi(\xi) \tag{2.40}$$

due to the symmetry of the density function about zero. Similarly, we can find that

$$\xi_p = \Phi^{-1}(p) = -\Phi^{-1}(1-p) \tag{2.41}$$

Example 2.2

> If a cantilever beam supports two random loads with means and standard deviations of $\mu_1 = 20$ kN, $\sigma_1 = 4$ kN and $\mu_2 = 10$ kN, $\sigma_2 = 2$ kN as shown in the accompanying drawing, the bending moment (M) and the shear force (V) at the fixed end due to the two loads are $M = L_1F_1 + L_2F_2$ and $V = F_1 + F_2$, respectively.

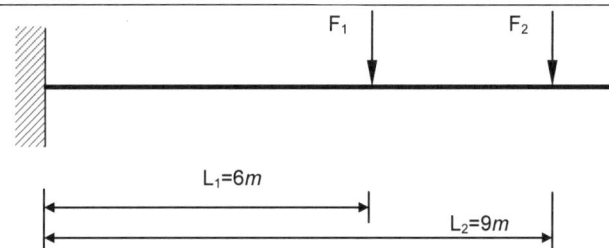

(a) If two loads are independent, what are the mean and the standard deviation of the shear and the bending moment at the fixed end?
(b) If two random loads are normally distributed, what is the probability that the bending moment will exceed 235 kNm?
(c) If two loads are independent, what is the correlation coefficient between V and M?

Solution:

(a) From the properties of the expected value operator (Equation 2.16 and Equation 2.18), the mean and the standard deviation of V and M can be obtained as

$$V = F_1 + F_2 \rightarrow E[V] = \mu_V = E[F_1] + E[F_2] = 20 + 10 = 30 \text{ kN}$$
$$Var[V] = Var[F_1] + Var[F_2] + 2Cov(F_1, F_2)$$
$$= 4^2 + 2^2 + 0 = 20 \text{ kN}^2$$
$$\therefore \sigma_V = \sqrt{20} = 4.47 \text{ kN}$$

$$M = L_1 F_1 + L_2 F_2 \rightarrow$$
$$E[M] = \mu_M = L_1 E[F_1] + L_2 E[F_2] = 6 \times 20 + 9 \times 10 = 210 \text{ kNm}$$
$$Var[M] = L_1^2 Var[F_1] + L_2^2 Var[F_2] + 2 L_1 L_2 Cov(F_1, F_2)$$
$$= 6^2 \times 4^2 + 9^2 \times 2^2 + 0 = 900 \text{ kNm}^2$$
$$\therefore \sigma_M = \sqrt{900} = 30 \text{ kNm}$$

(b) From the calculated results of (a), $\mu_M = 210$ kNm and
$$\sigma_M = \sqrt{900} = 30 \text{ kNm}$$

Therefore, the probability that the bending moment exceeds 235 kNm is

$$P(M > 235) = P\left(\xi > \frac{235 - 210}{30}\right)$$
$$= P(\xi > 0.8333) = 1 - \Phi(0.83) = 0.2023$$

(c) Regardless of the independence of F_1 and F_2, V and M can be correlated.

From Equation 2.22,

$$\begin{aligned}
\sigma_{VM} &= Cov(V,M) \\
&= E[(V-\mu_V)(M-\mu_M)] = E[VM] - \mu_V \mu_M \\
&= E[(F_1+F_2)(L_1 F_1 + L_2 F_2)] - (\mu_1+\mu_2)(L_1\mu_1+L_2\mu_2) \\
&= L_1 E[F_1^2] + L_2 E[F_2^2] + (L_1+L_2)E[F_1 F_2] \\
&\quad - L_1 \mu_1^2 - L_2 \mu_2^2 - (L_1+L_2)\mu_1 \mu_2 \\
&= L_1 \sigma_1^2 + L_2 \sigma_2^2 \quad (\because \text{Equation 2.19})
\end{aligned}$$

Thus, the correlation coefficient is obtained as

$$\rho_{VM} = \frac{\sigma_{VM}}{\sigma_V \sigma_M} = \frac{L_1 \sigma_1^2 + L_2 \sigma_2^2}{\sqrt{\sigma_1^2 + \sigma_2^2}\sqrt{L_1^2 \sigma_1^2 + L_2^2 \sigma_2^2}} = 0.98387$$

Example 2.3

Consider a cantilever beam structure subjected to a force P.

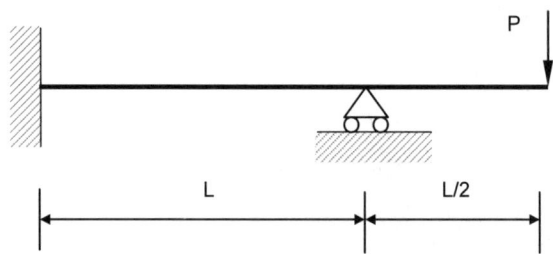

The displacement at the tip is given by

$$u = \frac{5PL^3}{48EI}$$

where E is Young's modulus and I is the area moment of the cross section. If E has a Gaussian distribution with $\mu_E = 10 \text{ kN}$, $\sigma_E = 2$ kN, derive the PDF of the displacement.

Solution:

According to Equation 2.4, $h_i^{-1}(u)$ and $\left|\dfrac{dh_i^{-1}}{du}\right|$ must be obtained. From the given formulation of the displacement:

$$E = \frac{5PL^3}{48I}\frac{1}{u} = \frac{c}{u} = h^{-1}(u)$$

where $c = 5PL^3/48I$.

Thus,

$$\frac{dE}{du} = -\frac{c}{u^2} \;\rightarrow\; \left|\frac{dh^{-1}}{du}\right| = \left|\frac{dE}{du}\right| = \frac{c}{u^2}$$

Finally, the derived density of the displacement is calculated as

$$f_U(u) = \frac{1}{\sigma_E\sqrt{2\pi}}\exp\left[-\frac{1}{2}\left(\frac{\frac{c}{u}-\mu_E}{\sigma_E}\right)^2\right]\frac{c}{u^2}$$

$$= \frac{c}{u^2\sigma_E\sqrt{2\pi}}\exp\left[-\frac{1}{2}\left(\frac{c-u\mu_E}{u\sigma_E}\right)^2\right]$$

Lognormal Distribution

The *lognormal distribution* plays an important role in probabilistic design because negative values of engineering phenomena are sometimes physically impossible. Typical uses of the lognormal distribution are found in descriptions of fatigue failure, failure rates, and other phenomena involving a large range of data. Examples are cycles to failure, material strength, loading variables, etc.

A situation may arise in reliability analysis where a random variable X is the product of several random variables $x_i : x = x_1 x_2 x_3 .... x_n$. Taking the natural logarithm of both sides,

$$ln\; x = ln\; x_1 + ln\; x_2 + ... + ln\; x_n$$

if no one term on the right side dominates, then by Equation 2.33, $ln\, x$ should be normally distributed. In the equation $Y = ln\, X$, the random variable X is said to follow lognormal distribution (Figure 2.5), and Y follows a normal distribution.

Thus the PDF of y is given by

$$f_Y(y) = \frac{1}{\sqrt{2\pi}\sigma_Y} \exp\left[-\frac{1}{2}\left(\frac{y-\mu_Y}{\sigma_Y}\right)^2\right], \quad -\infty < y < \infty \tag{2.42}$$

Since $Y = \ln X$, the above equation can be rewritten in terms of X as

$$f_X(x) = \frac{1}{\sqrt{2\pi}x\sigma_Y} \exp\left[-\frac{1}{2}\left(\frac{\ln x-\mu_Y}{\sigma_Y}\right)^2\right], \quad 0 \leq x < \infty \tag{2.43}$$

where

$$\sigma_Y^2 = \ln\left[\left(\frac{\sigma_X}{\mu_X}\right)^2 + 1\right] \tag{2.44}$$

and

$$\mu_Y = \ln \mu_X - \frac{1}{2}\sigma_Y^2 \tag{2.45}$$

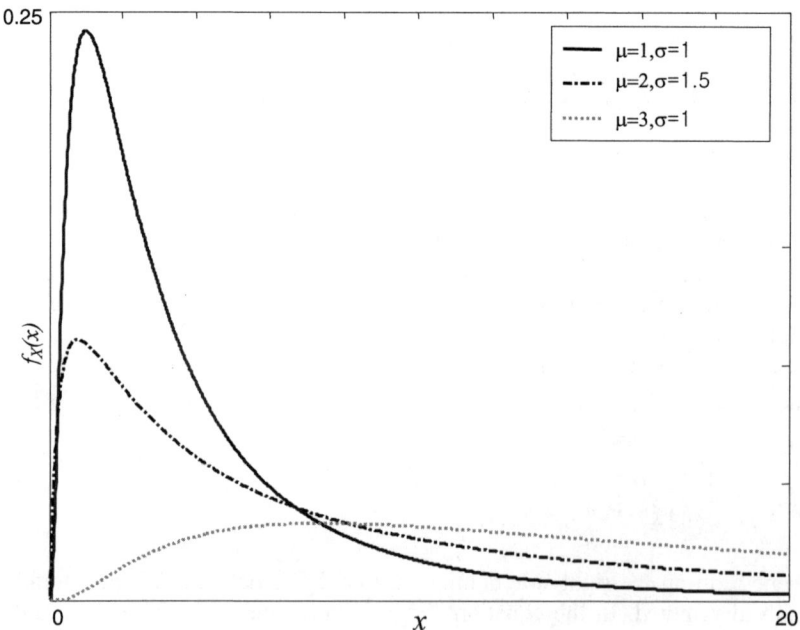

Figure 2.5. Lognormal Density Function

The CDF of the lognormal distribution is given by

$$F_X(x) = \frac{1}{\sigma_Y \sqrt{2\pi}} \int_0^x \frac{1}{x} \exp\left[-\frac{1}{2}\left(\frac{(\ln x - \mu_Y)^2}{2\sigma_Y^2}\right)\right] \qquad (2.46)$$

Example 2.4

If $\ln X$ has a Gaussian distribution, what is the distribution of X?

Solution:

This problem is the proof of the derived density of the lognormal distribution. Let $Y = \ln X$. Then $X = e^Y$. From Equation 2.12 and Equation 2.32, the mean of X is

$$\mu_X = E[X] = E[e^Y] = \frac{1}{\sigma_Y \sqrt{2\pi}} \int_{-\infty}^{\infty} \exp[y] \exp\left[-\frac{1}{2}\left(\frac{y - \mu_Y}{\sigma_Y}\right)^2\right] dy$$

$$= \left[\frac{1}{\sigma_Y \sqrt{2\pi}} \int_{-\infty}^{\infty} \exp\left\{-\frac{1}{2}\left(\frac{y - (\mu_Y + \sigma_Y^2)}{\sigma_Y}\right)^2\right\} dy\right] \exp(\mu_Y + \frac{1}{2}\sigma_Y^2)$$

Since the quantity inside the bracket of the above equation is the unit area of the Gaussian density function $\sim N(\mu_Y + \sigma_Y^2, \sigma_Y)$, we have

$$\mu_X = \exp(\mu_Y + \frac{1}{2}\sigma_Y^2) \;\;\rightarrow\;\; \mu_Y = \ln \mu_X - \frac{1}{2}\sigma_Y^2$$

This result is the same as Equation 2.45.

Similarly,

$$E[X^2] = \frac{1}{\sigma_Y \sqrt{2\pi}} \int_{-\infty}^{\infty} \exp[2y] \exp\left[-\frac{1}{2}\left(\frac{y - \mu_Y}{\sigma_Y}\right)^2\right] dy$$

$$= \left[\frac{1}{\sigma_Y \sqrt{2\pi}} \int_{-\infty}^{\infty} \exp\left\{-\frac{1}{2}\left(\frac{y - (\mu_Y + 2\sigma_Y^2)}{\sigma_Y}\right)^2\right\} dy\right] \exp[2(\mu_Y + \sigma_Y^2)]$$

$$= \exp[2(\mu_Y + \sigma_Y^2)]$$

From Equation 2.19, the variance of X is

$$Var[X] = \sigma_X^2 = \exp[2(\mu_Y + \sigma_Y^2)] - \exp[2(\mu_Y + \frac{1}{2}\sigma_Y^2)]$$
$$= \mu_X^2(\exp(\sigma_Y^2) - 1)$$

Thus, we obtain

$$\sigma_Y^2 = \ln\left[\left(\frac{\sigma_X}{\mu_X}\right)^2 + 1\right]$$

From the given condition

$$\left|\frac{dy}{dx}\right| = \frac{1}{x}$$

According to Equation 2.4,

$$f_X(x) = \frac{1}{\sqrt{2\pi} x \sigma_Y} \exp\left[-\frac{1}{2}\left(\frac{\ln x - \mu_Y}{\sigma_Y}\right)^2\right]$$

Therefore, if $\ln X$ is normal, the random variable X has a lognormal distribution.

Gamma Distribution

The *gamma distribution* (Figure 2.6) consists of the gamma function, a mathematical function defined in terms of an integral. This distribution is important because it allows us to define two families of random variables, the exponential and chi-square, which are used extensively in applied engineering and statistics.

The density function associated with the gamma distribution is defined by

$$f_X(x) = \frac{1}{\beta^\alpha \Gamma(\alpha)} x^{\alpha-1} e^{-x/\beta}, \quad 0 \le x < \infty \qquad (2.47)$$

where the parameters α and β satisfy $\alpha > 0$ and $\beta > 0$, and the gamma function is $\Gamma(\alpha) = \int_0^\infty x^{\alpha-1} e^{-x} dx$.

Let X be a gamma random variable with parameters α and β. Then the mean and variance for X are given by

$$E[X] = \mu = \alpha\beta, \quad V[X] = \sigma^2 = \alpha\beta^2 \qquad (2.48)$$

The gamma CDF is

$$F_X(x) = \frac{1}{\beta^\alpha \Gamma(\alpha)} \int_0^x t^{\alpha-1} e^{-x/\beta} dt \qquad (2.49)$$

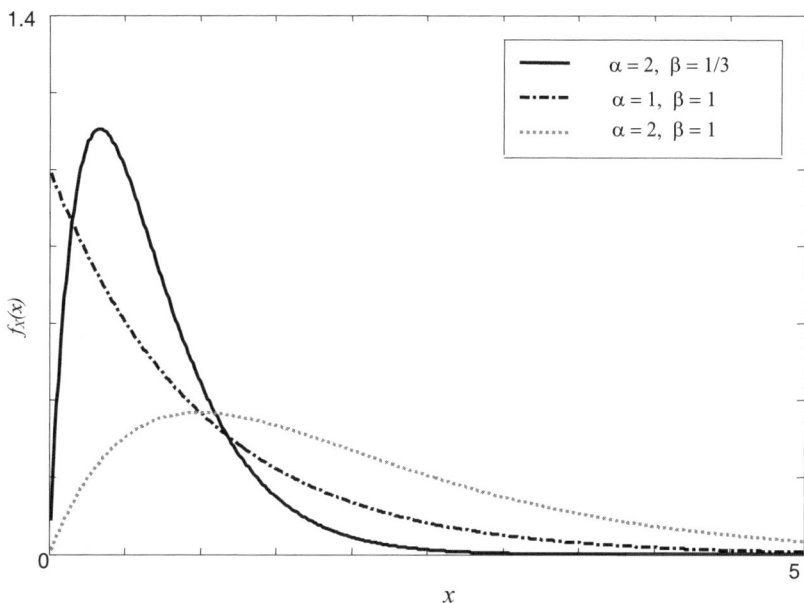

Figure 2.6. Gamma Density Functions

Extreme Value Distribution

The *extreme value distribution* is used to represent the maximum or minimum of a number of samples of various distributions. There are three types of extreme value distributions, namely Type I, Type II, and Type III. The *Type I extreme value distribution*, also referred to as the *Gumbel distribution*, is the distribution of the maximum or minimum of a number of samples of normally distributed data.

The density function of the Type I extreme value distribution is defined by

$$f_X(x) = \alpha \exp[-\exp(-\alpha(x-u))] \exp[-\alpha(x-u)], \qquad (2.50)$$
$$-\infty < x < \infty, \; \alpha > 0$$

where α and u are scale and location parameters, respectively.
The CDF of the extreme value distribution is given by

$$F_X(x) = \exp[-\exp(-\alpha(x-u))] \qquad (2.51)$$

Due to the functional form of Equation 2.51, it is also referred to as a doubly exponential distribution. Similar to the relationship between the Gaussian distribution and lognormal distribution, the *Type II extreme value distribution*, also referred to as the *Frechet distribution*, can be derived by using parameters $u = \ln v$, $\alpha = k$ in the Type I distribution. The PDF of the Type II extreme value distribution is

$$f_X(x) = \frac{k}{v}\left(\frac{v}{x}\right)^{k+1} \exp\left[-\left(\frac{v}{x}\right)^k\right], \quad 0 \leq x < \infty, \quad k \geq 2 \tag{2.52}$$

The corresponding CDF is

$$F_X(x) = \exp\left[-\left(\frac{v}{x}\right)^k\right] \tag{2.53}$$

The density functions of the Type I and Type II extreme value distributions are shown in Figure 2.7. The following subsection will discuss the last type of the extreme value distribution, the *Type III extreme value distribution*, also known as the *Weibull distribution*.

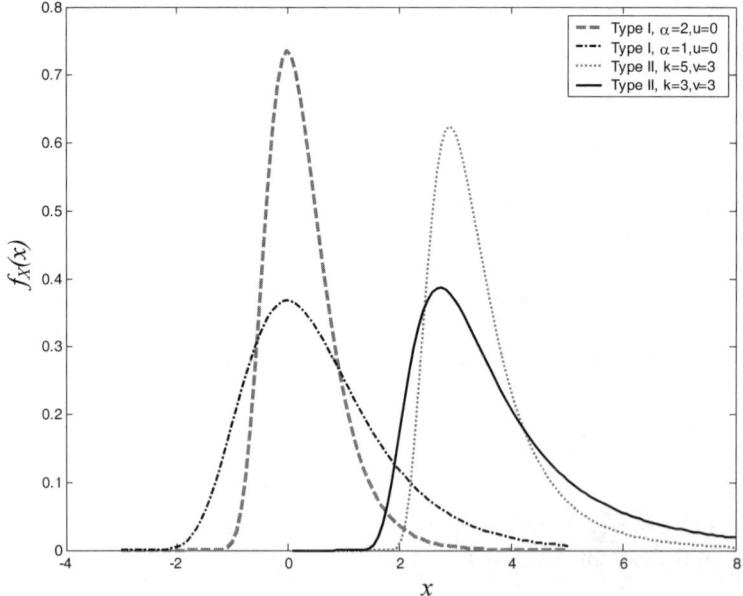

Figure 2.7. Type I and Type II Extreme Value Density Functions

Weibull Distribution

The Weibull distribution (Figure 2.8), also referred to as the Type III extreme value distribution, is well suited for describing the *weakest link* phenomena, or a situation where there are competing flaws contributing to failure. It is often used to describe fatigue, fracture of brittle materials, and strength in composites. The distribution of wind speeds at a given location on Earth can also be described with this distribution.

The probability density function is

$$f_X(x) = \frac{\alpha x^{\alpha-1}}{\beta^\alpha} \exp\left[-\left(\frac{x}{\beta}\right)^\alpha\right], \quad x \geq 0, \; \alpha > 0, \; \beta > 0 \tag{2.54}$$

and the CDF is

$$F_X(x) = 1 - \exp\left[-\left(\frac{x}{\beta}\right)^\alpha\right], \quad x > 0 \tag{2.55}$$

Every location is characterized by a particular shape and scale parameter. This is a two-parameter family, α and β. The moments in terms of the parameters are

$$E(X^n) = \beta^n \Gamma\left(\frac{n}{\alpha} + 1\right) \tag{2.56}$$

where $\Gamma(.)$ is the gamma function.

The mean and coefficient of variation are

$$\mu_X = \beta \, \Gamma\left(\frac{1}{\alpha} + 1\right) \tag{2.57}$$

$$COV_X = \left[\frac{\Gamma\left(\frac{2}{\alpha} + 1\right)}{\Gamma^2\left(\frac{1}{\alpha} + 1\right)} - 1\right]^{0.5} \tag{2.58}$$

The mean and standard deviation are complicated functions of the parameters α and β. However, the following simplified parameters, which provide very good accuracy over the range that is of interest to engineers, are recommended in [2]:

$$\alpha = COV_X^{-1.08}, \quad \beta = \frac{\mu_X}{\Gamma\left(\frac{1}{\alpha} + 1\right)} \tag{2.59}$$

To illustrate the use of the Weibull distribution in the weakest link phenomenon, suppose there is a chain with N links, each of which has a random strength X. The strength of the chain y based on the weakest link is

$$P(X > y) = P(x_1 > y \cap x_2 > y \cap x_3 > y \cap \ldots \cap x_N > y) \tag{2.60}$$

If all the link strengths are independent, then

$$P(X > y) = P(x_1 > y)P(x_2 > y)P(x_3 > y)\ldots P(x_N > y) \tag{2.61}$$

Furthermore, if all of the links follow the same strength distribution, a single CDF can express all the probabilities on the right side:

$$P(x_i < y) = 1 - P(x_i \leq y) = 1 - F_X(y) \tag{2.62}$$

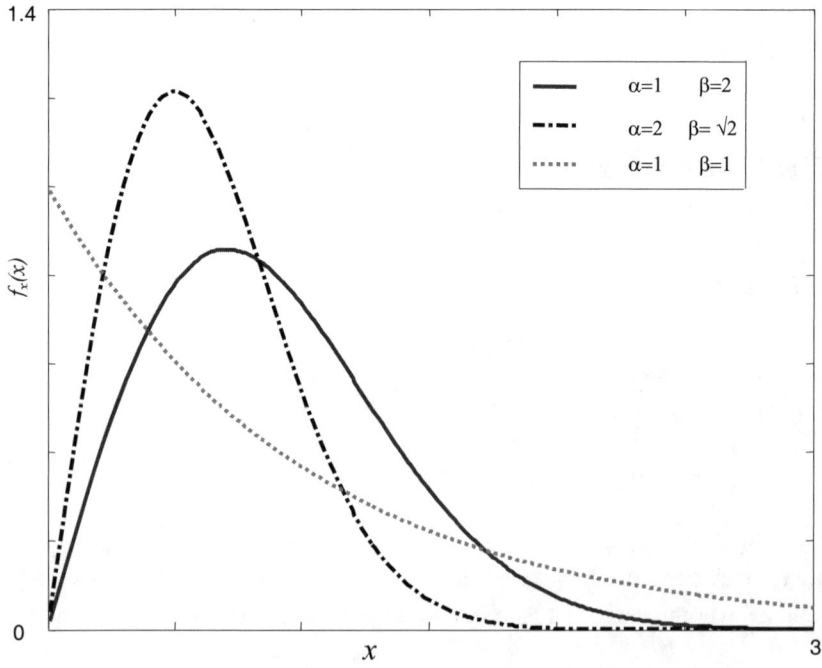

Figure 2.8. Weibull Density Function

So, the CDF for the chain can be expressed as

$$F_Y(y) = 1 - P(y < x_i) = 1 - [1 - F_X(y)]^N \tag{2.63}$$

Assuming the link strengths follow a Weibull distribution, Equation 2.55 can be substituted into the above expression:

$$F_Y(y) = 1 - \left[\exp\left(\frac{-y}{\beta}\right)^\alpha\right]^N = 1 - \exp\left[-N\left(\frac{y}{\beta}\right)^\alpha\right] \tag{2.64}$$

And so the strength of the entire chain is governed by the Weibull distribution:

$$F_Y(y) = 1 - \left[\exp\left(\frac{-y}{\beta'}\right)^\alpha\right] \tag{2.65}$$

with a scale parameter of

$$\beta' = N^{-\frac{1}{\alpha}} \beta \tag{2.66}$$

Example 2.5

A chain consists of welded links, each of which has Weibull-strength-distribution parameters of $\alpha = 5$ and $\beta = 5000$ N. Find (a) the mean strength of a link, (b) the mean strength of a 50-link and a 100-link chain, and (c) the load at which there is a 1% probability of failure for both chains.

Solution:

(a) From Equation 2.57, the mean strength is

$$\mu_X = 5000 \Gamma(1/5+1) = 5000(0.918) = 4590 \text{ N}$$

(b) From Equation 2.66

For a 50-link chain,

$$\beta' = 50^{-1/5}(5000) = 2287 \text{ N}$$
$$\mu_Y = 2287 \, \Gamma(1.2) = 2099 \text{ N}$$

For a 100-link chain,

$$\beta' = 100^{-1/5}(5000) = 1991 \text{ N}$$
$$\mu_Y = 1991\,\Gamma(1.2) = 1827 \text{ N}$$

(c) From Equation 2.65

$$0.01 = 1 - \exp\left(-\left[\frac{y}{\beta'}\right]^\alpha\right) \;\rightarrow\; \ln(0.99) = -\left[\frac{y}{\beta'}\right]^\alpha$$

$$y = \beta'[\ln(1/0.99)]^{1/\alpha}$$
$$y_{50} = 2287(0.3985) = 911 \text{ N}$$
$$y_{100} = 1991(0.5521) = 793 \text{ N}$$

Exponential Distribution

The *exponential distribution* is a special case of the Weibull distribution for $\alpha = 1$.

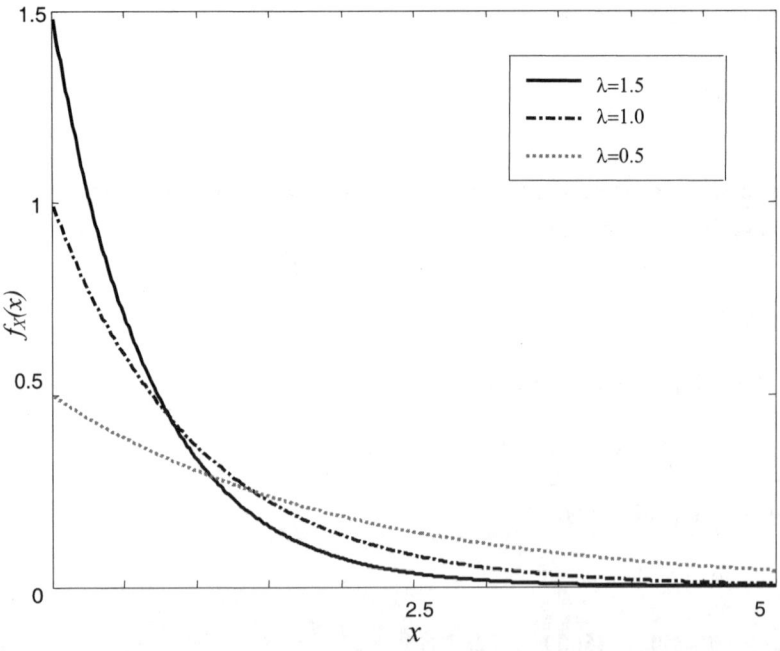

Figure 2.9. Exponential Density Function

The PDF is

$$f_X(x) = \lambda \exp[-\lambda x], \ 0 \le x < \infty \tag{2.67}$$

The CDF is given by

$$F_X(x) = 1 - \exp[-\lambda x], \ 0 \le x < \infty \tag{2.68}$$

The moments in terms of the parameter λ are

$$\mu_X = \frac{1}{\lambda}, \ \sigma_X^2 = \frac{1}{\lambda^2} \tag{2.69}$$

The exponential distribution is commonly used in reliability analysis. As shown in Figure 2.9, this distribution is well suited to represent the long flat portion of the *bathtub curve*, which is the phenomenon that probability failures are usually high early in the lifecycle, low in the middle, and rise strongly toward the end.

Example 2.6

If x_1 and x_2 are normally distributed with mean = 0 and standard deviation = 1, sketch the PDFs or histograms of (a) $x_1 + x_2$, (b) $\exp(x_1)$, (c) $x_1^2 + x_2^2$, and (d) $\sqrt{x_1^2 + x_2^2}$ by using numerical tools (*i.e.*, MATLAB® or Mathematica).

Solution:

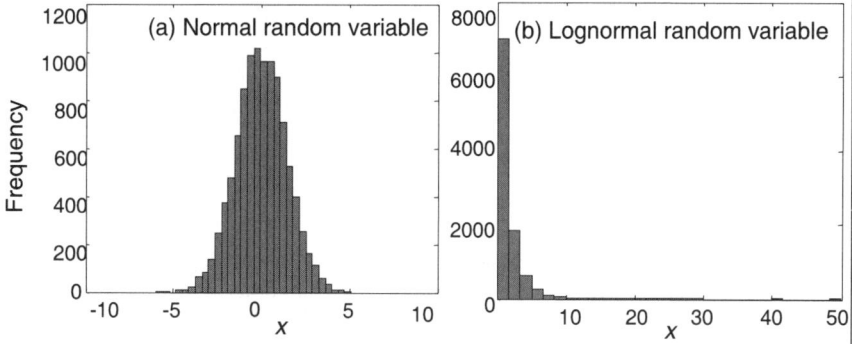

Figure 2.10. Probability Density Functions

When x_1 and x_2 are normally distributed, $x_1 + x_2$ is also normally distributed (Figure 2.10a), as mentioned earlier. This is also true for $ax_1 + bx_2$, where a and b are arbitrary constants. Furthermore, $\exp(x_1)$ is a lognormal random variable

(Figure 2.10b), and $x_1^2 + x_2^2$ is known as a Chi-squared random variable with two degrees of freedom (Figure 2.10c). And, $\sqrt{x_1^2 + x_2^2}$ is known as a Rayleigh random variable (Figure 2.10d). The properties of these distributions can be found in [6].

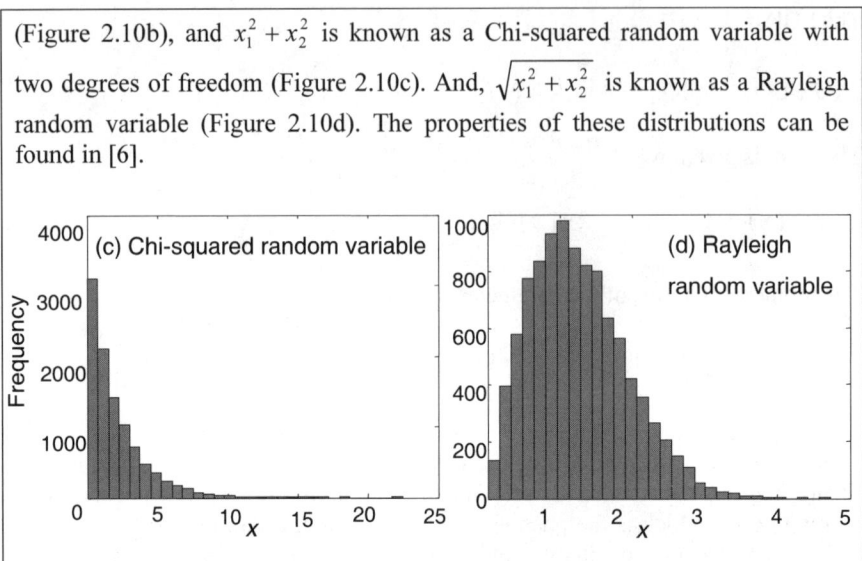

Figure 2.10. Probability Density Functions (contd.)

2.2 Random Field

A practical engineering structure has uncertainties in loads, geometry, material properties, manufacturing processes, and operational environments. These uncertainties fluctuate over space and time domains, and the responses of the structure are accordingly affected by these parameters. The estimation and representation of spatial- and time-variant data have become extremely important to realistic simulations that incorporate uncertainty analysis. The mathematical model of the variability, parameterized by the correlation between different locations, can be characterized by means of random field. This section briefly reviews the concept of the random field.

2.2.1 Random Field and Its Discretization

A *random field* is a random function of one or more variables. Many distributed properties in structural problems are random. For example, structural mechanical problems involve random fields, such as loads and stiffness properties. Efficient and realistic representation of the inputs will facilitate accurate estimations of random responses' statistics. Therefore, engineers should be able to handle these random field inputs and assess the allowable bounds for corresponding random responses, *i.e.*, stress and deflections, to determine the safety of structures. However, traditional deterministic analysis, such as the finite element method, uses a single design point, considering it sufficient to represent the response (Figure 2.11a). This simulation of a single design point is inadequate and unrealistic when characterizing systems under varying loads and material properties. For instance, in studying the response of an aircraft to gust loads, we cannot cover all types of gusts and speeds in a single simulation.

The mathematical model of the spatial variability, parameterized by the correlation between different locations, can be characterized by means of random field. Generally, the terminologies of a random field and a *random process* are used interchangeably in the literature, but the random field treats multidimensional variations, while the random process is used for a single coordinate, usually time [12], [13]. The basic idea of the random process is that the outcome of each experiment is a function over an interval of the domain rather than a single value (Figure 2.11b).

Thus, analysis of the random process is a realistic approach that can produce a whole design space instead of just a one-point result. The resulting function, which is generated for all the points ($\omega_1,\ldots, \omega_n$) in the sample space Ω, is known as a *realization* of a random process, and the collection of realizations is referred to as an *ensemble* [10],[12]. When we consider a set of samples in the interval, [t_0 t_n] (Figure 2.11b), the joint probability distributions of n random variables X can specify the particular random process. Thus, the moments of the random process $X(t)$ can be defined by similar formulas in accordance with the definition of the moments of the random variable.

The mean of the random process $X(t)$ is

$$\mu_X(t) = E[X(t)] = \int_{-\infty}^{\infty} x \, f_X(x,t)dx \qquad (2.70)$$

the *autocovariance* is

$$C_{XX}(t_1,t_2) = E[(X(t_1)-\mu_X(t_1))(X(t_2)-\mu_X(t_2))] \qquad (2.71)$$
$$= \int_{-\infty}^{\infty}\int_{-\infty}^{\infty}(x_1-\mu_X(t_1))(x_2-\mu_X(t_2))f_{XX}(x_1,x_2;t_1,t_2)dx_1dx_2$$

and the *autocorrelation* is

$$R_{XX}(t_1,t_2) = E[X(t_1)X(t_2)] \qquad (2.72)$$
$$= \int_{-\infty}^{\infty}\int_{-\infty}^{\infty} x_1 x_2 f_{XX}(x_1,x_2;t_1,t_2)dx_1dx_2$$

The autocorrelation function describes the correlation between all realizations at points t_1 and t_2. The prefix "auto" indicates that the integrand is composed of the same function at two points. Thus, the *cross-covariance* indicates a second-moment of two different functions. In particular, if the PDF $f_X(x,t)$ of the random process $X(t)$ is independent of t, namely $f_X(x,t) = f_X(x)$, then the process is referred to as a *stationary* process; otherwise, it is called a *nonstationary* process. Accordingly, all the moments of the stationary process are independent of t. If only the mean and the autocorrelation function of a random process are independent of t, then the process is said to be a *weakly stationary*. This process is a special case of homogenous processes that shows some symmetry in the domain. If the ensemble average of a stationary process is equal to the corresponding time average, the process is called *ergodic*. Figure 2.12 shows the classification of random processes. As seen in the figure, all ergodic processes are stationary, but not all stationary process are ergodic.

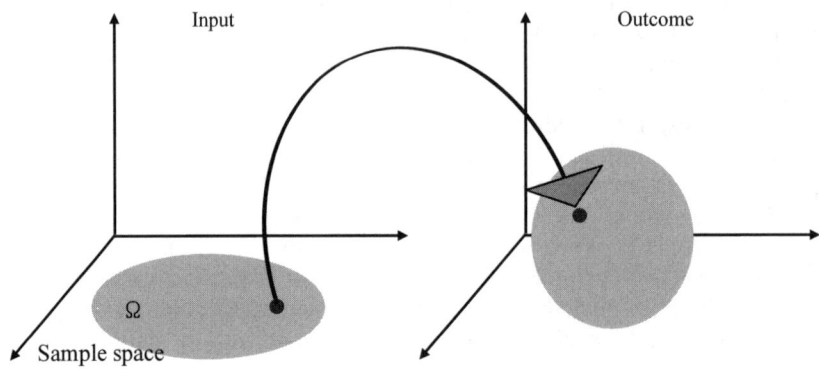

(a) Deterministic Concept

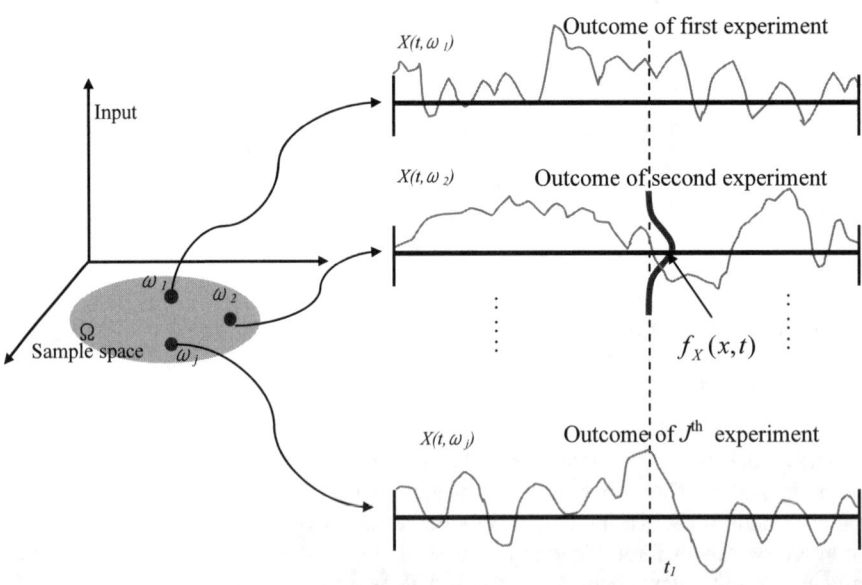

(b) Random Process Concept

Figure 2.11. Deterministic and Random Process Concepts

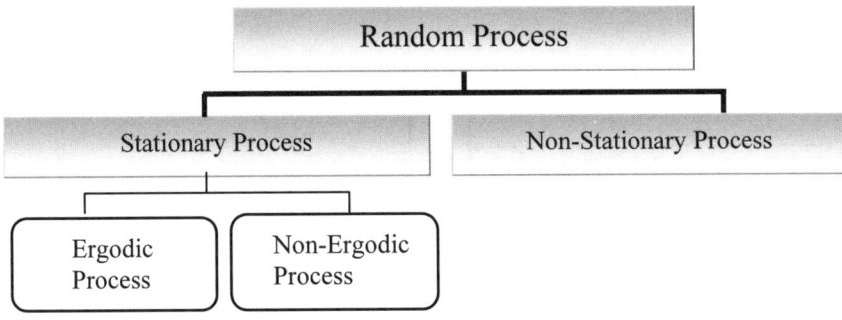

Figure 2.12. Classification of a Random Process

As previously stated, the random process can be thought of as a random function. Because manipulating random variables is easier than using the random function directly, a series of deterministic functions with random coefficients is frequently used to replace the random function. After its discretization, the continuous random process is an indexed set of an infinite number of random variables. Various methods, including the use of orthogonal polynomials and Taylor series representations, have been devised to replace random functions with random variables, depending on how the functions and the random coefficients are chosen. In the *Taylor series* representation [7],

$$U(x) \approx U_0 + U_1 x + U_2 x^2 + \ldots + U_n x^n \tag{2.73}$$

where $U(x)$ is a random function, and U_i are random variables with distributions determined by the distributions of $U(x)$. The deficiency of the Taylor series is that it requires many terms to get accurate results at points far from the origin. Thus, the use of orthogonal polynomials, which have a constant accuracy over the whole valid range of the approximation, is more suitable for the estimation of large fluctuations over domains. After first discussing the fundamentals of random field discretization, we provide complete details about the orthogonal polynomials, including the polynomial chaos expansion and the Karhunen-Loeve expansion [5].

Consider a simple cantilever beam, as illustrated in Figure 2.13a, with its Young's modulus, E, fluctuating over the length of the beam (Figure 2.13b). Obviously, the fluctuation of the Young's modulus should be considered in the analysis process. To do this, the *random field discretization* is used to describe the spatial variability of the stochastic structural properties over the structure. First, the randomness of the Young's modulus can be split into two parts, the mean part (Figure 2.13c) and the fluctuation part (Figure 2.13d), in order to reduce the bias and to better facilitate analysis. The discretization of the random field is similar to the finite element discretization of structures. In the discretization procedure, the particular value of E_n is assumed to have the same value for the entire n^{th} segment, and its accuracy depends on the size of the segments. After the discretization procedure, the random field can be replaced by a set of correlated random variables.

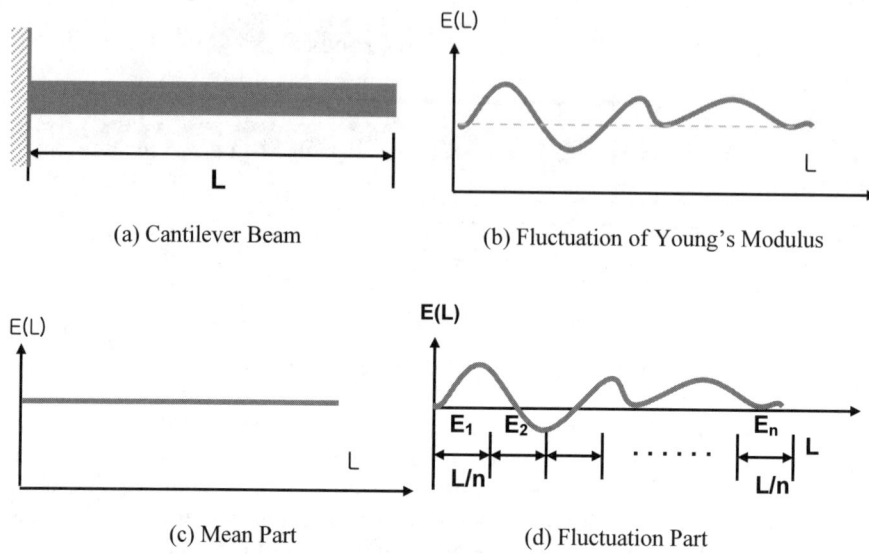

Figure 2.13. Random Field Discretization

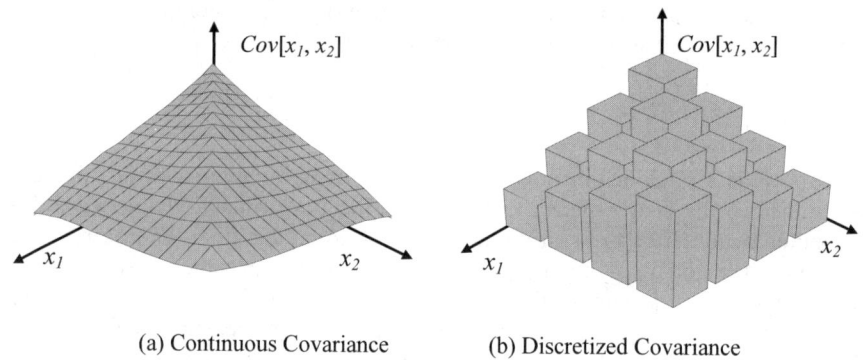

Figure 2.14. Discretization of Covariance

Several methods have been suggested to produce the random field discretization [8]. The utilization of orthogonal polynomials specifically will be discussed in Chapters 5 and 6. In order to properly discretize the random fields, the characterization and representation of the statistical correlation of each random variable (*i.e.*, E_n) are critical. Since the most widely used characterizations of the random fields are the first- and second-moment characterizations, the random-field discretization involves the discretization of its covariance function. The degree of correlation between the random process at nearby points can be specified by covariance functions. Figure 2.14b shows an illustration of the approximate covariance function for four elements along each direction in the finite element method. The structural properties of each element are modeled as random variables so that they have correspondingly different covariance values. Increasing the number of elements facilitates an accurate approximation of the actual covariance (Figure 2.14a). Thus, to ensure accurate analysis results, the engineer should consider the stochastic and modeling complexities of the problem before determining the size and number of elements.

2.2.2 Covariance Function

The degree of correlation between the random process at nearby points can be specified by *covariance functions*. If the variability of the random field is entirely random, the covariance function will decay asymptotically to zero. The points close together yield high correlation, and the lag points, vice versa. The following descriptions explain three well-known models of covariance functions, and the corresponding covariance functions are shown in Figure 2.15.

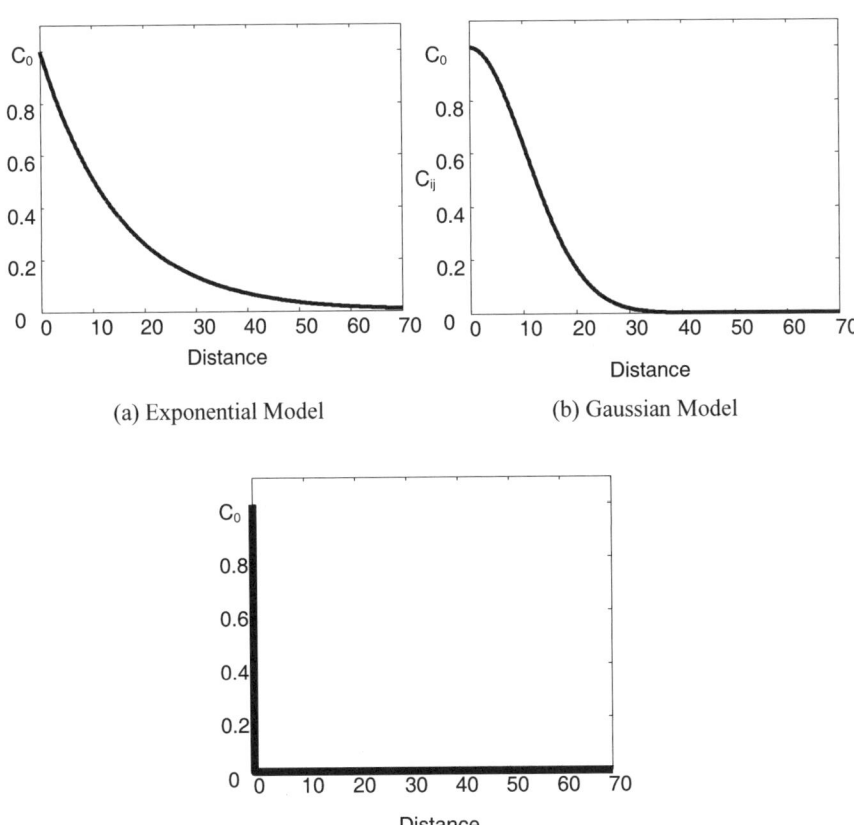

(a) Exponential Model

(b) Gaussian Model

(c) Nugget-effect Model

Figure 2.15. Covariance Functions

Exponential Model

The *exponential models* (Figure 2.15a) drop exponentially with increasing distance:

$$C_{ij} = C_0 \exp\left[-\left|\frac{x_{ij}}{l}\right|\right] \tag{2.74}$$

where C_0 is a variance, x_{ij} is the distance between two points (x_i and x_j), and l is the correlation length of the field. The correlation length indicates how quickly the covariance falls off. If the distance between two separated points is greater than the correlation length, the points are, statistically, nearly independent.

Gaussian Model

The *Gaussian model* (Figure 2.15b), also called the *squared exponential model*, is a commonly-used covariance function in random field analysis:

$$C_{ij} = C_0 \exp\left[-\left(\frac{x_{ij}}{l}\right)^2\right] \tag{2.75}$$

In contrast to the exponential model, the covariance in the Gaussian model remains flat at the origin. After the inflection point, it decays exponentially. Thus, this model provides a smoother stochastic process than the exponential model.

Nugget-effect Model

The *nugget-effect model*, also called the *delta-correlated model*, (Figure 2.15c) is an appropriate tool for discontinuous systems, which abruptly change values from one location to another:

$$C_{ij} = \begin{cases} C_0 & for \ |x_{ij}| = 0 \\ 0 & for \ |x_{ij}| > 0 \end{cases} \tag{2.76}$$

This model describes purely uncorrelated variation in population density, which is an equivalent concept to white noise in signal processing, and shows contributions to variability without spatial continuity.

The eigenfunctions and eigenvalues are extracted from the covariance function (C_{ij}) so that the collection of eigenfunctions represents the random fields. The extraction procedure of eigenfunctions and eigenvalues is known as *spectral decomposition* in continuous systems, and it is also known as the orthogonal transform or the Karhunen-Loeve transform. Chapters 5 and 6 describe the details of the Karhunen-Loeve transform and the new procedures developed for representing the random field.

Example 2.7

Suppose we have four points on a line, $x = [1, 2, 3, 4]$. (a) When a Gaussian covariance is given with a variance $C_0 = 1$, find a correlation length (l) which yields C_{14} or $C_{41} = 0.65$, (b) Use the correlation length obtained in step (a) to construct a Gaussian covariance matrix.

Solution:

(a) From Equation 2.75

$$l = |x_{ij}| / \sqrt{-\ln(C_{ij}/C_0)}, \quad \therefore \; l = |4-1| / \sqrt{-\ln(0.65)} = 4.5708$$

(b) The distance matrix is

$$x_{ij} = \begin{bmatrix} 0 & 1 & 2 & 3 \\ 1 & 0 & 1 & 2 \\ 2 & 1 & 0 & 1 \\ 3 & 2 & 1 & 0 \end{bmatrix}$$

Again, from Equation 2.75 with $l = 4.5708$, $C_0 = 1$

$$C_{ij} = \begin{bmatrix} 1 & 0.95 & 0.83 & 0.65 \\ 0.95 & 1 & 0.95 & 0.83 \\ 0.83 & 0.95 & 1 & 0.95 \\ 0.65 & 0.83 & 0.95 & 1 \end{bmatrix}$$

Note: Before proceeding, it is necessary to check the positive definiteness of the covariance matrix (all positive eigenvalues). If the matrix is not positive definite, there are several possible reasons: a) linear dependency – some of the covariances are linear function of other covariances, b) small sample size – the small sample size yields mere sampling fluctuations which cause the non-positive definite matrix, and c) missing data or input data error.

2.3 Fitting Regression Models

Regression analysis is the investigation of the functional relationship between two or more variables. Some specific aspects related to the proposed framework of

Chapters 5 and 6 are presented in this section. More complete details of the regression procedures are available in [9].

2.3.1 Linear Regression Procedure

In many situations, two or more variables are inherently related, and the investigation of the functional relationship between these measured variables (*independent variables*) and predicted variables (*dependent variables*) is the basic idea behind regression analysis. There are two types of regression analysis: linear and nonlinear. If the relation of the dependent and independent variable is assumed to be a linear function of some parameters, the regression model is called *linear*; otherwise, it is referred to as *nonlinear*. The earliest form of regression analysis for linear problems was studied by Gauss and Legendre in the 17th century. Their method, known as the least squares method, is a technique which minimizes the sum of squares of the residuals (differences between fitted function and given data) to find a best fit.

Consider the linear regression model:

$$y(x) = \beta_0 + \beta_1 f_1(x) + ... + \beta_k f_k(x) + \varepsilon \tag{2.77}$$

where β_i, $i = 0,1,2,...,k$, are the *regression coefficients* and ε, the error of the model equation, is assumed to be normally distributed with mean zero and variance σ_e^2.

Equation 2.77 can be written in matrix notation for n sample values of x and y as

$$Y = X\hat{\beta} + e \tag{2.78}$$

where

$$Y = \begin{bmatrix} y_1 \\ y_2 \\ \vdots \\ y_n \end{bmatrix} \quad X = \begin{bmatrix} 1 & f_1(x_1) & f_2(x_1) & \cdots & f_k(x_1) \\ 1 & f_1(x_2) & f_2(x_2) & \cdots & f_k(x_2) \\ \vdots & \vdots & \vdots & \vdots & \vdots \\ 1 & f_1(x_n) & f_2(x_n) & \cdots & f_k(x_n) \end{bmatrix} \quad \hat{\beta} = \begin{bmatrix} \beta_0 \\ \beta_1 \\ \vdots \\ \beta_k \end{bmatrix} \text{ and } e = \begin{bmatrix} \varepsilon_1 \\ \varepsilon_2 \\ \vdots \\ \varepsilon_n \end{bmatrix}$$

Generally, the method of least squares is used to obtain the regression coefficients:

$$\hat{\beta} = (X^T X)^{-1} X^T Y \tag{2.79}$$

The fitted model and the residuals are

$$\hat{Y} = X\hat{\beta} \text{ and } e = Y - \hat{Y} \tag{2.80}$$

The covariance matrix of $\hat{\beta}$ is

$$Cov(\hat{\beta}) = E\{(\hat{\beta} - E[\hat{\beta}])(\hat{\beta} - E[\hat{\beta}])^T\} = \sigma_e^2 (X^T X)^{-1} \qquad (2.81)$$

where $E(.)$ denotes the expected value.

The *total sum of squares* (Equation 2.82), *regression sum of squares* (Equation 2.83), and *error (residual) sum of squares* (Equation 2.84) are given as

$$SS_t = Y^T Y \qquad (2.82)$$

$$SS_r = \hat{Y}^T \hat{Y} = \hat{\beta}^T X^T Y \qquad (2.83)$$

$$SS_e = e^T e \text{ or } SS_e = SS_t - SS_r \qquad (2.84)$$

SS_t is the total variation in y, SS_r is the variation due to regression and SS_e is the part of the variation in y that cannot be explained by the regression. The smaller SS_e is as a fraction of SS_t is, the better is the quality of the regression model.

2.3.2 Linear Regression with Polynomial Fit

The linear model implicitly assumes that a plot of the dependent versus independent variables lies on a straight line before the addition with some random noise (Figure 2.16a). However, appropriate perception of the nonlinearity will provide better accuracy for quantifying nonlinear relationships (Figure 2.16b). The current section discusses the fitting of polynomials to data in some detail, and the next section presents statistical tests to help determine the appropriate regression model to use.

An understanding of polynomial approximation provides the foundation for most numerical analysis since polynomials are frequently used to approximate numerical solutions. In addition, solutions of many physical problems resemble polynomials, and in turn, polynomial approximation sometimes produces exact answers. Generally, the polynomial provides nonlinear relationships between response variables and explanatory variables. The response can be measured with fitted coefficients and selected ordered polynomials, and fitting errors. Another usage of polynomials is to approximate probability density or distribution functions.

Similar to the previous linear regression case (Equation 2.77), the linear regression of polynomial model of a one-dimensional case can be written as

$$y(x) = \beta_0 p_0(x) + \beta_1 p_1(x) + ... + \beta_m p_m(x) + \varepsilon \qquad (2.85)$$

where the degree of $p_i(x)$ is $i = 0,...,m$, and the polynomial approximation is said to be of order m for this case. The simplest polynomial model is the monomials of x^m (i.e., $p_0(x) = 1$, $p_1(x) = x$, ..., $p_m(x) = x^m$).

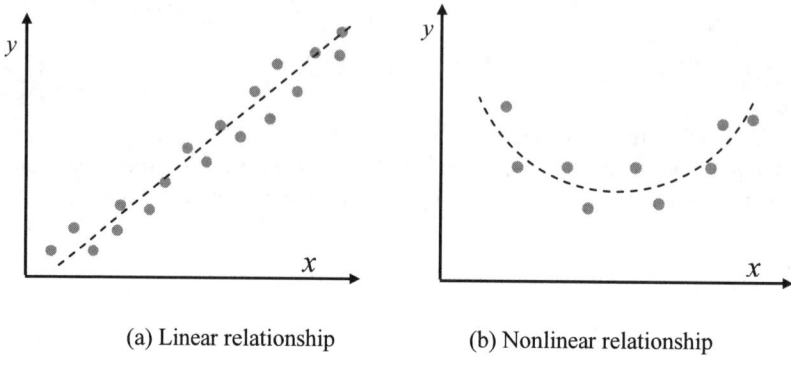

(a) Linear relationship   (b) Nonlinear relationship

Figure 2.16. Appropriate Regression Models

This polynomial regression model can be solved in precisely the same manner as in the previous section (Equation 2.79), and the X matrix can be written as

$$X = \begin{bmatrix} p_0(x_1) & p_1(x_1) & p_2(x_1) \dots & p_m(x_1) \\ p_0(x_2) & p_1(x_2) & p_2(x_2) \dots & p_m(x_2) \\ \vdots & \vdots & \vdots \ \vdots & \vdots \\ p_0(x_n) & p_1(x_n) & p_2(x_n) \dots & p_m(x_n) \end{bmatrix} \quad (2.86)$$

For the linear regression of polynomial model of the monomials of x^m, the columns of X can be nearly collinear. This causes an ill-conditioned problem, because negative values of x produce negative values for all odd powers, while positive values of x produce large positive values for the entire function. Hence, small changes in $p(x)$ lead to relatively large changes in the coefficients, β_i. More satisfactory solutions can be obtained when orthogonal polynomials (*i.e.*, Hermite, Laguerre, and Legendre) are used, reducing the effect of the ill-conditioned problem.

An important issue is to determine an appropriate order of polynomials. We can use high-order polynomials to closely fit representative sets of data, but the high-order model is generally less accurate in the area between selected sampling points. A relatively low-order polynomial fit to a large number of sampling points, provides a smooth interpolation that better fits the entire domain. The following section discusses the related topics of determining an appropriate order and checking the model adequacy of the regression.

2.3.3 ANOVA and Other Statistical Tests

Testing significance of regression involves *ANalysis Of VAriance* (ANOVA). ANOVA can determine the significant contributors of the model and can estimate the lack of fit and the confidence interval on the mean response. The test procedure is usually summarized in an analysis of variance table such as Table 2.2. In Table

2.2, n and k indicate the number of sample values or observations, and the number of treatments or regressors, respectively. When we omit the mean effect (β_o), the degrees of freedom of the total should be n-1 and the source of variance should be labeled as "Total, corrected" in Table 2.2. However, in the literature, the label "Total" is also used for the case of n-1 degrees of freedom. The test statistic F_o in Table 2.2 contributes to the significance test of the regression model. If the observed value of F_o is larger than the F-statistic, $F_o > F_{\alpha,1,df_e}$, then the coefficient is judged to have a significant effect on the regression model. The F-statistic has two parameters in this case, denoted by α and df_e. The df_e is the degrees of freedom of the residual, and α indicates the 100^{th}(1-α) percentile of the F distribution. The percentage of the F distribution for specific degrees of freedom can be calculated and tabulated (Appendix D).

Plots of the residuals, e versus the corresponding fitted values $\hat{Y}$, or the observed values, Y versus $\hat{Y}$, are good measures for determining model adequacy. These graphical plots and other statistical tests (e.g., normal probability plot [9]) yield the residual analysis, which can detect model inadequacies with little additional effort. Visual inspections of residuals are preferable to understand certain characteristics of the regression results. Analysts can easily construct the plots, which organize the data to reveal useful information. Example patterns of residual plots, including satisfactory, funnel, double bow, and nonlinear cases, are available in [4] and [9]. Abnormality of the residual plots indicates that the selected model is inadequate or that an error exists in the analysis. When the residual analysis detects these common types of model inadequacies, the analysts require or consider extra terms in the regression model (e.g., higher order or interaction terms).

Table 2.2. Analysis of Variance for Significance of Regression

Source of variance	Sum of squares	Degrees of freedom (df)	Mean square	F_0
Regression	SS_r	$df_r = k$	$MS_r = SS_r / df_r$	MS_r/MS_e
Residual	SS_e	$df_e = n-k$	$MS_e = SS_e / df_e$	
Total	SS_t	$df_t = n$		

To select the appropriate order of approximation polynomials, we can proceed using either of two strategies. One approach is a *forward selection procedure*, which involves increasing the polynomial order until the highest-order term is nonsignificant according to the significance test. The other approach is the *backward elimination procedure*, which fits a response model using the highest-order term and then deletes terms one at a time. Thus, the α value of F-statistics can indicate the acceptance and rejection levels of the regressors. Typically, the α values of 0.05 and 0.10 are common choices for both the acceptance and rejection levels, but these values can be adjusted according to the analyst's experience. Some researchers prefer to set a larger value for the rejection level α than for the acceptance level α to protect the rejection of regressors that are already admitted. Alternative methods that involve R^2, s^2, and C_p statistics select the best regression equations, and further discussions of their uses and advantages can be found in [4].

Example 2.8

Suppose we have this data:

x	0.8	1.0	1.2	1.4	1.6	1.8	2.0	2.2	2.4	2.6
y	24	20	10	13	12	6	5	1	1	0

(a) Fit the linear model ($y = \beta_0 + \beta_1 x + \varepsilon$) to the data above, and compute ANOVA.

(b) Fit the polynomial model ($y = \beta_0 + \beta_1 x + \beta_2 x^2 + \varepsilon$) to the data above, compute ANOVA, and check for the significance of the nonlinear term.

Solution:

In matrix notation, the coefficients can be obtained from Equation 2.79:

$$[X^T X] = \begin{bmatrix} 10 & 17 \\ 17 & 32.2 \end{bmatrix}, \quad [X^T X]^{-1} = \frac{1}{33}\begin{bmatrix} 32.2 & -17 \\ -17 & 10 \end{bmatrix},$$

$$[X^T Y] = \begin{bmatrix} 92 \\ 114 \end{bmatrix}, \quad \hat{\beta} = [X^T X]^{-1} X^T Y = \begin{bmatrix} 31.04 \\ -12.85 \end{bmatrix}$$

$\hat{y} = [\,20.76\ \ 18.19\ \ 15.62\ \ 13.05\ \ 10.48\ \ 7.91\ \ 5.34\ \ 2.77\ \ 0.20\ \ -2.36\,]^T$

From Equation 2.82 ~ 2.84,

$SS_t = Y^T Y = 1452.0$

$SS_r = \hat{Y}^T \hat{Y} = \hat{\beta}^T X^T Y = 1391.16$

$SS_e = e^T e = 60.84$

ANOVA is obtained from Table 2.2:

Source of variance	Sum of squares	Degrees of freedom (df)	Mean square	F_0
Regression	1391.16	2	695.58	91.40
Residual	60.84	8	7.61	
Total	1452.0	10		

The same procedure can be applied to the polynomial regression:

$$[X^TX] = \begin{bmatrix} 10 & 17 & 32.2 \\ 17 & 32.2 & 65.96 \\ 32.2 & 65.96 & 142.68 \end{bmatrix}, [X^TY] = \begin{bmatrix} 92 \\ 114 \\ 156 \end{bmatrix}$$

From Equation 2.79,

$$\hat{\beta} = [X^TX]^{-1}X^TY = \begin{bmatrix} 42.96 \\ -28.68 \\ 4.66 \end{bmatrix}$$

$$\hat{y} = [22.99\ 18.94\ 15.25\ 11.94\ 8.99\ 6.43\ 4.23\ 2.40\ 0.95\ -1.14]^T$$

From Equation 2.82 ~ 2.84,

$$SS_t = Y^TY = 1452.0$$
$$SS_r = \hat{Y}^T\hat{Y} = \hat{\beta}^TX^TY = 1409.35$$
$$SS_e = e^Te = 42.65$$

To check for the significance of the added model term, x^2, we need to break up the regression sum of squares (SS_r) into the component:

SS_{x^2} = SS_r of the added/current model - SS_r of the reduced/previous model
= 1409.35 - 1391.16 = 18.19

From Table 2.2

Source of variance	Sum of squares	Degrees of freedom (df)	Mean square	F_0
Regression	1409.35	3	469.78	77.14
$\beta_0 + \beta_1 x$	1391.16	2	695.58	114.22
$\beta_2 x^2$	18.19	1	18.19	2.99
Residual	42.65	7	6.09	
Total	1452.0	10		

Since 5% and 10% points of the F distribution (Appendix D) are $F_{.05,1,7} = 5.59$ and $F_{.10,1,7} = 3.59$, respectively, the coefficient, β_2, of the nonlinear term is not significant. Thus, we can assume that the exploration of higher-order models (third, fourth, etc.) are not necessary, and their effects are negligible.

2.4 References

[1] Ang, A.H.S., and Tang, W.H., *Probability Concept in Engineering Planning and Design*, Vol. I, John Wiley and Sons, NY, 1975.
[2] Cornell, C.A., "Bounds on the Reliability of Structural Systems," *Journal of Structural Division, ASCE*, Vol. 93 (1), February 1967, pp. 171-200.
[3] D'Errico, J. and Zaino, N., "Statistical Tolerancing Using a Modification of Taguchi's Method," *Technometrics*, Vol. 30, No. 4, Nov. 1988, pp. 397-405.
[4] Draper, N.R., and Smith, H., *Applied Regression Analysis*, Wiley, New York, 1981.
[5] Ghanem, R., and Spanos, P.D., *Stochastic Finite Elements: A Spectral Approach*, Springer-Verlag, NY, 1991.
[6] Johnson, N., Kotz, S., and Balakrishnan, N., *Continuous Univariate Distributions*, Volume 1, Wiley, NY, 1994
[7] Liu, W. K., and Belytschko, T., *Computational Mechanics of Probabilistic and Reliability Analysis*, Elmepress International, Lausanne, Switzerland, 1989.
[8] Matthies, H.G., Brenner, C.E., Bucher, C.G. and Soares, C.G., "Uncertainties in Probabilistic Numerical Analysis of Structures and Solids-Stochastic Finite Elements," *Structural Safety*, Vol.19, (3), 1997, pp. 283-336.
[9] Montgomery, D.C, and Peck E.A., *Introduction to Linear Regression Analysis*, Wiley, New York, 1992.
[10] Papoulis, A., *Probability, Random Variables, and Stochastic Processes*, Third Edition. McGraw-Hill, New York, 1991.
[11] Van Trees, H., *Detection, Estimation, and Modulation Theory Part I*, Wiley, New York, 1971.
[12] Vanmarcke, E., *Random Fields, Analysis and Synthesis*, MIT Press, Cambridge, MA., 1983.
[13] Vanmarcke, E., Shinozuka, M., Nakagiri, S., Schuëller, G. I., Grigoriu, M., "Random Fields and Stochastic Finite Elements," *Journal of Structural Safety*, Special Issue, No. 3, 1986, pp. 143-166.

3
Probabilistic Analysis

This chapter presents several probabilistic analysis methods, including the first- and second-order reliability methods, Monte Carlo simulation, Importance sampling, Latin Hypercube sampling, and stochastic expansions. Strengths and weaknesses of each method are discussed. Specifically, the ideas underlying Monte Carlo simulation and Latin Hypercube sampling are demonstrated, with intuitive illustrations of simple problems. A brief introduction of the stochastic finite element method, including the perturbation method, the Neumann expansion method, and the weighted integral method, is also presented.

3.1 Solution Techniques for Structural Reliability

Reliability analysis evaluates the probability of structural failure by determining whether the limit-state functions are exceeded. However, reliability analysis is not limited to calculation of the probability of failure. Evaluation of various statistical properties, such as probability distribution functions and confidence intervals of structural responses, plays an important role in reliability analysis. There are many ways to characterize the statistical properties of structural systems. In this section, we briefly discuss several methods for characterizing structural reliability, including the fundamentals of reliability analysis, sampling methods, stochastic finite element methods, and stochastic expansion. More specific details of each method will be discussed in Sections 3.2 and 3.3 and in Chapters 4, 5, and 6.

3.1.1 Structural Reliability Assessment

If, when a structure (or part of a structure) exceeds a specific limit, the structure (or part of the structure) is unable to perform as required, then the specific limit is called a *limit-state*. The structure will be considered unreliable if the failure probability of the structure limit-state exceeds the required value. For most structures, the limit-state can be divided into two categories:

Ultimate limit-states are related to a structural collapse of part or all of the structure. Examples of the most common ultimate limit-states are corrosion, fatigue, deterioration, fire, plastic mechanism, progressive collapse, fracture, etc. Such a limit-state should have a very low probability of occurrence, since it may risk the loss of life and major financial losses.

Serviceability limit-states are related to disruption of the normal use of the structures. Examples of serviceability limit-states are excessive deflection, excessive vibration, drainage, leakage, local damage, etc. Since there is less danger than in the case of ultimate limit-states, a higher probability of occurrence may be tolerated in such limit-states. However, people may not use structures that yield too much deflections, vibrations, etc.

Generally, the limit-state indicates the margin of safety between the resistance and the load of structures. The limit-state function, $g(\cdot)$, and probability of failure, P_f, can be defined as

$$g(X) = R(X) - S(X) \tag{3.1}$$

$$P_f = P\,[g(\cdot) < 0] \tag{3.2}$$

where R is the resistance and S is the loading of the system. Both $R(\cdot)$ and $S(\cdot)$ are functions of random variables X. The notation $g(\cdot) < 0$ denotes the *failure region*. Likewise, $g(\cdot) = 0$ and $g(\cdot) > 0$ indicate the *failure surface* and *safe region*, respectively.

The mean and standard deviation of the limit-state, $g(\cdot)$, can be determined from the elementary definition of the mean and variance. The mean of $g(\cdot)$ is

$$\mu_g = \mu_R - \mu_S \tag{3.3}$$

where, μ_R and μ_S are the means of R and S, respectively. And the standard deviation of $g(\cdot)$ is

$$\sigma_g = \sqrt{\sigma_R^2 + \sigma_S^2 - 2\rho_{RS}\sigma_R\sigma_S} \tag{3.4}$$

where, ρ_{RS} is the correlation coefficient between R and S, and σ_R and σ_S are the standard deviations of R and S, respectively.

The *safety index* or *reliability index*, β, is defined as

$$\beta = \frac{\mu_g}{\sigma_g} = \frac{\mu_R - \mu_S}{\sqrt{\sigma_R^2 + \sigma_S^2 - 2\rho_{RS}\sigma_R\sigma_S}} \tag{3.5}$$

If the resistance and the loading are uncorrelated ($\rho_{RS} = 0$), the safety index becomes

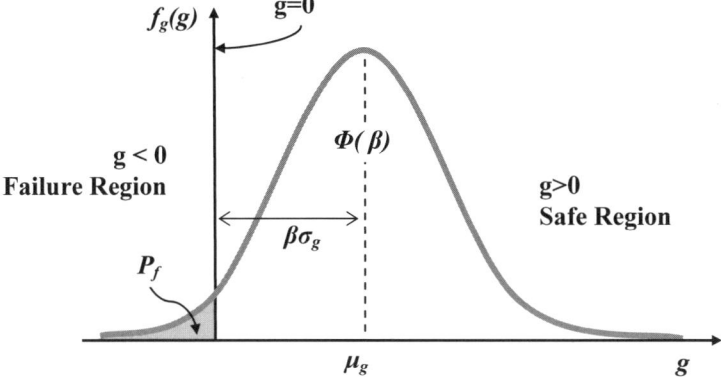

Figure 3.1. Probability Density for Limit-state $g(.)$

$$\beta = \frac{\mu_g}{\sigma_g} = \frac{\mu_R - \mu_S}{\sqrt{\sigma_R^2 + \sigma_S^2}} \tag{3.6}$$

The safety index indicates the distance of the mean margin of safety from $g(\cdot) = 0$. Figure 3.1 shows a geometrical illustration of the safety index in a one-dimensional case. The idea behind the safety index is that the distance from location measure μ_g to the limit-state surface provides a good measure of reliability. The distance is measured in units of the uncertainty scale parameter σ_g. The shaded area of Figure 3.1 identifies the probability of failure.

For a special case, the resistance, R, and loading, S, are assumed to be normally distributed and uncorrelated. The limit-state function is also normally distributed, since $g(\cdot)$ is a linear function of R and S. Thus, the probability density function of the limit-state function in this case is

$$f_g(g) = \frac{1}{\sigma_g \sqrt{2\pi}} \exp\left[-\frac{1}{2}\left(\frac{g - \mu_g}{\sigma_g}\right)^2\right] \tag{3.7}$$

The *probability of failure* is

$$P_f = \int_{-\infty}^{0} f_g(g)\, dg \tag{3.8}$$

When the normally distributed $g(\cdot) = 0$, the probability of failure is computed as

$$P_f = \int_{-\infty}^{0} \frac{1}{\sigma_g \sqrt{2\pi}} \exp\left[-\frac{1}{2}\left(\frac{0-\mu_g}{\sigma_g}\right)^2\right] dg \qquad (3.9)$$

$$= \int_{-\infty}^{0} \frac{1}{\sigma_g \sqrt{2\pi}} \exp\left(-\frac{1}{2}\beta^2\right) dg$$

$$= 1 - \Phi(\beta) = \Phi(-\beta)$$

where $\Phi(\cdot)$ is the standard normal cumulative distribution function.

For the multidimensional case, the generalization of Equation 3.8 becomes

$$P_f = P[g(X) \leq 0] = \int \cdots \int f_X(x_1,\ldots,x_n)\, dx_1 \ldots dx_n \qquad (3.10)$$

where $g(X)$ is the *n*-dimensional limit-state function and $f_X(x_1,\ldots,x_n)$ is the joint probability density function of all relevant random variables X.

Another well-known definition of reliability analysis is the *safety factor*, F:

$$F = \frac{R}{S} \qquad (3.11)$$

Failure occurs when $F = 1$, and if the safety factors are assumed to be normally distributed, the safety index is given by

$$\beta = \frac{\mu_F - 1}{\sigma_F} \qquad (3.12)$$

Example 3.1

> This example is taken from [17]. The figure below shows a simply-supported beam loaded at the midpoint by a concentrated force P. The length of the beam is L, and the bending moment capacity at any point along the beam is WT, where W is the plastic section modulus and T is the yield stress. All four random variables P, L, W, and T are assumed to be independent normal distributions. The mean values of P, L, W, and T are 10 kN, 8 m, 100×10^{-6} m³, and 600×10^3 kN/m² respectively. The standard deviations of P, L, W, and T are 2 kN, 0.1 m, 2×10^{-5} m³, and 10^5 kN/m², respectively. The limit-state function is given as
>
> $$g(\{P,L,W,T\}) = WT - \frac{PL}{4}$$
>
> Solve for the safety index, β, and the probability of failure, P_f, for this problem.

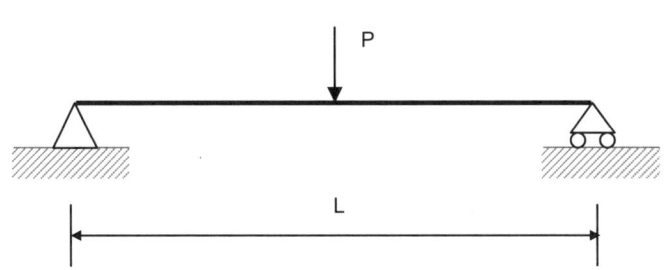

Solution:

From the given limit-state,

$$\mu_g = \mu_W \mu_T - \frac{1}{4}\mu_P \mu_L = 40$$

$$\mu_g^2 = \left(\mu_W \mu_T - \frac{1}{4}\mu_P \mu_L\right)^2 = 1600$$

Using Equation 2.19,

$$\sigma_g = \sqrt{E[g^2] - \mu_g^2}$$
$$= \frac{1}{4}\sqrt{\mu_P^2 \sigma_L^2 + \mu_L^2 \sigma_P^2 + \sigma_L^2 \sigma_P^2 + 16(\mu_W^2 \sigma_T^2 + \mu_T^2 \sigma_W^2 + \sigma_T^2 \sigma_W^2)}$$
$$= 16.2501$$

$$\because E[g^2] = E\left[\left(WT - \frac{PL}{4}\right)^2\right]$$
$$= (\sigma_W^2 + \mu_W^2)(\sigma_T^2 + \mu_T^2) - \frac{1}{2}\mu_W \mu_T \mu_P \mu_L + \frac{1}{16}(\sigma_p^2 + \mu_p^2)(\sigma_L^2 + \mu_L^2)$$

Using Equation 3.6, the safety index is calculated as

$$\beta = \frac{\mu_g}{\sigma_g} = 2.46153$$

In Chapter 4, this example will be solved by another approximation method, namely MVFOSM.

3.1.2 Historical Developments of Probabilistic Analysis

As shown in Section 1.2 (Figure 1.3), the probabilistic methods include the stochastic finite element method, the first- and second-order reliability method, sampling methods, the utilization of stochastic expansions based on the random process concept, etc. Each method requires different computational effort and provides different insight into the response variability and levels of accuracy. In this section, we briefly discuss historical developments of the most widely used methods in probabilistic analysis and discuss the advantages and disadvantages of each method. The subsequent sections, 3.2 and 3.3, describe the details of Monte Carlo simulation, Latin Hypercube sampling methods, and the stochastic finite element method. Chapters 4 and 5 provide details of the first- and second-order reliability methods and the state-of-the-art method of stochastic expansion, which incorporates several probabilistic approaches.

First- and Second-order Reliability Method

Due to the curse of dimensionality in the probability-of-failure calculation (Equation 3. 9), numerous methods are used to simplify the numerical treatment of the integration process. The Taylor series expansion is often used to linearize the limit-state $g(X) = 0$. In this approach, the first- or second-order Taylor series expansion is used to estimate reliability. These methods are referred to as the First Order Second Moment (FOSM) and Second Order Second Moment (SOSM) methods, respectively. FOSM is also referred to as the Mean Value First Order Second Moment method (MVFOSM), since it is a point expansion method at the mean point and the second moment is the highest-order statistical result used in this analysis. Although the implementation of FOSM is simple, it has been shown that the accuracy is not acceptable for low probability of failure ($P_f < 10^{-5}$) or for highly nonlinear responses [1]. In SOSM, the addition of a second-order term increases computational effort significantly, yet the improvement in accuracy is often minimal.

The safety index approach to reliability analysis given in the previous section is actually a mathematical optimization problem for finding the point on the structural response surface (limit-state approximation) that has the shortest distance from the origin to the surface in the standard normal space. Hasofer and Lind [15] provide a geographic interpretation of the safety index and improve the FOSM method by introducing the *Hasofer and Lind* (HL) *transformation*. In the transformation procedure, the design vector X is transformed into the vector of standardized, independent Gaussian variables, U. Because of rotational symmetry and the HL transformation, the design point in U-space represents the point of greatest probability density or maximum likelihood as shown in Figure 3.2. Because it makes the most significant contribution to the nominal failure probability $P_f = \Phi(-\beta)$, this design point is called the *Most Probable failure Point* (MPP).

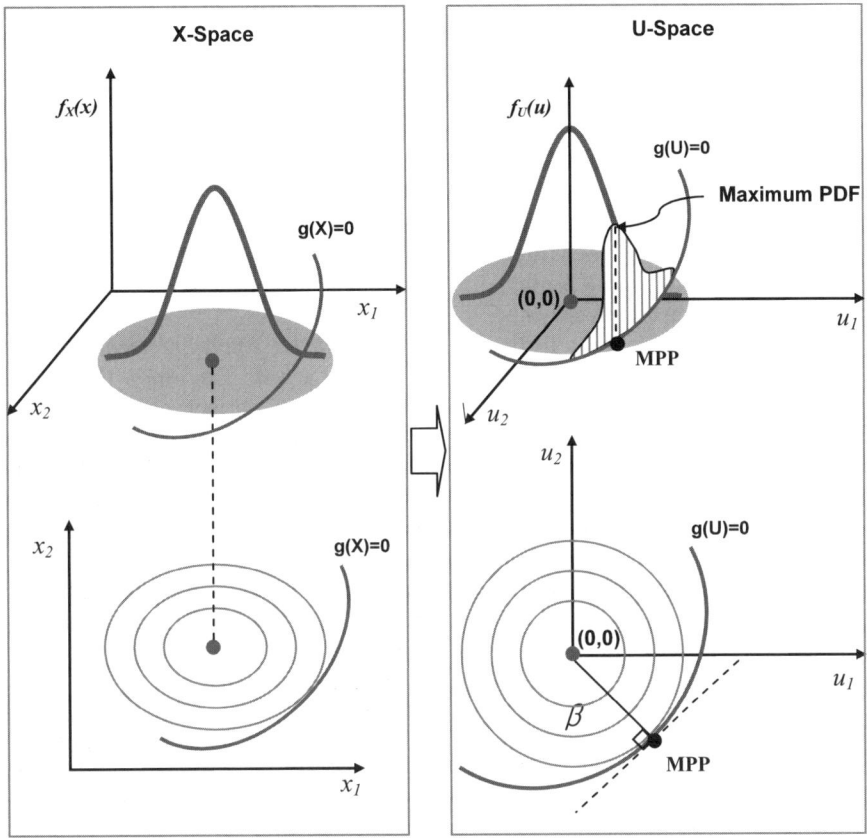

Figure 3.2. Transformation and MPP

Different approximate response surfaces $g(U)=0$ correspond to different methods for failure probability calculations. If the response surface is approached by a first-order approximation at the MPP, the method is called the *first-order reliability method* (FORM); if the response surface is approached by a second-order approximation at the MPP, the method is called the *second-order reliability method* (SORM). Furthermore, if the response surface is approached by a higher order approximation at the MPP, the method is called the *higher-order reliability Method* (HORM). Historically, the HL transformation method is often referred to as FORM. It is also referred to as the advanced FOSM or extended FOSM, but the acronym FORM is more prevalent for the HL method, since probability distributions are no longer approximated by their first and second moments [21].

In FORM, the limit-state is approximated by a tangent plane at the MPP. The approximate FORM results are used to specify a bound based on the probability of failure. If the approximations of the limit-state at the most probable failure point are accurate, then the bounds will produce satisfactory results; otherwise, this method may result in large error. FORM gives inaccurate results when the failure surface is highly nonlinear. Thus, FORM sometimes oscillates and converges on unreasonable values for probability of failure. The details of FORM and other common reliability methods are discussed in Chapter 4.

Stochastic Expansions

Stochastic expansion is an efficient tool for reliability analysis because the direct use of stochastic expansion, which is based on the concept of a random process, provides analytically appealing convergence properties for the stochastic analysis [4]. The purpose of the stochastic expansion is to better represent uncertainties of systems by introducing a series of polynomials aimed at characterizing the stochastic system being investigated.

Since the introduction of the *Spectral Stochastic Finite Element Method* (SSFEM) by Ghanem and Spanos [9], *Polynomial Chaos Expansion* (PCE) has been successfully used to represent uncertainty in a variety of applications, including structural response. PCE employs orthogonal polynomials of random variables. Most commonly, the random variables are standard-normal, and Hermite polynomials are used in SSFEM. PCE is convergent in the mean-square sense, and any order PCE consists of orthogonal polynomials. This property can simplify the calculation of moments in statistical procedures.

Tatang [30] introduced the *probabilistic collocation method* in which the responses of stochastic systems are projected onto the PCE. Delta functions at each collocation point serve as the test functions in a Galerkin method. Tatang obtained coefficients of the PCE by using the model outputs at selected collocation points (roots of the polynomials). Isukapalli [16] pointed out the limitation of the probabilistic collocation method for large-scale models and suggested a *stochastic response surface method* that uses the partial derivatives of model outputs with respect to model inputs. To obtain the partial derivatives of model outputs, ADIFOR, a FORTRAN programming library, was used in the stochastic response surface method. Recently, PCE was applied to the buckling eigenproblem by evaluating coefficients of the PCE through Monte Carlo Simulation [26]. Xiu *et al.* [32], [33] extended PCE to represent different distribution functions by using the *Askey scheme*. The Askey scheme, discussed in Chapter 5, classifies the hypergeometric orthogonal polynomials and indicates the limit transition relations between them. For instance, the Laguerre polynomials can be obtained from the Jacobi polynomials and can also be used to generate the Hermite polynomials.

Each of these methods has some limitations. In the case of the probabilistic collocation method, especially for many PCE degrees of freedom, the collocation points increase exponentially. Therefore, many collocation points are not sampled. Consequently, the collocation points selected to obtain unknown coefficients of the PCE do not guarantee a space filling design (Figure 3.3a), one that fills up the available design space with specified sampling points according to suitably-defined design criteria (such as maximize minimum distance between points [25]). If we are interested in the tail regions of a probability density function (Figure 3.4), then the data selection procedure should be reconsidered when applying the probabilistic collocation method, because the selected design points are concentrated in the high probability region.

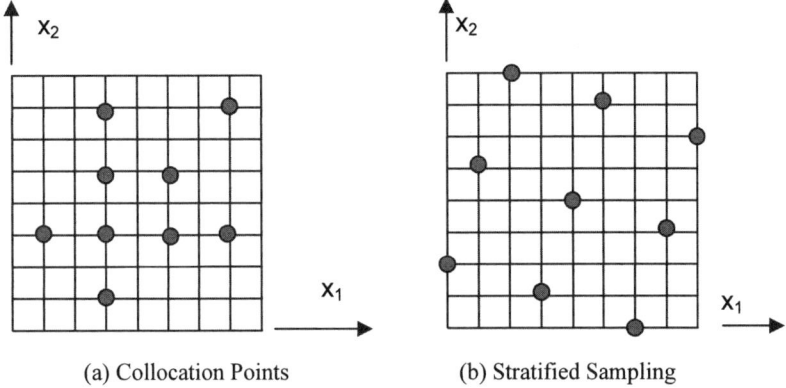

Figure 3.3. Comparison of Design Points of Probabilistic Collocation Method and Stratified Sampling

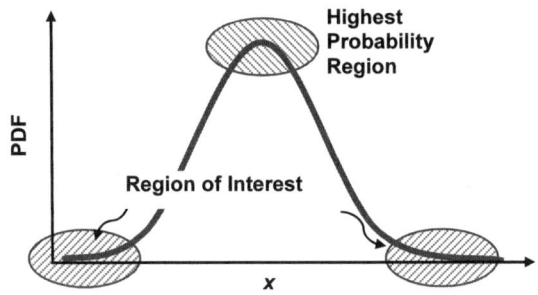

Figure 3.4. Regions of Interest in Probability Density Function

According to the preceding methods, we can classify the usage of stochastic expansions, including PCE and Karhunen-Loeve (KL) expansions, into two approaches: the *non-intrusive* and *intrusive formulation* procedures, as shown in Figure 3.5. An intrusive formulation is one in which the representation of uncertainty is expressed explicitly within the analysis of the system. Conversely, the results from structural analysis, in which uncertainty is not explicitly represented, are used in a non-intrusive formulation to characterize stochastic system behavior. In practice, this means that intrusive methods require access to modification of the analysis codes, whereas non-intrusive methods may treat the analysis code as a "black box." In one type of non-intrusive formulation, PCE is used to create the response surface without interfering with the finite element analysis procedure. Thus, this type of non-intrusive analysis is sometimes called the *stochastic response surface method*. Another type of non-intrusive formulation is the probabilistic collocation method. By contrast, the intrusive formulation method uses PCE and KL expansions to directly modify the stiffness matrix of a finite element analysis procedure. SSFEM and the *stochastic Galerkin FEM* [2] are both intrusive formulations.

The KL expansion can be applied to represent the characteristics of an uncertain system when its covariance function is known. However, if we do not have the information to form the covariance function, in the case of structural responses, then PCE can be used to represent this kind of uncertainty instead of the KL expansion. Chapter 6 includes a novel procedure using the non-intrusive formulation with an ANOVA and the Latin Hypercube Sampling (LHS) method. This procedure can guarantee that each of the input variables has all portions of its range represented (Figure 3.3b).

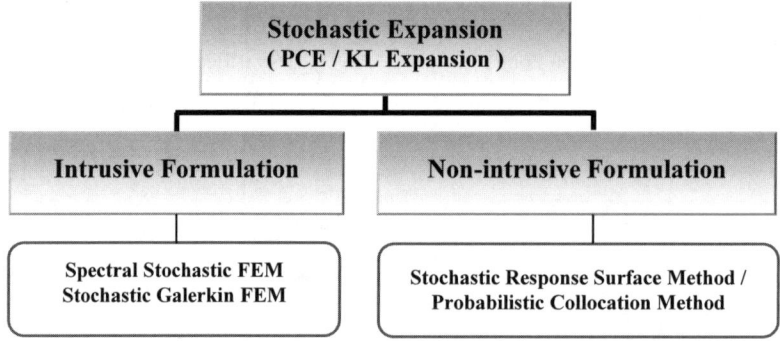

Figure 3.5. Intrusive and Non-Intrusive Formulation

3.2 Sampling Methods

A basic advantage of sampling methods is their direct utilization of experiments to obtain mathematical solutions or probabilistic information concerning problems whose system equations cannot be solved easily by known procedures. The following descriptions help to clarify the basic ideas and limitations of three representative sampling methods: Monte Carlo Simulation, Importance Sampling, and LHS.

3.2.1 Monte Carlo Simulation (MCS)

Monte Carlo Simulation, named after the casino games of Monte Carlo, Monaco, originates from the research work of Neumann and Ulam in 1949 [29]. Random behavior is found in games of chance such as slots, roulette wheels, and dice. Monte Carlo Simulation (MCS) is known as a *simple random sampling method* or *statistical trial method* that make realizations based on randomly generated sampling sets for uncertain variables. Application of the Monte Carlo method to probabilistic structural analysis problems is comparatively recent, becoming practical only with the advent of digital computers. It is a powerful mathematical tool for determining the approximate probability of a specific event that is the outcome of a series of stochastic processes. The Monte Carlo method consists of digital generation of random variables and functions, statistical analysis of trial outputs, and variable reduction techniques. These are discussed briefly in this section.

The computation procedure of MCS is quite simple:

1) Select a distribution type for the random variable
2) Generate a sampling set from the distribution
3) Conduct simulations using the generated sampling set

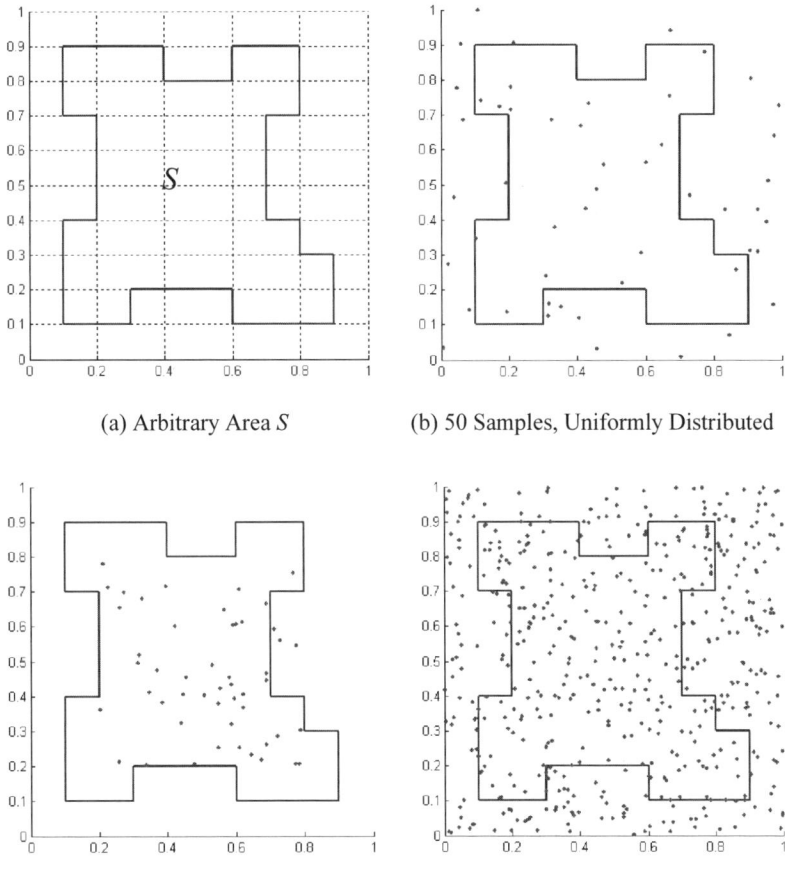

(a) Arbitrary Area S (b) 50 Samples, Uniformly Distributed

(c) 50 Samples, Aimed at the Center (d) 500 Samples, Uniformly Distributed

Figure 3.6. Example of "Hit or Miss" Method

The example shown in Figure 3.6, slightly modified from Sobol's book [29], clearly shows the basic idea and the critical aspects of MCS. Simple random sampling is a possible tool to measure the arbitrarily selected area S from the unit square plane (Figure 3.6a). The area can be approximately calculated by the ratio of m/n where n is the number of sampling points and m is the number of points that fall inside the area S. The exact area is geometrically calculated as 0.47. The obtained area, determined by using 50 random samples based on uniform distribution, yields 0.42, which means that 21 points appear inside the area S. It is obvious that the distribution type and the boundary limit are decisive parts of the sampling procedure.

When the sampling scheme is aimed at the center of the unit area, a distorted result might be obtained, as shown in Figure 3.6b. In this case, 47 points appear inside, and the ratio 47/50 yields the overestimated area of 0.94. Another important factor in the accuracy of the sampling method is the number of sampling points. When the number of sampling points is increased to 500 uniformly distributed points, the accurate result of 0.476 is obtained (Figure 3.6d).

The same idea can be extended to the analysis of structural reliability, which was described in Section 3.1. First, the sampling set of the corresponding random variables are generated according to the probability density functions. Next, we set the mathematical model of $g(\cdot)$, namely the limit-state, which can determine failures for the drawing samples of the random variables. Then, after conducting simulations using the generated sampling set, we can easily obtain the probabilistic characteristics of the response of the structures. In the above example, "Hit or Miss" of the area S represents the function $g(\cdot)$. If the limit-state function $g(\cdot)$ is violated, the structure or structural element has "failed." The trial is repeated many times to guarantee convergence of the statistical results. In each trial, sample values can be digitally generated and analyzed. If N trials are conducted, the probability of failure is given approximately by

$$P_f = \frac{N_f}{N} \tag{3.13}$$

where N_f is the number of trials for which $g(\cdot)$ is violated out of the N experiments conducted.

Generation of Random Variables

One of the key features in Monte Carlo Sampling is the generation of a series of values of one or more random variables with specified probability distributions. The most commonly used generation method is the *inverse transform method*. Let $F_X(x_i)$ be the CDF of random variable x_i. By definition, the numerical value of $F_X(x_i)$ is a value in the interval of [0,1]. Assuming that v_i is the generated uniformly distributed random number $(0 \leq v_i \leq 1)$, the inverse transform method is used to equate v_i to $F_X(x_i)$ as follows:

$$F_X(x_i) = v_i \text{ or } x_i = F_X^{-1}(v_i) \tag{3.14}$$

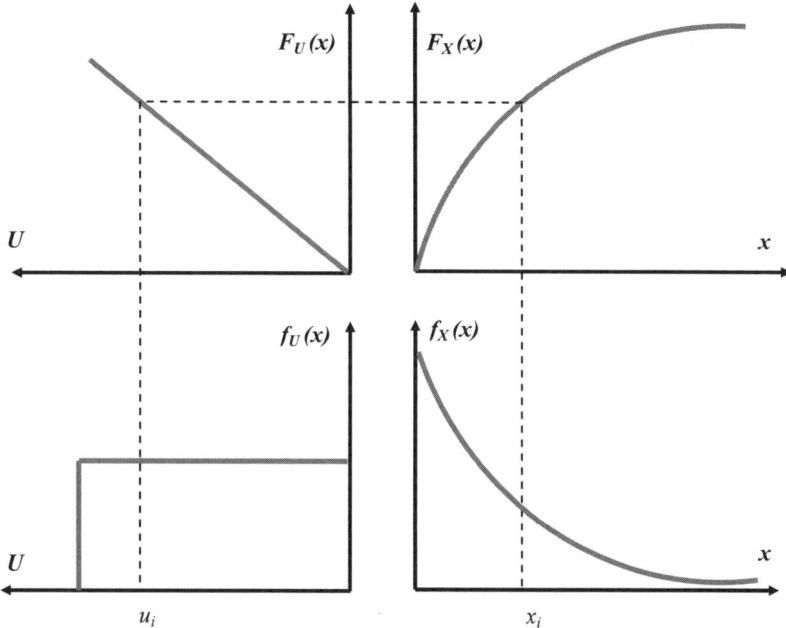

Figure 3.7. Inverse CDF Method for Exponential Distribution

This method can be applied to variables for which a cumulative distribution function has been obtained from direct observation, or where an analytic expression for the inverse cumulative function, $F^{-1}(.)$, exists. The inverse transform technique is graphically summarized in Figure 3.7. The random number generator produces uniform random numbers between 0 and 1 based on arbitrarily selected seed values. From the generated uniform random number, the corresponding CDF value of the uniform distribution and target distribution can easily be obtained. The final step is to obtain the random number for the target PDF using Equation 3.14.

Example 3.2

Generate exponential random variables whose PDF and CDF are given in Section 2.1.2 (Equation 2.67 and 2.68), and sketch the probability density function or histogram by using the Equation obtained.

Solution:

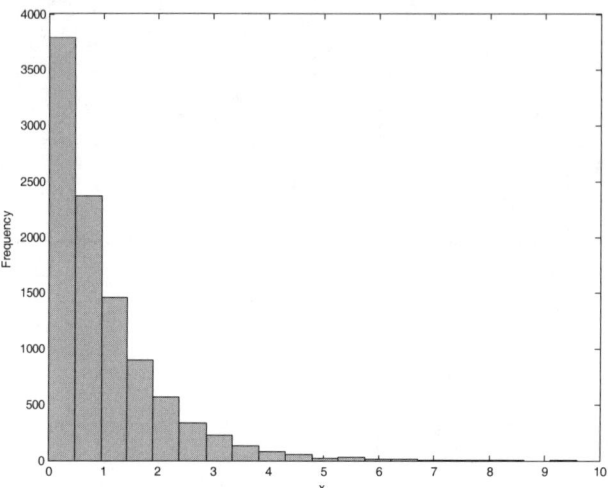

Figure 3.8. Exponential Distribution (Example 3.2)

The probability density function (Equation 2.67) of the exponential distribution is

$$f_X(x) = \lambda \exp[-\lambda x] \quad (x > 0)$$

The cumulative distribution function (Equation 2.68) is

$$F_X(x) = 1 - \exp[-\lambda x]$$

Let v_i be the random numbers from uniform distribution over the interval [0,1].

From Equation 3.14

$$F_X(x_i) = v_i \quad \rightarrow \quad 1 - \exp[-\lambda x_i] = v_i$$

Solve for x_i

$$x_i = -\frac{1}{\lambda} \ln(1 - v_i)$$

Hence, for any v_i, the exponential random variables can be generated with the above equation. The corresponding sketch of the histogram (Figure 3.8) is obtained after generating 10,000 samples of v_i with $\lambda = 1$.

Calculation of the Probability of Failure

Equation 3.13 gives the simplest Monte Carlo approach for reliability problems; it may be the most widely used, but it is not the most efficient, especially in complex systems. As defined in Section 3.1, failure occurs when the limit-state $g(X) \leq 0$, and the failure probability

$$P_f = P[g(X) \leq 0] = \int \cdots \int_{g(X) \leq 0} f_X(X) dX \qquad (3.15)$$

is written as

$$P_f = \int \cdots \int I[X] f_X(X) dX \qquad (3.16)$$

where $I[X]$ is an indicator function, which equals 1 if [$g(X) \leq 0$] is "true" and 0 if [$g(X) \leq 0$] is "false." It can be seen that the expectation of the indicator random variable for the failure event is just the probability that failure occurs. Hence,

$$P_f = P[g(X) \leq 0] = E[I(X)] = \mu_I = \mu_{P_f} \qquad (3.17)$$

and its variance is

$$Var[I(X)] = E[I(X)^2] - \{E[I(X)]\}^2 \qquad (3.18)$$

$$= E[I(X)] - \{E[I(X)]\}^2$$

$$= E[I(X)]\{1 - E[I(X)]\} = P_f(1 - P_f)$$

In order to evaluate P_f by the Monte Carlo method, a sample value for basic variable x_i with a cumulative distribution $F_x(x_i)$ must be drawn. The inverse transform method can be used to obtain the random variate, in which a uniformly distributed random number $u_i (0 \leq u_i \leq 1)$ is generated and equated to $F_x(x_i)$, i.e., $x_i = F^{-1}_x(u_i)$.

Hence, independent random numbers $u_1, u_2, \ldots, u_n$ are drawn from the density $f_x(x_i)$, and the estimate of P_f is obtained:

$$\hat{P}_f = \frac{1}{n} \sum_{i=1}^{\hat{n}} I(u_i) \qquad (3.19)$$

where $\hat{P}_f$ represents the Monte Carlo estimator of μ_{P_f} and $\hat{n}$ is the number of independent random numbers.

The variance of the sample mean is computed as

$$Var(\hat{P}_f) = \frac{\sigma_I^2}{\hat{n}} = \sigma_{P_f}^2 \qquad (3.20)$$

So the sample variance is given as

$$S_I^2 = \frac{1}{\hat{n}-1}\{\sum_{i=1}^{\hat{n}}I^2(u_i) - \hat{n}[\frac{1}{\hat{n}}\sum_{i=1}^{\hat{n}}I(u_i)]^2\} \qquad (3.21)$$

In principle, the Monte Carlo simulation is only worth exploiting when the number of trials of simulation is less than the number of integration points required in numerical integration. This is achieved for higher dimensions by replacing the systematic selection of points by a random selection, under the assumption that the options selected will be in some way unbiased in their representation of the function being integrated. To produce defensible results for a case with a large number of random variables, a large number of sampling sets is required. Thus, there are limitations to obtaining high accuracy for large-scale problems, which require tremendous computational time and effort. To improve the rate of convergence, several modifications can be introduced to MCS using quasi-random points [31]. The next two sections describe well-known approaches in variance reduction methods.

Example 3.3

Estimate the safety index β of Example 3.1 by using the MCS method with the same limit-state function, mean values, standard deviations, and distributions of the random variables.

Solution:

No. of Simulations	1,000	10,000	50,000	100,000	200,000	500,000	1,000,000	2,000,000
β	3.0902	2.9290	2.8228	2.8175	2.8574	2.8642	2.8507	2.8499
P_f	0.001	0.0017	0.00238	0.00242	0.00214	0.00209	0.002181	0.002187

As shown in the table, the MCS results (using MATLAB®) are converged to three digits after 200,000 runs ($\beta = 2.85, P_f = 0.0021$).

Example 3.4

Find solutions of x_1 and x_2 for the following problem by using a sampling method.

$$\text{Minimize } f = x_1^2 + 2x_2^2 - 4x_1x_2 + 8$$
$$\text{subject to } -x_1^2 + x_2 + 900 \leq 0$$
$$x_1 x_2 - 1000 \leq 0$$
$$0 \leq x_1 \leq 999, \ 0 \leq x_2 \leq 999$$

Hint: x_1 and x_2 are uniformly distributed in the interval [0, 999]

Solution:

This problem is called a constrained nonlinear optimization problem. Although efficient numerical methods exist, *i.e.*, Sequential Quadratic Programming (SQP), to solve these kinds of problems, we are obtaining the solution by using a sampling method. The basic strategy is very simple: Check all possible combinations of x_1 and x_2 with corresponding f values. Then, we seek a minimum value of f and a corresponding sampling set of x_1 and x_2. If the sampling set satisfies the given constraints, the data set is our solution. Accordingly, the procedure requires large simulation numbers to obtain accurate results through numerical simulation tools. The following solutions are obtained by using the MCS and SQP methods in MATLAB®:

(a) MCS Results

No. of Simulations	10,000	20,000	50,000	100,000	200,000	500,000	1,000,000	2,000,000
x_1	32.63	39.00	34.91	36.51	36.91	36.89	36.98	38.11
x_2	28.65	21.71	27.40	27.05	25.82	26.97	27.01	26.21
f	-1025.08	-915.22	-1098.19	-1145.89	-1108.14	-1155.95	-1160.90	-1161.30

(b) SQP result with 11 iterations of optimization

$$x_1 = 37.61, x_2 = 26.59, \text{ and } f = -1163.57$$

As shown in the MCS result of Example 3.4, the number of simulations required to provide optimum designs similar to the SQP result is quite large. But the SQP also has drawbacks; for instance, inappropriate initial points may result in local optima.

3.2.2 Importance Sampling

Variance reduction techniques have a dual purpose: to reduce the computational cost of a sample run and to increase accuracy using the same number of runs. In structural reliability analysis, where the probability of failure is generally relatively small, the direct (crude) Monte Carlo simulation procedure becomes inefficient. For example, in many pressure vessel technology problems, the probability of failure could be as small as 10^{-5} or 10^{-10}; this implies that at least a million simulation repetitions are required to predict this behavior. If the limit-state function $g(X)$ represents the mathematical model for structural simulation problems, the tail of the distribution of $g(X)$ is the most important factor. In order to predict the risk accurately and to increase the efficiency of the simulation by expediting execution times and minimizing computer storage requirements, the simulated iteration must concentrate the sample points in this part. Slow convergence is a severe difficulty associated with the direct Monte Carlo method and has led to the development of several variance reduction techniques. The importance sampling method, systematic sampling method, stratified sampling method, split sampling method, LHS method, conditional expectation method, and antithetic variates method are some of the popular variance reduction techniques [31]. Here, the importance sampling method is briefly introduced as an illustration of the concept of variance reduction techniques.

The importance sampling method is a modification of Monte Carlo simulation in which the simulation is biased for greater efficiency. In importance sampling, the sampling is done primarily in the tail of the distribution, rather than spreading it out evenly, in order to ensure that sufficient simulated failures occur.

The failure probability of Equation 3.16 can be written as

$$P_f = P[g(X) \leq 0] = \int \cdots \int \frac{I[X] f_X(X)}{f_X^*(X)} f_X^*(X) dX \tag{3.22}$$

where $f_X^*(X)$ is the importance sampling probability density function. The expectation function of the indicator function in Equation 3.22 can be written in the form

$$P_f = E[\frac{I(X) f_X(X)}{f_X^*(X)}] = \mu_{P_f} \tag{3.23}$$

Let $x_1^*, x_2^*, \ldots, x_{\hat{n}}^*$ denote random observations from the importance sampling function, $f_X^*(\cdot)$. Then an unbiased estimate of P_f is given by

$$\hat{P}_f^* = \frac{1}{\hat{n}} \sum_{i=1}^{\hat{n}} \frac{I(x_i^*) f_x(x_i^*)}{f_x^*(x_i^*)} \tag{3.24}$$

The choice of $f_x^*(\cdot)$ is quite important. If the density $f_x^*(\cdot)$ has been chosen so that there is an abundance of observations for which $I(x_i^*) = 1$, and if the ratio $f_x(x_i^*)/f_x^*(x_i^*)$ does not change much with different values of x_i^*, then $Var[\hat{P}_f^*]$ will be much less than $Var[\hat{P}_f]$ (Equation 3.20). Consequently, $\hat{P}_f^*$ requires many fewer observations than $\hat{P}_f$ (Equation 3.19) to achieve the same degree of precision.

The variance of $\hat{P}_f^*$ is given by

$$Var[\hat{P}_f^*] = \frac{1}{\hat{n}} \{ \int \cdots \int [\frac{I(x)f_x(x)}{f_x^*(x)}]^2 f_x^*(x)dx - \mu_{P_f}^2 \} \tag{3.25}$$

$$= \frac{1}{\hat{n}} \{ \int \cdots \int \frac{[I(x)]^2 [f_x(x)]^2}{f_x^*(x)} dx - \mu_{P_f}^2 \}$$

The ideal choice of $f_x^*(x)$ is obtained using calculus.

$$\frac{\partial}{\partial [f_x^*(x)]} \{Var[\hat{P}_f^*] + \lambda [\int \cdots \int f_x^*(x)dx - 1]\} = 0 \tag{3.26}$$

where λ is a Lagrange multiplier. This can be solved by using the calculus of variations. We obtain

$$f_X^*(x) = \frac{|I(x)f_x(x)|}{\int \cdots \int |I(x)f_x(x)|dx} \tag{3.27}$$

Substituting into Equation 3.25, it is easily found that

$$Var[\hat{P}_f^*] = \frac{1}{\hat{n}} \{ [\int \cdots \int |I(x)f_x(x)|dx]^2 - \mu_{P_f}^2 \} \tag{3.28}$$

If $|I(x)f_x(x)|$ is positive everywhere, the multiple integral is identical with μ_{P_f} and $Var[\hat{P}_f^*] = 0$. In this case the optimal function $f_x^*(\cdot)$ is

$$f_x^*(x) = \frac{I(x)f_x(x)}{\mu_{P_f}} \tag{3.29}$$

It can be seen that a good choice for $f_x^*(\cdot)$ can produce zero variance. Since μ_{P_f} is unknown, this is impossible. However, it demonstrates that if more effort is put into obtaining a close initial estimate of P_f, then the $Var[\hat{P}_f^*]$ will be much less than the variance of $\hat{P}_f$ in Equation 3.19. Conversely, the variance can actually be increased using a very poor choice for $f_x^*(\cdot)$. Thus, the application of importance sampling is sometimes referred to as an art that must be used with caution.

3.2.3 Latin Hypercube Sampling (LHS)

If there is an array of symbols or numbers and each occurs just once, the specific array is called a "Latin Square." The term "Hypercube" represents the extension of this concept to higher dimensions for many design variables. Therefore, LHS method, also known as the "Stratified Sampling Technique," represents a multivariate sampling method that guarantees non-overlapping designs. LHS, which has been successfully used to generate multivariate samples of statistical distributions, was first proposed by McKay, et al. [20]. In LHS, the distribution for each random variable can be subdivided into n equal probability intervals or bins. Each bin has one analysis point. There are n analysis points, randomly mixed, so each of the n bins has $1/n$ of the distribution probability. Figure 3.9 shows the basic steps for the general LHS method, which are:

    1) Divide the distribution for each variable into n non-overlapping intervals on the basis of equal probability.
    2) Select one value at random from each interval with respect to its probability density.
    3) Repeat steps 1) and 2) until you have selected values for all random variables, such as $x_1, x_2, ..., x_k$.
    4) Associate the n values obtained for each x_i with the n values obtained for the other $x_{j \neq i}$ at random.

The regularity of probability intervals on the probability distribution function ensures that each of the input variables has all portions of its range represented, resulting in relatively small variance in the response. At the same time, the analysis is much less computationally expensive to generate. The LHS method also provides flexible sample sizes while ensuring stratified sampling; i.e., each of the input variables is sampled at n levels.

To illustrate LHS, consider the following example. Suppose a design of LHS is desired to a sample size n of 4 from a normal distribution with a mean of 3.0, a standard deviation of 0.2. Let U_m be the random number of the standard uniform distribution (uniformly distributed between 0 and 1), where $m = 1, 2, 3,$ and 4. Each of the random numbers U_m should be scaled to obtain a corresponding cumulative probability, P_m:

$$P_m = \left(\frac{1}{n}\right) U_m + \left(\frac{m-1}{n}\right) \quad\quad\quad (3.30)$$

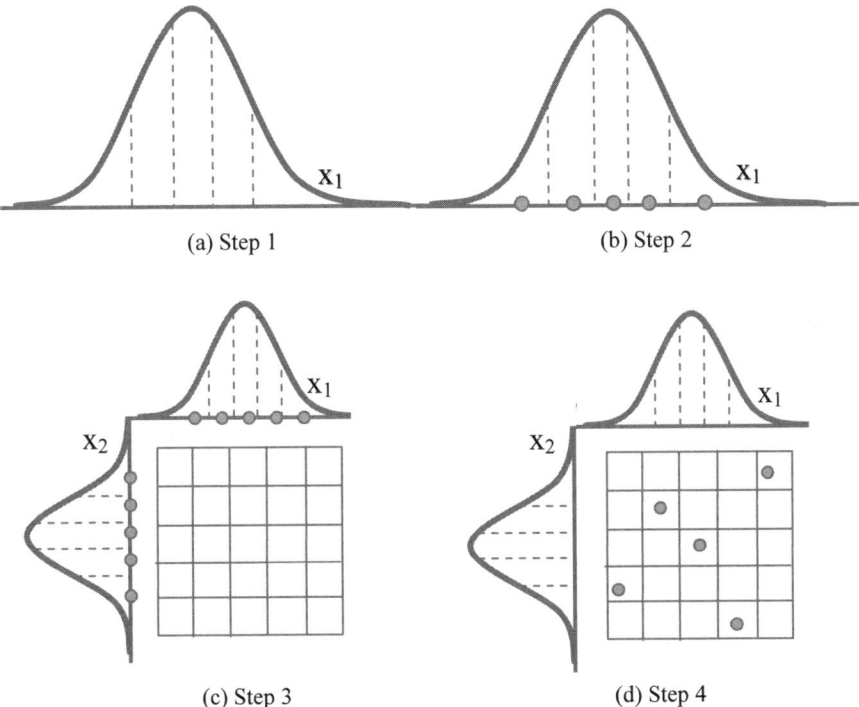

Figure 3.9. Basic Concept of LHS: Two Variables and Five Realizations

Table 3.1. LHS Design of Size 4, $N(3, 0.2)$

Interval No. (m)	Uniform Random No. (U_m)	Scaled Probability (P_m)	Corresponding Standard Normal Variables	Corresponding Normal Values $N(3, 0.2)$
1	0.2844	0.0711	-1.4676	2.7605
2	0.4692	0.3673	-0.3390	2.9322
3	0.0648	0.5162	0.0406	3.0081
4	0.9883	0.9971	2.7571	3.5514

Thus, each P_m lies within the mth interval: P_1, P_2, P_3, and P_4 will fall within each of the four intervals (0, 0.25), (0.25, 0.5), (0.5, 0.75), and (0.75, 1.0). The values P_m will be used to obtain the corresponding standard normal values (ξ) from the inverse normal distribution function. For instance, we generate a uniform random variable, namely $U_1 = 0.2844$. Then, the corresponding $P_1 = 0.0711$, $\xi_1 = \Phi^{-1}(P_1) = -1.4676$, and $x_1 = \mu_x + \sigma_x \xi_1 = 2.7605$. The complete results for the four intervals are summarized in Table 3.1.

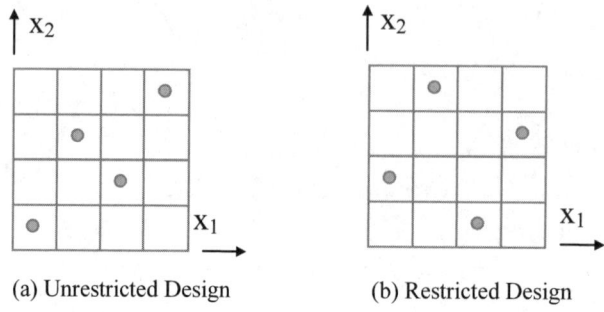

Figure 3.10. Two-dimensional LHS Designs for Four Realizations

The same procedure can be readily conducted for n-dimensional space. Therefore, the LHS method provides flexible sample sizes while ensuring stratified sampling (*i.e.*, each of the input variables is sampled at n levels). The remaining step of the LHS method is the pairing procedure for the generated n-dimensional variables. It should be noted that the random permutation of the pairing procedure (unrestricted pairing procedure) to generate an n-dimensional sampling set does not guarantee well-stratified designs. Figure 3.10a shows an unrestricted pairing result of the LHS design for two variables with four realizations, *i.e.*, (1, 1), (2, 3), (3, 2), and (4, 4), generated by the random number generator of a numerical package, such as MATLAB® or Mathematica®. The generated design shown in Figure 3.10a is highly correlated, and it requires proper estimation of error. On the other hand, Figure 3.10b shows a restricted design of the LHS, *i.e.*, (1, 2), (2, 4), (3, 1), and (4, 3), generated by a optimal sampling procedure [25],[31]. The restricted pairing procedure requires structuring the sampling points across the whole domain, and some situations require controlling the correlation structure of the generated sampling set. These issues trigger much research work to improve stratified designs and to control sample correlations. Chapter 6 discusses a controlling correlation method of restricted pairing procedures using the Karhunen-Loeve transform [5], but a more detailed description of other restricted pairing methods can be found in [18], [23], and [24].

3.3 Stochastic Finite Element Method (SFEM)

The Finite Element Method (FEM) is used in various fields of structural engineering and deals with deterministic parameters, despite the fact that the problems to which they are applied involve uncertainties of considerable degree. When there are requirements for analyzing stochastic behavior of structural systems with random parameters, one of the applicable solutions is the Stochastic Finite Element Method (SFEM). SFEM, which combines probability theory with deterministic FEM procedures, is becoming robust enough to allow engineers to estimate the risk of structural systems. This section describes the basic concept and offers an overview of SFEM.

3.3.1 Background

In general, the deterministic system can be governed by a set of partial differential equations and associated boundary conditions. If we do not have closed-form solutions of such a system, numerical analysis, which discretizes the geometry into a set of nodes, can be a possible approach to analyze the system. The field variable is defined in terms of FEM mesh nodal quantities. Thus, the set of partial differential equations is transformed to a system of equations in terms of nodal responses. Deterministic FEM in linear elasticity yields $Ku = f$, where K is the global stiffness matrix, u is the response vector of the structure, and f is the possible excitation of the system.

SFEM is an extension of deterministic FEM for considering the fluctuation of structural properties, loads, and responses of stochastic systems. SFEM, which combines probability theory and statistics within FEM, provides an efficient alternative to time-costly MCS and allows engineers to estimate the risk of structural designs. The main difference between SFEM and deterministic FEM is in the incorporation of randomness into the formulation. The basic formulation for SFEM is described in the literatures as [27], [28]

$$K = K_0 + \Delta K = \sum_{e=1}^{M_e} K_0^{(e)} + \sum_{e=1}^{M_e} \Delta K^{(e)} \tag{3.31}$$

The global stiffness matrix can be split into two parts: the deterministic part, K_0, and the fluctuation (uncertainty) part, ΔK. The global stiffness matrix K is obtained by assembling the element stiffness matrices $K^{(e)}$ of M_e elements.

From the linear finite element equation $Ku = f$, we get

$$(K_0 + \Delta K)u = f \tag{3.32}$$

where u is the random response of the structure and f is the possibly random excitation of the system.

There are four typical methods for formulating the fluctuation part of the SFEM procedure: the perturbation method, the Neumann expansion method, the weighted integral method, and the spectral stochastic finite element method. The following subsections describe corresponding details of each method.

3.3.2 Perturbation Method

The *perturbation method* is also known as the Taylor series expansion method and, generally, its effectiveness is restricted in that the random fluctuations must be small (*i.e.*, COV < 0.2). Hart and Collins [14] and Cambou [3] dealt with randomness in FEM modeling using first-order perturbation theory. Handa and

Anderson [13] obtained the variance of the structural response considering the first-order variance of force, displacement, and stiffness. Nakagiri and Hisada [22] indicate that the second-order perturbation is impractical for large-scale problems because of its huge computational effort in calculating the second-order function gradients.

Basic Formulations

The linear finite element equation is

$$Ku = f \tag{3.33}$$

where K, u, and f denote the stiffness matrix, nodal displacement vector, and nodal force vector, respectively. In general cases, the elements of K contain random parameters such as uncertain material properties. Therefore, the displacement or response vector u is random, since K contains random variables.

The displacement of the given system can be

$$u = K^{-1} f \tag{3.34}$$

And the derivative of displacement is obtained as

$$\left.\frac{\partial u}{\partial x_i}\right|_{x=\bar{x}} = \left.\frac{\partial (K^{-1} f)}{\partial x_i}\right|_{x=\bar{x}} = \frac{\partial K^{-1}}{\partial x_i} f(\bar{x}) + K^{-1}(\bar{x}) \left.\frac{\partial f}{\partial x_i}\right|_{x=\bar{x}} \tag{3.35}$$

where x_i are random variables and $\bar{x}$ is the mean of random variables.

To obtain $\dfrac{\partial K^{-1}}{\partial x_i}$, the relationship $KK^{-1} = I$ can be used:

$$\frac{\partial (KK^{-1})}{\partial x_i} = \frac{\partial I}{\partial x_i} = 0, \quad \frac{\partial K^{-1}}{\partial x_i} = -K^{-1} \frac{\partial K}{\partial x_i} K^{-1} \tag{3.36}$$

Substituting Equation 3.36 into Equation 3.35 yields

$$\left.\frac{\partial u}{\partial x_i}\right|_{x=\bar{x}} = -K^{-1}(\bar{x}) \frac{\partial K}{\partial x_i} u(\bar{x}) + K^{-1}(\bar{x}) \left.\frac{\partial f}{\partial x_i}\right|_{x=\bar{x}} \tag{3.37}$$

$$= K^{-1}(\bar{x}) \left(-\frac{\partial K}{\partial x_i} u(\bar{x}) + \left.\frac{\partial f}{\partial x_i}\right|_{x=\bar{x}} \right) = K^{-1}(\bar{x}) \bar{P}_i$$

where $\overline{P}_i = -\dfrac{\partial K}{\partial x_i} u(\overline{x}) + \dfrac{\partial f}{\partial x_i}\bigg|_{x=\overline{x}}$ (3.38)

The derivative of the total stiffness matrix $\dfrac{\partial K}{\partial x_i}$ is obtained by assembling the derivative of the element stiffness matrices.

According to the first-order Taylor series, the displacement vector, u, can be expanded about $\overline{x}$:

$$u - u(\overline{x}) = \sum_{i=1}^{n} \dfrac{\partial u}{\partial x_i}(x - \overline{x})$$ (3.39)

The covariance matrix of the displacement vector can be obtained from Equation 3.39 and Equation 3.37. Finally, we get

$$Cov(u,u) = E[(u-\overline{u})(u-\overline{u})^T] = \overline{K}^{-1}\overline{P}C_x\overline{P}^T\overline{K}^{-1}$$ (3.40)

where C_x is the covariance matrix of the random variable x.

3.3.3 Neumann Expansion Method

The *Neumann expansion method* has been used by several researchers within the framework of Monte Carlo simulation [27]. The main advantage of the Neumann expansion method is that the matrix K_0 has to be decomposed only once for all samples in conjunction with the Monte Carlo simulation. Due to this single matrix decomposition, computing time can be greatly reduced. However, determining the covariance matrix among all elements of the fluctuation part of the stiffness matrix requires extremely high computational effort [19].

Basic Procedure

Solving the stochastic problem given by Equation 3.32,

$$(K_0 + \Delta K)u = f$$ (3.41)

yields

$$u = (K_0 + \Delta K)^{-1} f$$ (3.42)

In the Neumann series expansion, the inverse of $K_0 + \Delta K$ yields

$$(K_0 + \Delta K)^{-1} = K_0^{-1} - K_0^{-1}\Delta K \ K_0^{-1} + (K_0^{-1}\Delta K)^2 K_0^{-1} - ... \qquad (3.43)$$

It is well-known that the Neumann expansion converges if the absolute values of all the eigenvalues of $P = K_0^{-1}\Delta K$ are less than 1 [19].

The inverse of the global stiffness matrix can be expanded by the Neumann series

$$(K_0 + \Delta K)^{-1} = (I - P + P^2 - P^3 + ...)K_0^{-1} \qquad (3.44)$$

$$= \sum_{k=0}^{\infty} (-K_0^{-1}\Delta K)^k K_0^{-1}$$

The displacement vector can also be represented by the series

$$u = (I - P + P^2 - P^3 + ...)K_0^{-1}f \qquad (3.45)$$

$$= u_0 - Pu_0 + P^2 u_0 - P^3 u_0 + ...$$

$$= u_0 - u_1 + u_2 - u_3 + ...$$

The series solution is equivalent to the following recursive formula

$$u_i = Pu_{i-1} \qquad (3.46)$$
$$K_0 u_i = \Delta K u_{i-1}, \ i = 1,2,......,n$$

The mean value and the covariance matrix using a first-order Neumann expansion, $u = (I - P)u_0$, are given by

$$E[u] = u_0 \qquad (3.47)$$
$$Cov[u,u] = E[Pu_0 u_0^T P^T]$$

C_{ij} of the covariance matrix can be calculated as

$$C_{ij} = \sum_{k=1}^{M}\sum_{l=1}^{M}\sum_{m=1}^{M}\sum_{n=1}^{M} K_{0ik}^{-1}(u_{0l}u_{0m}{}^T)K_{0nj}^{-1}E[(\Delta K_{kl})(\Delta K_{mn})] \qquad (3.48)$$

where M is the degree of freedom of the stiffness matrix.

3.3.4 Weighted Integral Method

The *weighted integral method* was proposed by Deodatis and Shinozuka [6], [7]. The main feature of this method is that it does not require discretization of the random field. Instead, everything is determined by the mesh chosen for the deterministic part of the stiffness matrix. The weighted integral method requires about ten times more computational effort than the perturbation scheme method [19].

Formulation of Weighted Integral Method

Suppose the Young's modulus has randomness and fluctuates over the element length according to

$$E(x) = E_0[1 + f(x)] \tag{3.49}$$

where E_0 is the mean value of the Young's modulus and $f(x)$ is a zero-mean homogeneous Gaussian random field.

In order to avoid the possibility of obtaining a negative value of the Young's modulus, $f(x)$ is assumed to be bounded as, say

$$-0.8 < f(x) < 0.8 \tag{3.50}$$

The stochastic stiffness matrix $K^{(e)}$ can be expressed as

$$K^{(e)} = K_0^{(e)} + \Delta K^{(e)} \tag{3.51}$$

where

$$K_0^{(e)} = \int_{\Omega e} B^T D_0 B \, d\Omega^e \quad \text{and} \tag{3.52}$$

$$\Delta K^{(e)} = \int_{\Omega e} f(x) B^T D_0 B \, d\Omega^e \tag{3.53}$$

Here, D is the elasticity matrix and B is the matrix that relates the strains and displacement in the deterministic FEM procedure.

The B matrix can be decomposed as

$$B = B_i P_i; \quad i = 1, 2, \ldots, N_P \tag{3.54}$$

where B_i is the constant matrix and P_i is the independent polynomial [28].

The weighted integral is defined as

$$X_i = \int_{\Omega_e} f(x) P_i d\Omega^e \ ; \quad i = 1, 2, \ldots, N_P \tag{3.55}$$

Using weighted integrals, the element stiffness matrix reads

$$K^{(e)} = K_0^{(e)} + \sum_{i=1}^{N_P} X_i^{(e)} \Delta K_i^{(e)} \tag{3.56}$$

By means of a first-order Taylor series expansion, the global approximation of u yields

$$u \approx u_0 + \sum_{e=1}^{Me} \sum_{i=1}^{N_P} \left. \frac{\partial u}{\partial X_i^{(e)}} \right|_E (X_i^{(e)} - \overline{X}_i^{(e)}) \tag{3.57}$$

where $\overline{X}_i^{(e)}$ denotes the mean value of the random variables $X_i^{(e)}$ and $|_E$ means evaluation at $\overline{X}_i^{(e)}$.

The partial derivative of the displacement vector $\dfrac{\partial u}{\partial X_i^{(e)}}$ can be derived by differentiating $Ku = f$

$$\frac{\partial u}{\partial X_i^{(e)}} = -K_0^{-1} \frac{\partial K}{\partial X_i^{(e)}} u_0 \tag{3.58}$$

Substituting Equation 3.58 into Equation 3.57, and assuming zero mean random variables, it follows that

$$u = u_0 + \sum_{e=1}^{Me} \sum_{i=1}^{N_P} K_0^{-1} \left. \frac{\partial K}{\partial X_i^{(e)}} \right|_E u_0 X_i^{(e)} \tag{3.59}$$

From Equation 3.59, the mean and covariance of u are readily determined:

$$E(u) = u_0 \tag{3.60}$$

$$Cov(u,u) = \sum_{e_1=1}^{Me} \sum_{e_2=1}^{Me} \sum_{i_1=1}^{N_P} \sum_{i_2=1}^{N_P} K_0^{-1} \left. \frac{\partial K}{\partial X_i^{(e)}} \right|_E u_0 u_0^T X_i^{(e)} \left. \frac{\partial K}{\partial X_i^{(e)}} \right|_E^T K_0^{-1} E[X_{i_1}^{(e_1)} X_{i_2}^{(e_2)}] \tag{3.61}$$

3.3.5 Spectral Stochastic Finite Element Method

As discussed in Section 3.1.2, the SSFEM, suggested by Ghanem and Spanos, is the intrusive formulation of the stochastic expansions (*i.e.*, KL expansion and PCE). In Chapter 6, a brief review of SSFEM follows a discussion of the details and important properties of KL expansion and PCE.

3.4 References

[1] Aerospace Information Report 5080, *Integration of Probabilistic Methods into the Design Process*, Society of Automotive Engineers, 1997.
[2] Babuska, I., Tempone, R., and Zouraris, G.E., "Galerkin Finite Element Approximations of Stochastic Elliptic Differential Equations," *SIAM Journal on Numerical Analysis*, Vol. 42, 2004, pp. 800-825.
[3] Cambou, B., "Application of First Order Uncertainty Analysis in the Finite Element Method in Linear Elasticity," *Proceedings of Second International Conference on Application of Statistics and Probability in Soil and Structural Engineering*, London, England, 1971.
[4] Cameron, R.H. and Martin, W.T., "The Orthogonal Development of Nonlinear Functionals in Series of Fourier-Hermite Functionals," *Annals of Mathematics*, Vol. 48, 1947, pp. 385-392.
[5] Choi, S., Canfield, R.A, and Grandhi, R.V., "Estimation of Structural Reliability for Gaussian Random Fields," *Structure & Infrastructure Engineering*, Nov. 2005 (In press).
[6] Deodatis, G., and Shinozuka, M., "Stochastic FEM Analysis of Nonlinear Dynamic Problems," *Stochastic Mechanics*, Vol. 3, Princeton University, Princeton, N.J., 1988, pp. 27-54.
[7] Deodatis, G., and Shinozuka, M., "Weighted Integral Method. II: Response Variability and Reliability," *Journal of Engineering Mechanics*, ASCE, Vo.117, (8), 1991, pp.1865–1877.
[8] Elishakoff I., *Safety Factors and Reliability: Friends or Foes?*, Kluwer Academic Publishers, Boston, 2004.
[9] Ghanem, R. and Spanos, P.D., *Stochastic Finite Elements: A Spectral Approach*, Springer-Verlag, NY, 1991.
[10] Ghanem, R. and Spanos, P.D., "Stochastic Finite Element Expansion for Random Media," *Journal of Engineering Mechanics*, Vol. 115, No. 5, 1989, pp. 1035-1053.
[11] Grandhi, R.V. and Wang, L.P., *Structural Reliability Analysis and Optimization: Use of Approximations*, NASA CR-1999-209154, 1999.
[12] Haldar, A., and Mahadevan, S., *Reliability Assessment Using Stochastic Finite Element Analysis*, John Wiley & Sons, NY, 2000.
[13] Handa, K., and Anderson, K., "Application of Finite Element Method in the Statistical Analysis of Structures," *International Conference on Sturctural Safety and Reliability (ICOSSAR)*, Elsevier, 1981.
[14] Hart, G. C. and Collins, J. D., "The Treatment of Randomness in Finite Element Modeling," *SAE Shock and Vibrations Symposium*, Los Angeles, CA, Oct. 1970, pp. 2509-2519.
[15] Hasofer, A. M, and Lind, N. C., "Exact and Invariant Second-Moment Code Format," *Journal of the Engineering Mechanics Division*, ASCE 100, EM1, 1974, pp.111-121.
[16] Isukapalli, S.S., *Uncertainty Analysis of Transport-Transformation Models*, Ph.D. Dissertation, Rutgers, the State University of New Jersey, New Brunswick, NJ, 1999.

[17] Madsen, H.O., Krenk, S., and Lind, N. C., *Methods of Structural Safety*, Prentice-Hall, Englewood Cliffs, New Jersey, 1986.
[18] Manteufel, R.D., "Distributed Hypercube Sampling Algorithm." *Proceedings of 42nd AIAA Structures, Structural Dynamics, and Materials Conference*. AIAA-01-1673. Apr. 2001.
[19] Matthies, H.G., Brenner, C.E., Bucher, C.G., and Soares, C.G., "Uncertainties in Probabilistic Numerical Analysis of Structures and Solids-Stochastic Finite Elements," *Structural Safety*, Vol.19, (3), 1997, pp. 283-336.
[20] McKay, M.D., Beckman, R.J., and Conover, W.J., "A Comparison of Three Methods for Selecting Values of Input Variables in the Analysis of Output from a Computer Code," *Technometrics*, Vol. 21, (2), 1979, pp. 239-245.
[21] Melchers, R. E., *Structural Reliability Analysis and Prediction*, Ellis Horwood Limited, UK., 1987.
[22] Nakagiri, S., and Hisada, T., "Stochastic Finite Element Method Applied to Structural Analysis with Uncertain Parameters," *Proceedings of the International Conference on FEM*, August 1982, pp. 206-211.
[23] Novák, D., Lawanwisut, W., Bucher, C., "Simulation of Random Fields Based on Orthogonal Transform of Covariance Matrix and Latin Hypercube Sampling," *Proceedings of International Conference on Monte Carlo Simulation MC 2000*, Monte Carlo, Monaco, June 2000, pp. 129-136.
[24] Owen A.B., "Controlling Correlations in Latin Hypercube Samples," *Journal of the American Statistical Association*, Vol. 89, (428), 1994, pp.1517-1522.
[25] Park, J. S., "Optimal Latin-Hypercube Designs for Computer Experiments," *Journal of Statistical Planning and Inference*, Vol. 39, (1), 1994, pp. 95-111.
[26] Pettit, C.L., Canfield, R.A., and Ghanem, R., "Stochastic Analysis of an Aeroelastic System," presented at *15th ASCE Engineering Mechanics Conference*, Columbia University, New York, NY, June 2-5, 2002.
[27] Schuëller G.I. (Ed.), "A State-of-the-Art Report on Computational Stochastic Mechanics," *Journal of Probabilistic Engineering Mechanics*, Vol. 12, (4), 1997, pp. 197-313.
[28] Shinozuka, M., and Deodatis, G., "Response Variability of Stochastic Finite Element Systems," *Stochasitc Mechanics*, Vol. 1, Department of Civil Engineering and Engineering Mechanics, Columbia University, New York, NY, 1986.
[29] Sobol I.M., *A Primer for the Monte Carlo Method*, CRC Press, 1994.
[30] Tatang, M.A., Direct Incorporation of Uncertainty in Chemical and Environmental Engineering Systems, Ph.D. Dissertation, Massachusetts Institute of Technology, Cambridge, MA, 1995.
[31] Wyss, G.D., and Jorgensen, K.H., "A User's Guide to LHS: Sandia's Latin Hypercube Sampling Software," SAND98-0210, Sandia National Lab. PO Box 5800, Albuquerque, New Mexico, 1998.
[32] Xiu, D., and Karniadakis, G., "The Wiener-Askey Polynomial Chaos for Stochastic Differential Equations," *SIAM Journal on Scientific Computing*, Vol. 24, (2), 2002, pp. 619-644.
[33] Xiu, D., Lucor, D. Su, C., Karniadakis, G., "Stochastic Modeling of Flow-Structure Interactions Using Generalized Polynomial Chaos," *Journal of Fluids Engineering*, Vol. 124, No. 51, 2002, pp. 51-59.

4

Methods of Structural Reliability

This chapter presents methods for two significant reliability measures: safety index and probability of failure. Because of the iterative nature of calculating these measures, use of limit-state function approximations is a necessary aspect. However, efficient selection of suitable approximations at different stages of reliability analysis makes these tools practical for many large-scale engineering problems. Also, the physical interpretation of sensitivity factors as used in design is discussed. At the end of the chapter, several engineering problems are presented with corresponding results for use as test cases.

4.1 First-order Reliability Method (FORM)

In principle, random variables are characterized by their first moment (mean), second moment (variance), and higher moments. Different ways of approximating the limit-state function form the basis for different reliability analysis algorithms (*i.e.*, FORM, SORM, *etc.*). In this section, we first discuss the first-order second moment method, and then the details of FORM, because the development of FORM can be traced to FOSM.

4.1.1 First-order Second Moment (FOSM) Method

The FOSM method, also referred to as the Mean Value FOSM (MVFOSM), simplifies the functional relationship and alleviates the complexities of the probability-of-failure calculation. The name "first-order" come from the first-order expansion of the function. As implied, inputs and outputs are expressed as the mean and standard deviation. Higher moments, which might describe skew and flatness of the distribution, are ignored.

In the MVFOSM method, the limit-state function is represented as the first-order Taylor series expansion at the mean value point. Assuming that the variables X are statistically independent, the approximate limit-state function at the mean is written as

$$\tilde{g}(X) \approx g(\mu_X) + \nabla g(\mu_X)^T (X_i - \mu_{X_i}) \tag{4.1}$$

where, $\mu_X = \{\mu_{x_1}, \mu_{x_2}, \ldots \mu_{x_n}\}^T$, and $\nabla g(\mu_X)$ is the gradient of g evaluated at μ_X,

$$\nabla g(\mu_X) = \left\{ \frac{\partial g(\mu_X)}{\partial x_1}, \frac{\partial g(\mu_X)}{\partial x_2}, \ldots, \frac{\partial g(\mu_X)}{\partial x_n} \right\}^T.$$

The mean value of the approximate limit-state function $\tilde{g}(X)$ is

$$\mu_{\tilde{g}} \approx E[g(\mu_X)] = g(\mu_X) \tag{4.2}$$

Because

$$\text{Var}[g(\mu_X)] = 0, \quad \text{Var}[\nabla g(\mu_X)] = 0 \tag{4.3}$$

$$\begin{aligned} \text{Var}[\nabla g(\mu_X)^T (X - \mu_X)] &= \text{Var}[\nabla g(\mu_X)^T X] - \mu_X \text{Var}[\nabla g(\mu_X)] \\ &= \text{Var}[(\nabla g(\mu_X)^T X] \\ &= [\nabla g(\mu_X)^T]^2 \text{Var}(X) \end{aligned} \tag{4.4}$$

The variance of the approximate limit-state function $\tilde{g}(X)$ is

$$\text{Var}[\tilde{g}(X)] \approx \text{Var}[g(\mu_X)] + \text{Var}[\nabla g(\mu_X)^T (X - \mu_X)] \tag{4.5}$$

Therefore, the standard deviation of the approximate limit-state function is

$$\begin{aligned} \sigma_{\tilde{g}} &= \sqrt{\text{Var}[\tilde{g}(X)]} = \sqrt{[\nabla g(\mu_X)^T]^2 \text{Var}(X)} \\ &= \left[\sum_{i=1}^{n} \left(\frac{\partial g(\mu_X)}{\partial x_i} \right)^2 \sigma_{x_i}^2 \right]^{\frac{1}{2}} \end{aligned} \tag{4.6}$$

The reliability index β is computed as:

$$\beta = \frac{\mu_{\tilde{g}}}{\sigma_{\tilde{g}}} \tag{4.7}$$

Equation 4.7 is the same as Equation 3.6 if the limit-state function is linear. If the limit-state function is nonlinear, the approximate limit-state surface is obtained by linearizing the original limit-state function at the mean value point. Therefore, this method is called the mean-value method, and the β given in Equation 4.7 is called a MVFOSM reliability index.

In a general case with independent variables of n-dimensional space, the failure surface is a hyperplane and can be defined as a linear-failure function:

$$\tilde{g}(X) = c_0 + \sum_{i=1}^{n} c_i x_i \qquad (4.8)$$

The reliability index given in Equation 4.7 can still be used for this n-dimensional case, in which

$$\mu_{\tilde{g}} = c_0 + c_1 \mu_{x_1} + c_2 \mu_{x_2} + \ldots + c_n \mu_{x_n} \qquad (4.9)$$

$$\sigma_{\tilde{g}} = \sqrt{\sum_{i=1}^{n} c_i^2 \sigma_{x_i}^2} \qquad (4.10)$$

The MVFOSM method changes the original complex probability problem into a simple problem. This method directly establishes the relationship between the reliability index and the basic parameters (mean and standard deviation) of the random variables via Equation 4.7. However, there are two serious drawbacks in the MVFOSM method:

1) Evaluation of reliability by linearizing the limit-state function about the mean values leads to erroneous estimates for performance functions with high nonlinearity, or for large coefficients of variation. This can be seen from the following mean value calculation of $\tilde{g}(X)$, which assumes that truncation of the Taylor series expansion for a case of only one random variable at the first three terms is

$$\tilde{g}(X) \approx g(\mu_X) + (X - \mu_X)\nabla g(\mu_X) + \frac{(X - \mu_X)^2}{2} \nabla g^2(\mu_X) \qquad (4.11)$$

The mean value of the approximate limit-state function $\tilde{g}(X)$ can be calculated as

$$\mu_{\tilde{g}} \approx E[g(\mu_X)] + E[(X - \mu_X)\nabla g(\mu_X)] + E\left[\frac{(X - \mu_X)^2}{2} \nabla g^2(\mu_X)\right] \qquad (4.12)$$

Because

$$E[g(\mu_X)] = g(\mu_X) \qquad (4.13a)$$

$$E[(X - \mu_X)\nabla g(\mu_X)] = E[(X\nabla g(\mu_X)] - E[\mu_X \nabla g(\mu_X)]$$

$$= \nabla g(\mu_X)E(X) - \mu_X \nabla g(\mu_X) = 0 \quad (4.13b)$$

$$E\left[\frac{(X-\mu_X)^2}{2}\nabla^2 g(\mu_X)\right] = \frac{1}{2}\nabla^2 g(\mu_X)E[(X-\mu_X)^2]$$

$$= \frac{1}{2}\nabla g^2(\mu_X)Var(X) \quad (4.13c)$$

From Equation 4.13c, it is obvious that the third term on the right side of Equation 4.11 depends on the variance of X and the second-order gradients of the limit-state function. If the variance of X is small or the limit-state function is close to linear, the third term of Equation 4.11 can be ignored and the mean value of $\tilde{g}(X)$ is the same as Equation 4.2. Otherwise, large errors in the mean value estimation will result.

2) The MVFOSM method fails to be invariant with different mathmatically equivalent formulations of the same problem. This is a problem not only for nonlinear forms of $g(\cdot)$, but also for certain linear forms. Example 4.2 shows that two different equivalent formulations of the limit-state function for the same problem result in different safety indices.

Example 4.1

The performance function is
$$g(x_1, x_2) = x_1^3 + x_2^3 - 18$$
in which x_1 and x_2 are the random variables with normal distributions (mean $\mu_{x_1} = \mu_{x_2} = 10$, standard deviation $\sigma_{x_1} = \sigma_{x_2} = 5$). Find the safety-index β by using the mean-value FOSM method, and check the accuracy of the obtained result with the MCS.

Solution:

The mean of the linearized performance function is

$$\mu_{\tilde{g}} = g(\mu_{x_1}, \mu_{x_2}) = 1982.0$$

From Equation 4.6, the standard deviation of the linearized performance function is

$$\sigma_{\tilde{g}} = \sqrt{\left(\frac{\partial g(\mu_{x_1}, \mu_{x_2})}{\partial x_1}\sigma_{x_1}\right)^2 + \left(\frac{\partial g(\mu_{x_1}, \mu_{x_2})}{\partial x_2}\sigma_{x_2}\right)^2}$$

$$= \sqrt{(3\times 10^2 \times 5.0)^2 + (3\times 10^2 \times 5.0)^2} = 2121.32$$

From Equation 4.7, the safety-index β is

$$\beta = \frac{\mu_{\tilde{g}}}{\sigma_{\tilde{g}}} = \frac{1982.0}{2121.32} = 0.9343$$

We can expect that the accuracy of the MVFOSM method is not acceptable, since the given limit-state function is highly nonlinear. The result of MCS (1,000,000 runs) yields $P_f = 0.005524$ and $\beta = 2.5412$.

Example 4.2

Consider Example 3.1 with the same structural and statistical properties. Investigate the invariant property of MVFOSM for two different formulations of the same limit-state function. Two different formulations of the limit-state function can be given as:

$$g_1(P, L, W, T) = WT - \frac{PL}{4}$$

$$g_2(P, L, W, T) = T - \frac{PL}{4W}$$

Solution:

The safety-index for the g_1 function is

$$\beta_1 = \frac{\mu_{g_1}}{\sigma_{g_1}}$$

$$= \frac{100 \times 10^{-6} \times 600 \times 10^3 - \frac{10 \times 8}{4}}{\sqrt{(-2 \times 2)^2 + (-2.5 \times 0.1)^2 + (600 \times 10^3 \times 2 \times 10^{-5})^2 + (100 \times 10^{-6} \times 10^5)^2}}$$

$$= 2.48$$

and the safety index for the function g_2 is

$$\beta_2 = \frac{\mu_{g_2}}{\sigma_{g_2}}$$

$$= \frac{600 \times 10^3 - \frac{10 \times 8}{4 \times 100 \times 10^{-6}}}{\sqrt{(-2 \times 10^4 \times 2)^2 + (-2.5 \times 10^3)^2 + (4 \times 10^4)^2 + (1 \times 10^5)^2}}$$

$$= 3.48$$

> β_1 and β_2 are different even though the above two limit-state equations are equivalent. This lack of invariance was overcome by the Hasofer and Lind method.

4.1.2 Hasofer and Lind (HL) Safety-index

Searching for the MPP on the limit-state surface is a key step in the HL method. The improvement of the HL method compared with the MVFOSM also comes from changing the expansion point from the mean value point to the MPP. In Section 3.1.1, Figure 3.1 shows how the reliability index could be interpreted as the measure of the distance from the origin to the failure surface. In the one-dimensional case, the standard deviation of the safety margin was conveniently used as the scale. To obtain a similar scale in the case of multiple variables, Hasofer and Lind [8] proposed a linear mapping of the basic variables into a set of normalized and independent variables, u_i.

Consider the fundamental case with the independent variables of strength, R, and stress, S, which are both normally distributed. First, Hasofer and Lind introduced the standard normalized random variables:

$$\hat{R} = \frac{R - \mu_R}{\sigma_R}, \quad \hat{S} = \frac{S - \mu_s}{\sigma_s} \tag{4.14}$$

where μ_R and μ_S are the mean values of random variables R and S, respectively, and σ_R and σ_S are the standard deviations of R and S, respectively.

Next, transform the limit-state surface $g(R, S) = R - S = 0$ in the original (R, S) coordinate system into the limit-state surface in the standard normalized $(\hat{R}, \hat{S})$ coordinate system,

$$g(R(\hat{R}), S(\hat{S})) = \hat{g}(\hat{R}, \hat{S}) = \hat{R}\sigma_R - \hat{S}\sigma_S + (\mu_R - \mu_S) = 0 \tag{4.15}$$

Here, the shortest distance from the origin in the $(\hat{R}, \hat{S})$ coordinate system to the failure surface $\hat{g}(\hat{R}, \hat{S}) = 0$ is equal to the safety-index, $\beta = \hat{O}P^* = (\mu_R - \mu_S)/\sqrt{\sigma_R^2 + \sigma_S^2}$, as shown in Figure 4.1. The point $P^*(\hat{R}^*, \hat{S}^*)$ on $\hat{g}(\hat{R}, \hat{S}) = 0$, which corresponds to this shortest distance, is often referred to as the MPP.

In a general case with normally distributed and independent variables of n-dimensional space, the failure surface is a nonlinear function:

$$g(X) = g(\{x_1, x_2, ... x_n\}^T) \tag{4.16}$$

Transform the variables into their standardized forms:

$$u_i = \frac{x_i - \mu_{x_i}}{\sigma_{x_i}} \qquad (4.17)$$

where μ_{x_i} and σ_{x_i} represent the mean and the standard deviation of x_i, respectively. The mean and standard deviation of the standard normally distributed variable, u_i, are zero and unity, respectively.

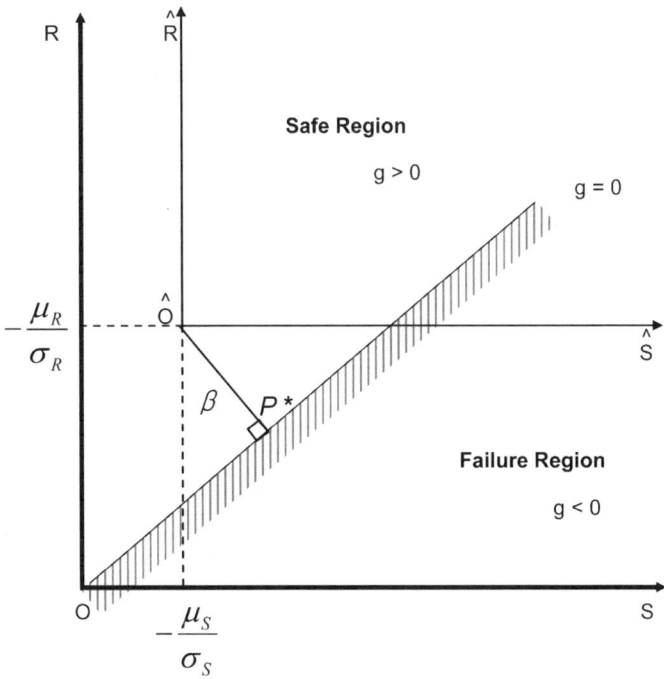

Figure 4.1. Geometrical Illustration of Safety-index

Any orthogonal distribution of standard normally distributed variables $U = \{u_1, u_2, ..., u_n\}^T$ results in a new set of normalized and uncorrelated variables. Therefore, the distributions of U are rotationally symmetric with respect to second moment distribution. Based on the transformation of Equation 4.17, the mean value point in the original space (X-space) is mapped into the origin of the normal space (U-space). The failure surface $g(X)=0$ in X-space is mapped into the corresponding failure surface $g(U)=0$ in U-space, as shown in Figure 3.2 and Figure 4.2. Due to the rotational symmetry of the second-moment representation of U, the geometrical distance from the origin in U-space to any point on $g(U)=0$ is simply the number of standard deviations from the mean value point in X-space to the corresponding point on $g(X)=0$. The distance to the failure surface can then be measured by the safety-index function:

$$\beta(U) = (U^T U)^{1/2} = \|U\|_2, U \in g(U) = 0 \tag{4.18}$$

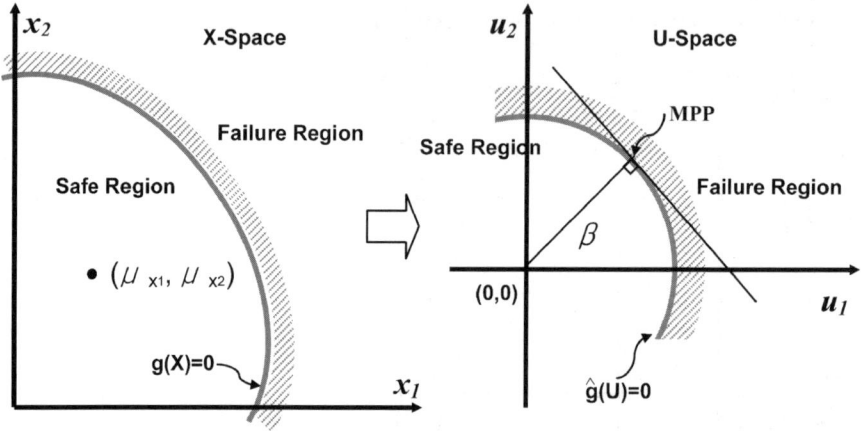

Figure 4.2. Mapping of Failure Surface from X-space to U-space

The safety-index β is the shortest distance from the origin to the failure surface $g(U) = 0$, i.e.,

$$\beta = \min_{U \in g(U) = 0} (U^T U)^{\frac{1}{2}} \tag{4.19}$$

This safety-index is also called the Hasofer and Lind (HL) safety-index, β_{HL}. The point $U^*(u_1^*, u_2^*, ..., u_n^*)$ on $g(U)=0$ is the design point. The values-of-safety indices given in Equations 4.7 and 4.19 are the same when the failure surface is a hyperplane. The Hasofer and Lind reliability index can also be interpreted as a FOSM reliability index. The value of β_{HL} is the same for the true failure surface as well as for the approximate tangent hyperplane at the design point. The ambiguity in the value of the first-order reliability index is thus resolved when the design point is taken as the linearization point. The resultant reliability index is a sensible measure for the distance to the failure surface.

4.1.3 Hasofer and Lind Iteration Method

Equation 4.19 shows that safety-index, β, is the solution of a constrained optimization problem in the standard normal space.

Minimize: $\beta(U) = (U^T U)^{\frac{1}{2}}$ (4.20a)

Subject to: $g(U) = 0$ (4.20b)

There are many algorithms available that can solve this problem, such as mathematical optimization schemes or other iteration algorithms. In [6], several constrained optimization methods were used to solve this optimization problem, including primal methods (feasible directions, gradient, projection, reduced gradient), penalty methods, dual methods, and Lagrange multiplier methods. Each method has its advantages and disadvantages, depending upon the attributes of the method and the nature of the problem. In the following description, the most commonly used recursive algorithms, the HL and HL-RF methods, are introduced to solve the reliability problems.

The HL method was proposed by Hasofer and Lind. Rackwitz and Fiessler extended the HL method to include random variable distribution information, calling their extended method the HL-RF method. Assuming that the limit-state surface with n-dimensional normally distributed and independent random variables X is

$$g(X) = g(\{x_1, x_2, ..., x_n\}^T) = 0 \tag{4.21}$$

This limit-state function can be linear or nonlinear. Based on the transformation given in Equation 4.17, the limit-state function given in Equation 4.21 is transformed into

$$g(U) = g(\{\sigma_{x_1} u_1 + \mu_{x_1}, \sigma_{x_2} u_2 + \mu_{x_2}, ..., \sigma_{x_n} u_n + \mu_{x_n}\}^T) = 0 \tag{4.22}$$

The normal vector from the origin $\hat{O}$ to the limit-state surface $g(U)$ generates an intersection point P^* as shown in Figure 4.1 and Figure 4.2. The distance from the origin to the MPP is the safety-index β. The first-order Taylor series of expansion of $g(U)$ at the MPP U^* is

$$\tilde{g}(U) \approx g(U^*) + \sum_{i=1}^{n} \frac{\partial g(U^*)}{\partial U_i}(u_i - u_i^*) \tag{4.23}$$

where k denotes the iteration number of the recursive algorithm.

From the transformation of Equation 4.17, we have

$$\frac{\partial \hat{g}(U)}{\partial u_i} = \frac{\partial g(X)}{\partial x_i} \sigma_{x_i} \tag{4.24}$$

The shortest distance from the origin to the above approximate failure surface given in Equation 4.23 is

$$\hat{OP}^* = \beta = \frac{g(U^*) - \sum_{i=1}^{n} \frac{\partial g(U^*)}{\partial x_i} \sigma_{x_i} u_i^*}{\sqrt{\sum_{i=1}^{n} (\frac{\partial g(U^*)}{\partial x_i} \sigma_{x_i})^2}} \qquad (4.25)$$

The direction cosine of the unit outward normal vector is given as

$$\cos\theta_{x_i} = \cos\theta_{u_i} = -\frac{\frac{\partial g(U^*)}{\partial u_i}}{|\nabla g(U^*)|} \qquad (4.26)$$

$$= -\frac{\frac{\partial g(X^*)}{\partial x_i}\sigma_{x_i}}{\left[\sum_{i=1}^{n}(\frac{\partial g(X^*)}{\partial x_i}\sigma_{x_i})^2\right]^{1/2}} = \alpha_i$$

where α_i expresses the relative effect of the corresponding random variable on the total variation. Thus, it is called the *sensitivity factor*. More details about α_i will be given later in Section 4.1.4.

The coordinates of the point P^* are computed as

$$u_i^* = \frac{x_i^* - \mu_{x_i}}{\sigma_{x_i}} = \hat{OP}^* \cos\theta_{x_i} = \beta\cos\theta_{x_i} \qquad (4.27)$$

The coordinates corresponding to P^* in the original space are

$$x_i^* = \mu_{x_i} + \beta\sigma_{x_i}\cos\theta_{x_i}, (i=1,2,\ldots,n) \qquad (4.28)$$

Since P^* is a point on the limit-state surface,

$$g(\{x_1^*, x_2^*, \ldots x_n^*\}^T) = 0 \qquad (4.29)$$

The main steps of the HL iteration method are:

1) Define the appropriate limit-state function of Equation 4.21
2) Set the mean value point as an initial design point, i.e, $x_{i,k} = \mu_{x_i}$ $i=1,2,\ldots,n$, and compute the gradients $\nabla g(X_k)$ of the limit-state function at this point. Here, $x_{i,k}$ is the ith element in the vector X_k of the kth iteration

3) Compute the initial β using the mean-value method (Cornell safety-index), i.e., $\beta = \mu_{\tilde{g}}/\sigma_{\tilde{g}}$ and its direction cosine
4) Compute a new design point X_k and U_k (Equations 4.27 and 4.28), function value, and gradients at this new design point
5) Compute the safety-index β using Equation 4.25 and the direction cosine or sensitivity factor from Equation 4.26
6) Repeat steps 4)~6) until the estimate of β converges
7) Compute the coordinates of the design point X_k or most probable failure point (MPP), X^*

In some cases, the failure surface may contain several points corresponding to stationary values of the reliability-index function. Therefore, it may be necessary to use several starting points to find all the stationary values $\beta_1, \beta_2, ..., \beta_m$. This is called a *multiple MPP* problem.

The HL safety-index is

$$\beta_{HL} = \min\{\beta_1, \beta_2, ..., \beta_m\} \tag{4.30}$$

From Equations 4.7 and 4.25, the difference between the MVFOSM method and the HL method is that the HL method approximates the limit-state function using the first-order Taylor expansion at the design point $X^{(k)}$ or $U^{(k)}$ instead of the mean value point μ_X. Also, the MVFOSM method does not require iterations, while the HL method needs several iterations to converge for nonlinear problems. The HL method usually provides better results than the mean-value method for nonlinear problems. How well a linearized limit-state function, $\tilde{g}(U) = 0$, approximates a nonlinear function $g(U)$ in terms of the failure probability P_f depends on the shape of $g(U)=0$. If it is concave towards the origin, P_f is underestimated by the hyperplane approximation. Similarly, a convex function implies overestimation. However, there is no guarantee that the HL algorithm converges in all situations. Furthermore, the HL method only considers normally distributed random variables, so it cannot be used for non-Gaussian random variables.

Example 4.3a

Solve the safety-index β of Example 4.1 by using the HL method with the same performance function, mean values, standard deviations, and distributions of the random variables.

Solution:

(1) Iteration 1:

(a) Set the mean value point as an initial design point and set the required β convergence tolerance to $\varepsilon_r = 0.001$. Compute the limit-state function value and gradients at the mean value point:

$$g(X_1) = g(\mu_{x_1}, \mu_{x_2}) = \mu_{x_1}^3 + \mu_{x_2}^3 - 18$$
$$= 10.0^3 + 10.0^3 - 18 = 1982.0$$

$$\frac{\partial g}{\partial x_1}\bigg|_{\mu_X} = 3\mu_{x_1}^2 = 3 \times 10^2 = 300 \ , \ \frac{\partial g}{\partial x_2}\bigg|_{\mu_X} = 3\mu_{x_2}^2 = 3 \times 10^2 = 300$$

(b) Compute the initial β using the mean-value method and its direction cosine α_i

$$\beta_1 = \frac{\mu_{\tilde{g}}}{\sigma_{\tilde{g}}}$$

$$= \frac{g(X_1)}{\sqrt{(\frac{\partial g(\mu_{x_1}, \mu_{x_2})}{\partial x_1}\sigma_{x_1})^2 + (\frac{\partial g(\mu_{x_1}, \mu_{x_2})}{\partial x_2}\sigma_{x_2})^2}}$$

$$= \frac{1982.00}{\sqrt{(300 \times 5.0)^2 + (300 \times 5.0)^2}}$$

$$= 0.9343$$

$$\alpha_1 = -\frac{\frac{\partial g}{\partial x_1}\bigg|_{\mu_X}\sigma_{x_1}}{\sqrt{(\frac{\partial g(\mu_{x_1}, \mu_{x_2})}{\partial x_1}\sigma_{x_1})^2 + (\frac{\partial g(\mu_{x_1}, \mu_{x_2})}{\partial x_2}\sigma_{x_2})^2}}$$

$$= -\frac{300 \times 5.0}{\sqrt{(300 \times 5.0)^2 + (300 \times 5.0)^2}}$$

$$= -0.7071$$

$$\alpha_2 = -\frac{\frac{\partial g}{\partial x_2}\bigg|_{\mu_X}\sigma_{x_2}}{\sqrt{(\frac{\partial g(\mu_{x_1}, \mu_{x_2})}{\partial x_1}\sigma_{x_1})^2 + (\frac{\partial g(\mu_{x_1}, \mu_{x_2})}{\partial x_2}\sigma_{x_2})^2}}$$

$$= -\frac{300 \times 5.0}{\sqrt{(300 \times 5.0)^2 + (300 \times 5.0)^2}}$$

$$= -0.7071$$

(c) Compute a new design point X_2 from Equation 4.28

$$x_{1,2} = \mu_{x_1} + \beta_1 \sigma_{x_1} \alpha_1 = 10.0 + 0.9343 \times 5.0 \times (-0.7071) = 6.6967$$

$$x_{2,2} = \mu_{x_2} + \beta_1 \sigma_{x_2} \alpha_2 = 10.0 + 0.9343 \times 5.0 \times (-0.7071) = 6.6967$$

$$u_{1,2} = \frac{x_{1,2} - \mu_{x_1}}{\sigma_{x_1}} = \frac{6.6967 - 10.0}{5.0} = -0.6607$$

$$u_{2,2} = \frac{x_{2,2} - \mu_{x_2}}{\sigma_{x_2}} = \frac{6.6967 - 10.0}{5.0} = -0.6607$$

(2) Iteration 2:

(a) Compute the limit-state function and its gradient at X_2

$$g(X_2) = (x_{1,2})^3 + (x_{2,2})^3 - 18 = 6.6967^3 + 6.6967^3 - 18 = 582.63$$

$$\left.\frac{\partial g}{\partial x_1}\right|_{X_2} = 3 \times (x_{1,2})^2 = 3 \times 6.6967^2 = 134.5374$$

$$\left.\frac{\partial g}{\partial x_2}\right|_{X_2} = 3 \times (x_{2,2})^2 = 3 \times 6.6967^2 = 134.5374$$

(b) Compute β using Equation 4.25 and the direction cosine α_i

$$\beta_2 = \frac{g(X_2) - \sum_{i=1}^{2} \frac{\partial g(X_2)}{\partial x_i} \sigma_{x_i} u_{i,2}}{\sqrt{\sum_{i=1}^{2} (\frac{\partial g(X_2)}{\partial x_i} \sigma_{x_i})^2}}$$

$$= \frac{582.63 - 134.5374 \times 5.0 \times (-0.6607) - 134.5374 \times 5.0 \times (-0.6607)}{\sqrt{(134.5347 \times 5.0)^2 + (134.5347 \times 5.0)^2}}$$

$$= 1.5468$$

$$\alpha_1 = -\frac{\left.\frac{\partial g}{\partial x_1}\right|_{X_2} \sigma_{x_1}}{\sqrt{(\left.\frac{\partial g}{\partial x_1}\right|_{X_2} \sigma_{x_1})^2 + (\left.\frac{\partial g}{\partial x_2}\right|_{X_2} \sigma_{x_2})^2}}$$

$$= -\frac{134.5374 \times 5.0}{\sqrt{(134.5374 \times 5.0)^2 + (134.5374 \times 5.0)^2}}$$

$$= -0.7071$$

$$\alpha_1 = \alpha_2 = -0.7071$$

(c) Compute a new design point X_3

$$x_{1,3} = \mu_{x_1} + \beta_2 \sigma_{x_1} \alpha_1 = 10.0 + 1.5468 \times 5.0 \times (-0.7071) = 4.5313$$

$$x_{2,3} = x_{1,3} = 4.5313$$

$$u_{1,3} = \frac{x_{1,3} - \mu_{x_1}}{\sigma_{x_1}} = \frac{4.5313 - 10.0}{5.0} = -1.0937$$

$$u_{2,3} = u_{1,3} = -1.0937$$

(d) Check β convergence

$$\varepsilon = \frac{|\beta_2 - \beta_1|}{\beta_1} = \frac{1.5468 - 0.9343}{0.9343} = 0.6556$$

Since $\varepsilon > \varepsilon_r$, continue the process.

Table 4.1. Iteration Results in the HL Method (Example 4.3a)

Iteration No.	1	2	3	4	5	6	7
$g(X_k)$	1982.0	582.63	168.08	45.529	10.01	1.1451	0.023
$\frac{\partial g}{\partial x_1}\|_{X_k}$	300	134.5374	61.598	30.0897	17.43	13.5252	12.9917
$\frac{\partial g}{\partial x_2}\|_{X_k}$	300	134.5374	61.598	30.0897	17.43	13.5252	12.9917
β	0.9343	1.5468	1.9327	2.1467	2.2279	2.2398	2.2401
α_1	-0.7071	-0.7071	-0.7071	-0.7071	-0.7071	-0.7071	-0.771
α_2	-0.7071	-0.7071	-0.7071	-0.7071	-0.7071	-0.7071	-0.7071
$x_{1,k}$	6.6967	4.5313	3.1670	2.4104	2.1233	2.0810	2.0801
$x_{2,k}$	6.6967	4.5313	3.1670	2.4104	2.1233	2.0810	2.0801
$u_{1,k}$	-0.6607	-1.0937	-1.3666	-1.5179	-1.5753	-1.5838	-1.5840
$u_{2,k}$	-0.6607	-1.0937	-1.3666	-1.5179	-1.5753	-1.5838	-1.5840
ε	-	0.6556	0.2495	0.1107	0.036	0.005	0.0001

The same procedures can be repeated until the stopping criterion ($\varepsilon < \varepsilon_r$) is satisfied. The iteration results are summarized in Table 4.1. The safety-index β is 2.2401. Since the limit-state function value at the MPP, X^*, is close to zero, this safety-index can be considered as the shortest distance from the origin to the limit surface. Compared with the safety-index $\beta = 0.9343$ obtained from the MVFOSM method given in Example 4.1, the safety-index computed from the HL method is much more accurate for this highly nonlinear problem.

Example 4.3b

Solve the safety-index β of Example 4.1 by using the HL method and the mean value of $x_2 = 9.9$ instead of $x_2 = 10.0$. The other properties remain the same.

Solution:

In this example, the performance function, the mean value of x_1, the standard deviations, and the distributions of both random variables are the same as in Example 4.1. The only difference between Example 4.3a and 4.3b is that the mean value of x_2 is 9.9 instead of 10.0.

(1) Iteration 1:

(a) Set the mean value point as an initial design point and set the required β convergence tolerance to $\varepsilon_r = 0.001$. Compute the limit-state function value and gradient at the mean value point.

$$g(X_1) = g(\mu_{x_1}, \mu_{x_2}) = \mu_{x_1}^3 + \mu_{x_2}^3 - 18$$

$$= 10.0^3 + 9.9^3 - 18 = 1952.299$$

$$\frac{\partial g}{\partial x_1}\bigg|_{\mu_x} = 3\mu_{x_1}^2 = 3 \times 10^2 = 300$$

$$\frac{\partial g}{\partial x_2}\bigg|_{\mu_x} = 3\mu_{x_2}^2 = 3 \times 9.9^2 = 294.03$$

(b) Compute the initial β value using the mean-value method and its direction cosine α_i

$$\beta_1 = \frac{\mu_{\tilde{g}}}{\sigma_{\tilde{g}}}$$

$$= \frac{g(X_1)}{\sqrt{\left(\frac{\partial g(\mu_{x_1}, \mu_{x_2})}{\partial x_1}\sigma_{x_1}\right)^2 + \left(\frac{\partial g(\mu_{x_1}, \mu_{x_2})}{\partial x_2}\sigma_{x_2}\right)^2}}$$

$$= \frac{1952.299}{\sqrt{(300 \times 5.0)^2 + (294.03 \times 5.0)^2}}$$

$$= 0.9295$$

$$\alpha_1 = -\frac{\frac{\partial g}{\partial x_1}\bigg|_{\mu_x}\sigma_{x_1}}{\sqrt{\left(\frac{\partial g(\mu_{x_1}, \mu_{x_2})}{\partial x_1}\sigma_{x_1}\right)^2 + \left(\frac{\partial g(\mu_{x_1}, \mu_{x_2})}{\partial x_2}\sigma_{x_2}\right)^2}}$$

$$= -\frac{300 \times 5.0}{\sqrt{(300 \times 5.0)^2 + (294.03 \times 5.0)^2}}$$

$$= -0.7142$$

$$\alpha_2 = -\frac{\frac{\partial g}{\partial x_2}\big|_{\mu_X} \sigma_{x_2}}{\sqrt{(\frac{\partial g(\mu_{x_1},\mu_{x_2})}{\partial x_1}\sigma_{x_1})^2 + (\frac{\partial g(\mu_{x_1},\mu_{x_2})}{\partial x_2}\sigma_{x_2})^2}}$$

$$= -\frac{294.03 \times 5.0}{\sqrt{(300 \times 5.0)^2 + (294.03 \times 5.0)^2}}$$

$$= -0.7000$$

Table 4.2. Iteration Results in the HL Method (Example 4.3b)

Iteration No.	1	2	....	21	22	23
$g(X_k)$	1952.299	573.8398	....	678.9088	676.7346	677.655
$\frac{\partial g}{\partial x_1}\big|_{X_k}$	300	133.8982	....	218.0401	56.9049	217.6582
$\frac{\partial g}{\partial x_2}\big|_{X_k}$	294.03	132.5409	....	54.4352	216.2786	54.61056
β	0.9295	1.5387	....	1.1636	1.1650	1.1657
α_1	-0.7142	-0.7107	....	-0.9702	-0.2544	-0.9699
α_2	-0.7000	-0.7035	....	-0.2422	-0.9671	-0.2434
$x_{1,k}$	6.6808	4.5323	....	4.3553	8.5178	4.3468
$x_{2,k}$	6.6468	4.4877	....	8.4908	4.2666	8.4816
$u_{1,k}$	-0.6638	-1.0935	....	-1.1289	-0.2964	-1.1306
$u_{2,k}$	-0.6506	-1.0825	....	-0.2818	-1.1267	-0.2837
ε	-	0.6554	....	0.002	0.0012	0.0006

(c) Compute a new design point X_2 from Equation 4.28

$$x_{1,2} = \mu_{x_1} + \beta_1 \sigma_{x_1} \alpha_1 = 10.0 + 0.9295 \times 5.0 \times (-0.7142) = 6.6808$$

$$x_{2,2} = \mu_{x_2} + \beta_1 \sigma_{x_2} \alpha_2 = 9.9 + 0.9295 \times 5.0 \times (-0.7000) = 6.6468$$

$$u_{1,2} = \frac{x_{1,2} - \mu_{x_1}}{\sigma_{x_1}} = \frac{6.6808 - 10.0}{5.0} = -0.6638$$

$$u_{2,2} = \frac{x_{2,2} - \mu_{x_2}}{\sigma_{x_2}} = \frac{6.6468 - 9.9}{5.0} = -0.6506$$

The same procedures can be repeated until the stopping criterion ($\varepsilon < \varepsilon_r$) is satisfied. The iteration results are summarized in Table 4.2. The safety-index converges after 23 iterations; however, the MPP is not on the limit-state surface ($g(X^*) = 677.655$). Also, beginning with iterations 21, 22, and 23, the design point X^* oscillates. If a convergence check to determine whether or not the MPP is on the surface is added, the process will continue. However, no final MPP on the surface can be found, even after hundreds of iterations, due to the oscillation. From this example, it is clear that the HL method may not converge in some cases due to its linear approximation. A more efficient method will be used to deal with this problem, and the correct safety-index for this example will be given in Example 4.6.

4.1.4 Sensitivity Factors

As mentioned in Section 4.1.3, the direction cosine of the unit outward normal vector of the limit-state function α_i, given in Equation 4.26, is defined as the sensitivity factor. The sensitivity factor shows the relative importance of each random variable to the failure probability. The sensitivity of the failure probability or the safety index to small changes in the random variables can be examined, which usually provides information useful to studying the statistical variation of the response.

In Equation 4.26, the physical meaning of α_i implies the relative contribution of each random variable to the failure probability (Figure 4.3). For example, the larger the α_i value is, the higher the contribution towards the failure probability. This is due to

$$\alpha_1^2 + \alpha_2^2 + \ldots + \alpha_n^2 = 1 \tag{4.31}$$

In fact, α_i is the sensitivity of the safety-index β at the MPP. From the definition of β as the distance from the origin to the limit-state surface, $g(U) = 0$, it follows that

$$\frac{\partial \beta}{\partial u_i} = \frac{\partial}{\partial u_i} \sqrt{u_1^2 + u_2^2 + \ldots + u_n^2} = \frac{u_i}{\beta} = \alpha_i, \ (i = 1,2,\ldots,n) \tag{4.32}$$

The sensitivity factors for the failure probability P_f are

$$\frac{\partial \beta}{\partial u_i} = \frac{\partial}{\partial u_i} \Phi(-\beta) = \phi(-\beta) \frac{\partial \beta}{\partial u_i} \quad (4.33)$$

where $\phi(\cdot)$ represents the standard normal density function.

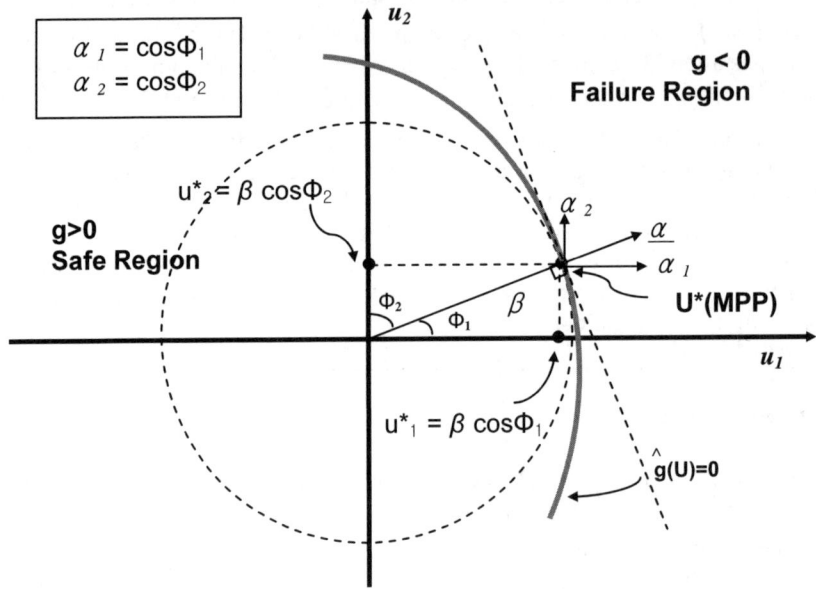

Figure 4.3. Sensitivity Factors

In Equation 4.26, $\partial g(X)/\partial x_i$ represents the sensitivity of the performance function $g(X)$, which measures the change in the performance function resulting from changes in the physical random variables. However, the sensitivity of the safety-index β represents both the change in the safety-index due to the change in random variables and the uncertainty. That is, it depends on the sensitivity of the performance function, $\partial g(X)/\partial x_i$ and on the standard deviation of random variable σ_i (from Equation 4.26).

In summary, computing α_i provides the sensitivity of the safety-index with respect to u_i, which has two major purposes. First, these sensitivity factors show the contributions of the random variables to the safety-index or failure probability. Second, the sign of the sensitivity factor shows the relationship between the performance function and the physical variables. A positive α_i means that the performance function $g(U)$ decreases as the random variable increases, and a negative factor means $g(U)$ increases as the random variable increases.

The sensitivity of the failure probability will have the same direction, but α_i is multiplied by the probability-density function value (Equation 4.33). For example, the limit-state function is

$$\overline{S} - S(X) = g(X) \tag{4.34}$$

where $\overline{S}$ is the allowable stress and $S(X)$ is the structural stress. If α_i is positive, it means $\partial g / \partial x_i$ is negative and $\partial S(X) / \partial x_i$ is positive. In very simplified terms, if the random variable value increases, both $S(X)$ and the failure probability increase.

4.1.5 Hasofer Lind - Rackwitz Fiessler (HL-RF) Method

In the Hasofer-Lind method, the random variables X are assumed to be normally distributed. In non-Gaussian cases, even when the limit-state function $g(X)$ is linear, the structural probability calculation given in Equation 3.9 is inappropriate. However, many structural reliability problems involve non-Gaussian random variables. It is necessary to find a way to solve these non-Gaussian problems. There are many methods available for conducting the transformations, such as Rosenblatt [16], and Hohenbichler and Rackwitz [10]. A simple, approximate transformation called the equivalent normal distribution, or the normal tail approximation, is described below. The main advantages of this transformation are:

1) It does not require the multi-dimensional integration
2) Transformation of non-Gaussian variables into equivalent normal variables has been accomplished prior to the solution of Equations 4.21 – 4.29
3) Equation 3.9 for calculation of the structural probability is retained
4) It often yields excellent agreement with the exact solution of the multi-dimensional integral of probability formula.

When the variables are mutually independent, the transformation is given as

$$u_i = \Phi^{-1}[F_{x_i}(x_i)] \tag{4.35}$$

where $\Phi^{-1}[.]$ is the inverse of $\Phi[.]$

One way to get the equivalent normal distribution is to use the Taylor series expansion of the transformation at the MPP X^*, neglecting nonlinear terms [13],

$$u_i = \Phi^{-1}[F_{x_i}(x_i^*)] + \frac{\partial}{\partial x_i}([\Phi^{-1}F_{x_i}(x_i)])|_{x_i^*} (x_i - x_i^*) \tag{4.36}$$

where

$$\frac{\partial}{\partial x_i}\Phi^{-1}[F_{x_i}(x_i)] = \frac{f_{x_i}(x_i)}{\phi(\Phi^{-1}[F_{x_i}(x_i)])} \qquad (4.37)$$

Upon substituting (4.37) into (4.36) and rearranging,

$$u_i = \frac{x_i^* - [x_i^* - \Phi^{-1}[F_{x_i}(x_i^*)]\phi(\Phi^{-1}[F_{x_i}(x_i^*)])/f_{x_i}(x_i^*)]}{\phi(\Phi^{-1}[F_{x_i}(x_i^*)])/f_{x_i}(x_i^*)} \qquad (4.38a)$$

which can be written as

$$u_i = \frac{x_i - \mu_{x_i'}}{\sigma_{x_i'}} \qquad (4.38b)$$

where $F_{x_i}(x_i)$ is the marginal cumulative distribution function, $f_{x_i}(x_i)$ is the probability density function, and $\mu_{x_i'}$ and $\sigma_{x_i'}$ are the equivalent means and standard deviations of the approximate normal distributions, and which are given as

$$\sigma_{x_i'} = \frac{\phi(\Phi^{-1}[F_{x_i}(x_i^*)])}{f_{x_i}(x_i^*)} \qquad (4.39a)$$

$$\mu_{x_i'} = x_i^* - \Phi^{-1}[F_{x_i}(x_i^*)]\sigma_{x_i'} \qquad (4.39b)$$

Another way to get equivalent normal distributions is to match the cumulative distribution functions and probability density function of the original, non-normal random variable distribution, and the approximate or equivalent normal random variable distributions at the MPP [16]. Assuming that x_i' is an equivalent normally distributed random variable, the cumulative distribution function values of x_i and x_i' are equal:

$$F_{x_i}(x_i^*) = F_{x_i'}(x_i^*) \qquad (4.40)$$

or $F_{x_i}(x_i^*) = \Phi(\frac{x_i^* - \mu_{x_i'}}{\sigma_{x_i'}}) \qquad (4.41)$

so

$$\mu_{x_i'} = x_i^* - \Phi^{-1}[F_{x_i}(x_i^*)]\sigma_{x_i'} \qquad (4.42)$$

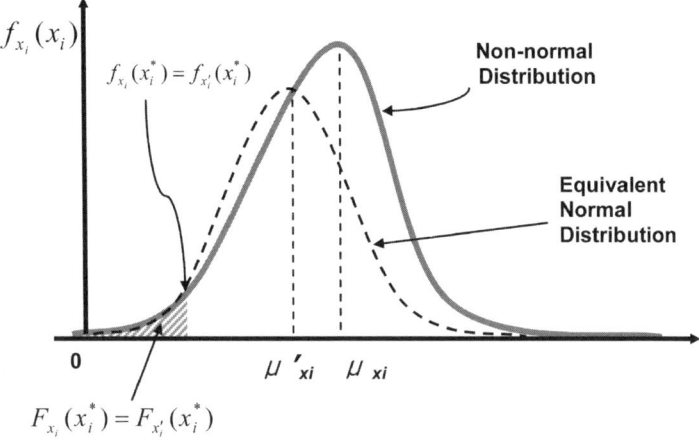

Figure 4.4. Normal Tail Approximation

The probability density function value of x and x'_i at x^*_i are equal:

$$f_{x_i}(x^*_i) = f_{x'_i}(x^*_i) \tag{4.43}$$

$$f_{x_i}(x^*_i) = \frac{1}{\sigma_{x'_i}} \phi \left(\frac{x^*_i - \mu_{x'_i}}{\sigma_{x'_i}} \right) \tag{4.44}$$

From Equations 4.42 and 4.44, the equivalent mean $\mu_{x'_i}$ and standard deviation $\sigma_{x'_i}$ of the approximate normal distributions are derived as Equations 4.39a and 4.39b. This normal-tail approximation is shown in Figure 4.4. Using Equations 4.39a and 4.39b, the transformation of the random variables from the X-space to the U-space can easily be performed, and the performance function $g(U)$ in U-space is approximately obtained.

The RF algorithm is similar to the Hasofer-Lind iteration method shown in Section 4.1.3, except that steps 2) and 4) are necessary to implement the calculation of the mean and standard deviation of the equivalent normal variables based on Equations 4.39a and 4.39b. The RF method is also called the HL-RF method, since the iteration algorithm was originally proposed by Hasofer and Lind and later extended by Rackwitz and Fiessler to include random variable distribution information. A computer program based on the HL-RF method can be readily developed to perform the reliability analysis. A flow chart of the algorithm is given in Figure 4.5. The following examples are given to illustrate the HL-RF method and the transformation of non-Gaussian variables.

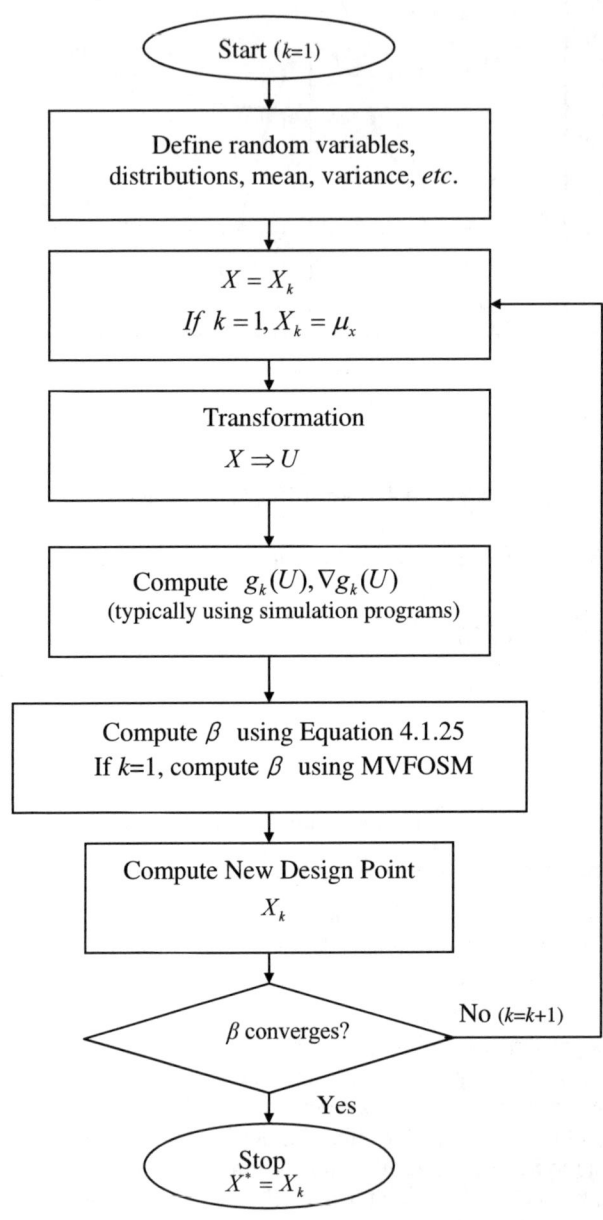

Figure 4.5. HL-RF Method Flow-chart

Example 4.4

Let x be a random variable having a lognormal distribution. Its mean value and standard deviation are $\mu_x = 120$ and $\sigma_x = 12$. Calculate the mean value and standard deviation of the equivalent normal distribution variable x' at $x^* = 80.0402$.

Solution:

(1) Compute the mean value, μ_y, and standard deviation, σ_y, of normally distributed variable y ($y = \ln x$) using Equations 2.44 and 2.45:

$$\sigma_y = \sqrt{\ln\left[\left(\frac{\sigma_x}{\mu_x}\right)^2 + 1\right]} = \sqrt{\ln\left[\left(\frac{12}{120}\right)^2 + 1\right]} = 0.09975$$

$$\mu_y = \ln\mu_x - \frac{1}{2}\sigma_y^2 = \ln 120 - \frac{1}{2} \times 0.09975^2 = 4.7825$$

(2) Compute the density function value at x^*:

$$f_x(x^*) = \frac{1}{\sqrt{2\pi}x^*\sigma_y}\exp\left[-\frac{1}{2}\left(\frac{\ln x^* - \mu_y}{\sigma_y}\right)^2\right]$$

$$= \frac{1}{\sqrt{2\pi} \times 80.0402 \times 0.09975}\exp\left[-\frac{1}{2}\left(\frac{\ln 80.0402 - 4.7825}{0.099975}\right)^2\right]$$

$$= 1.6114 \times 10^{-5}$$

(3) Compute the cumulative distribution function value at x^*:

From the density function above, it is obvious that the cumulative distribution function can be given as

$$F_x(x^*) = \Phi\left(\frac{\ln x^* - \mu_y}{\sigma_y}\right)$$

(4) Compute $\Phi^{-1}[F_x(x^*)]$:

$$\Phi^{-1}[F_x(x^*)] = \frac{\ln x^* - \mu_y}{\sigma_y} = \frac{\ln 80.0402 - 4.7835}{0.09975} = -4.0098$$

(5) Compute $\phi(\Phi^{-1}[F_x(x^*)])$:

$$\phi(\Phi^{-1}[F_x(x^*)]) = \frac{1}{\sqrt{2\pi}} \exp\left[-\frac{1}{2}\left(\frac{\ln x^* - \mu_y}{\sigma_y}\right)^2\right]$$

$$= \frac{1}{\sqrt{2\pi}} \exp\left[-\frac{1}{2}\left(\frac{\ln 80.0402 - 4.7825}{0.09975}\right)^2\right]$$

$$= 1.2865 \times 10^{-4}$$

(6) Compute the mean value and standard deviation of the equivalent normal distribution, $\mu_{x'}$ and $\sigma_{x'}$, using Equations 4.39a and 4.39b:

$$\sigma_{x'} = \frac{\phi(\Phi^{-1}[F_x(x^*)])}{f_x(x^*)} = \frac{1.2865 \times 10^{-4}}{1.6114 \times 10^{-5}}$$

$$= 7.9841$$

$$\mu_{x'} = x^* - \Phi^{-1}[F_x(x^*)]\sigma_{x'}$$

$$= 80.042 + 4.0098 \times 7.9841 = 112.0553$$

Using $\mu_{x'}$ and $\sigma_{x'}$, the standard normal variable of x can be computed from Equation 4.38b.

Example 4.5

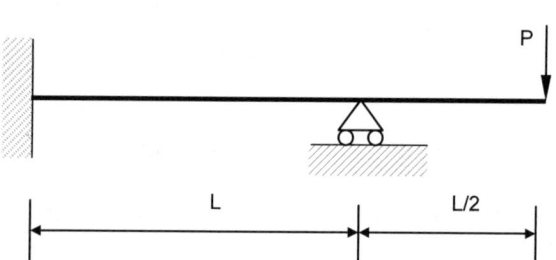

Consider the plane frame structure shown above. Evaluate the safety-index β and the coordinates x_i^* of the MPP. The limit state on the displacement is

$$d = \frac{5PL^3}{48EI} \leq \frac{L}{30} = d_{max}$$

where d_{max} is an allowable maximum displacement, E is Young's modulus, and I is the area moment of the cross-section. There are three random variables in this example. E and I are normally distributed, and the CDF and other parameters of the random variable P are defined by

$$F_P(x) = \exp[-[\exp\{-\alpha(x-\delta)\}]] \quad \ldots \ldots \ldots \ldots \ldots \ldots \text{(A)}$$

$$\mu_P = \delta + \frac{0.577}{\alpha}, \quad \sigma_P = \frac{1.283}{\alpha} \quad \ldots \ldots \ldots \ldots \ldots \ldots \text{(B)}$$

(*Note*: this distribution is known as the type-I extreme value distribution)

The mean values, μ_P, μ_L, μ_E, μ_I, of the load P, the beam length L, the Young's modulus E, and the area moment I are 4 kN, 5 m, 2.0×10^7 kN/m², and 10^{-4} m⁴, respectively. The corresponding standard deviations, σ_P, σ_L, σ_E, σ_I, are 1 kN, 0 m, 0.5×10^7 kN/m², and 0.2×10^{-4} m⁴. Here, L is a deterministic parameter ($L = 5$ m) because $\sigma_L = 0$.

Solution:

The limit-state function is given as

$$EI - 78.12P = 0$$

(1) Compute the scale and location parameters of the type-I extreme-value distribution using Equation B for the variable P.

$$\mu_P = \delta + 0.5772/\alpha, \quad \sigma_P = 1.2825/\alpha$$

Substituting $\mu_P = 4$ kN and $\sigma_P = 1$ kN into the above formulas, the scale and location parameters are obtained as

$$\delta = 3.5499, \quad \alpha = 1.2825$$

(2) Iteration 1:

(a) Compute the mean and standard deviation of the equivalent normal distribution for P:

First, assuming the design point, $X_1 = \{E_1, I_1, P_1\}^T$, as the mean value point, the coordinates of the initial design point are

$$E_1 = \mu_E = 2 \times 10^7, \quad I_1 = \mu_I = 10^{-4}, \quad P_1 = \mu_P = 4$$

The density function value at P_1 is

$$\begin{aligned}f_P(P_1) &= \alpha \exp\{-(P_1-\delta)\alpha - \exp[-(P_1-\delta)\alpha]\} \\ &= 1.2825 \exp\{-(4-3.5499) \times 1.2825 - \exp[-(-4-3.5499) \times 1.2825]\} \\ &= 0.4107\end{aligned}$$

The cumulative distribution value at P_1 is

$$\begin{aligned} F_P(P_1) &= \exp\{-\exp[-(P_1-\delta)\alpha]\} \\ &= \exp\{-\exp[-(-4-3.5499)1.2825\alpha]\} \\ &= 0.5704 \end{aligned}$$

Therefore, the standard deviation and mean value of the equivalent normal variable at P_1 from Equations 4.44a and 4.44b are

$$\sigma_{P'} = \frac{\phi(\Phi^{-1}[F_P(P_1)])}{f_P(P_1)} = \frac{\phi(\Phi^{-1}[0.5703])}{0.4107} = \frac{0.3927}{0.4107} = 0.9561$$

where $\Phi^{-1}[0.5704] = 0.177$ and $\phi[0.177] = 0.3927$.

$$\mu_{P'} = P_1 - \Phi^{-1}[F_P(P_1)]\sigma_{P'} = 4 - \Phi^{-1}[0.5704] \times 0.9561 = 3.8304$$

(b) Compute the function value and gradients of the limit-state function at the mean value point:

$$g(E_1, I_1, P_1) = EI - 78.12P = 2\times 10^7 \times 10^{-4} - 78.12 \times 4 = 1687.52$$

$$\frac{\partial g(X_1)}{\partial E} = 10^{-4}, \quad \frac{\partial g(X_1)}{\partial I} = 2\times 10^7, \quad \frac{\partial g(X_1)}{\partial P} = -78.12$$

(c) Compute the initial β using the mean-value method and its direction cosine α_i:

$$\beta_1 = \frac{\mu_{\tilde{g}}}{\sigma_{\tilde{g}}}$$

$$= \frac{g(E_1, I_1, P_1)}{\sqrt{(\frac{\partial g(E_1, I_1, P)}{\partial E}\sigma_E)^2 + (\frac{\partial g(E_1, I_1, P)}{\partial I}\sigma_I)^2 + (\frac{\partial g(E_1, I_1, P)}{\partial P}\sigma_P)^2}}$$

$$= \frac{1687.52}{\sqrt{(10^{-4}\times 0.5\times 10^7)^2 + (2\times 10^7 \times 0.2\times 10^{-4})^2 + (-78.12\times 0.9561)^2}}$$

$$= 2.6383$$

$$\alpha_E = -\frac{\frac{\partial g}{\partial E}\big|_{\mu_X}\sigma_E}{\sqrt{(\frac{\partial g(E_1, I_1, P)}{\partial E}\sigma_E)^2 + (\frac{\partial g(E_1, I_1, P)}{\partial I}\sigma_I)^2 + (\frac{\partial g(E_1, I_1, P)}{\partial P}\sigma_P)^2}}$$

$$= -\frac{10^{-4}\times 0.5\times 10^7}{\sqrt{(10^{-4}\times 0.5\times 10^7)^2 + (2\times 10^7 \times 0.2\times 10^{-4})^2 + (-78.12\times 0.9561)^2}}$$

$$= -0.7756$$

$$\alpha_I = -\frac{\frac{\partial g}{\partial I}\Big|_{\mu_x}\sigma_I}{\sqrt{(\frac{\partial g(E_1,I_1,P)}{\partial E}\sigma_E)^2+(\frac{\partial g(E_1,I_1,P)}{\partial I}\sigma_I)^2+(\frac{\partial g(E_1,I_1,P)}{\partial P}\sigma_P)^2}}$$

$$= -\frac{2\times10^7\times 0.2\times 10^{-4}}{\sqrt{(10^{-4}\times 0.5\times 10^7)^2+(2\times 10^7\times 0.2\times 10^{-4})^2+(-78.12\times 0.9561)^2}}$$

$$= -0.6205$$

$$\alpha_P = -\frac{\frac{\partial g}{\partial P}\Big|_{\mu_x}\sigma_P}{\sqrt{(\frac{\partial g(E_1,I_1,P)}{\partial E}\sigma_E)^2+(\frac{\partial g(E_1,I_1,P)}{\partial I}\sigma_I)^2+(\frac{\partial g(E_1,I_1,P)}{\partial P}\sigma_P)^2}}$$

$$= -\frac{-78.122\times 0.9561}{\sqrt{(10^{-4}\times 0.5\times 10^7)^2+(2\times 10^7\times 0.2\times 10^{-4})^2+(-78.12\times 0.9561)^2}}$$

$$= 0.1159$$

(d) Compute the coordinates of the new design point using Equation 4.28:

$$E_2 = \mu_E + \beta_1\sigma_E\alpha_E$$
$$= 2\times 10^7 + 2.6383\times 0.5\times 10^7\times(-0.7756)$$
$$= 9.7687\times 10^6$$

$$I_2 = \mu_I + \beta_1\sigma_I\alpha_I$$
$$= 10^{-4} + 2.6383\times 0.2\times 10^{-4}\times(-0.6205)$$
$$= 0.6726\times 10^{-4}$$

$$P_2 = \mu_{P'} + \beta_1\sigma_{P'}\alpha_P = 3.8304 + 2.6383\times 0.9561\times 0.1159$$
$$= 4.1227$$

$$u_{E,2} = \frac{E_2-\mu_E}{\sigma_E} = \frac{9.7687\times 10^6 - 2\times 10^7}{0.5\times 10^7} = -2.0463$$

$$u_{I,2} = \frac{I_2-\mu_I}{\sigma_I} = \frac{0.6726\times 10^{-4} - 10^{-4}}{0.2\times 10^{-4}} = -1.6370$$

$$u_{P,2} = \frac{P_2-\mu_P}{\sigma_P} = \frac{4.1227-3.8304}{0.9561} = 0.3057$$

(3) Iteration 2:

(a) Compute the mean and standard deviation of the equivalent normal distribution at P_2. The density function value at P_2 is

$$f_P(P_2) = \alpha\exp\{-P_2-\delta)\alpha-exp[-(P_2-\delta)\alpha]\}$$

$$= 1.2825 exp\{-(4.1227 - 3.5499) \times 1.2825 - exp[-(4.1227 - 3.5499) \times 1.2825]\}$$
$$= 0.3808$$

The cumulative distribution function value at P_2 is

$$F_P(P_2) = \exp\{-\exp[-(P_2 - \delta)\alpha]\}$$
$$= \exp\{-\exp[-(4.1227 - 3.5499) \times 1.2825]\} = 0.6189$$

The standard deviation and mean value of the equivalent normal variable at P_2 are

$$\sigma_{P'} = \frac{\phi(\Phi^{-1}[F_P(P_2)])}{f_P(P_2)} = \frac{\phi(\Phi^{-1}[0.6189])}{0.3808} = \frac{0.3811}{0.3808}$$
$$= 1.0007$$

where $\Phi^{-1}[0.6189] = 0.3028$ and $\phi[0.6189] = 0.3811$

$$\mu_{P'} = P_2 - \Phi^{-1}[F_P(P_2)]\sigma_{P'}$$
$$= 4.1227 - \Phi^{-1}[0.6189] \times 1.007 = 3.8197$$

(b) Compute the function value and gradients of the limit-state function at $X_2(E_2, I_2, P_2)$

$$g(E_2, I_2, P_2) = EI - 78.12P$$
$$= 97.6871 \times 10^5 \times 0.6726 \times 10^{-4} - 78.12 \times 4.1227$$
$$= 334.9737$$

$$\frac{\partial g(X_2)}{\partial E} = 6.726 \times 10^{-5}, \quad \frac{\partial g(X_2)}{\partial I} = 97.6871 \times 10^5,$$

$$\frac{\partial g(X_2)}{\partial P} = -78.12$$

(c) Compute the initial β using Equation 4.25 and the direction cosine α_i:

$$\beta_2 = \frac{g(X_2) - \frac{\partial g(X_2)}{\partial E}\sigma_E u_{E,2} - \frac{\partial g(X_2)}{\partial I}\sigma_I u_{I,2} - \frac{\partial g(X_2)}{\partial P}\sigma_P u_{P,2}}{\sqrt{(\frac{\partial g(X_2)}{\partial E}\sigma_E)^2 + (\frac{\partial g(X_2)}{\partial I}\sigma_I)^2 + (\frac{\partial g(X_2)}{\partial P}\sigma_P)^2}}$$

$$= \frac{334.9737 + 672.6 \times 0.5 \times 2.0463 + 976.87 \times 0.2 \times 1.6370 + 78.12 \times 1.0007 \times 0.3057}{\sqrt{(6.726 \times 10^{-5} \times 0.5 \times 10^7)^2 + (9.7687 \times 10^6 \times 0.2 \times 10^{-4})^2 + (-78.12 \times 1.0007)^2}}$$

$$= 3.4449$$

$$\alpha_E = -\frac{\frac{\partial g}{\partial E}\big|_{\mu_x} \sigma_E}{\sqrt{(\frac{\partial g(E_2,I_2,P_2)}{\partial E}\sigma_E)^2 + (\frac{\partial g(E_2,I_2,P_2)}{\partial I}\sigma_I)^2 + (\frac{\partial g(E_2,I_2,P_2)}{\partial P}\sigma_P)^2}}$$

$$= -\frac{6.726 \times 10^{-5} \times 0.5 \times 10^7}{\sqrt{(6.726 \times 10^{-5} \times 0.5 \times 10^7)^2 + (9.7687 \times 10^6 \times 0.2 \times 10^{-4})^2 + (-78.12 \times 1.0007)^2}}$$

$$= -0.8477$$

$$\alpha_I = -\frac{\frac{\partial g}{\partial I}\big|_{\mu_x} \sigma_I}{\sqrt{(\frac{\partial g(E_2,I_2,P_2)}{\partial E}\sigma_E)^2 + (\frac{\partial g(E_2,I_2,P_2)}{\partial I}\sigma_I)^2 + (\frac{\partial g(E_2,I_2,P_2)}{\partial P}\sigma_P)^2}}$$

$$= -\frac{9.7687 \times 10^6 \times 0.2 \times 10^{-4}}{\sqrt{(6.726 \times 10^{-5} \times 0.5 \times 10^7)^2 + (9.7687 \times 10^6 \times 0.2 \times 10^{-4})^2 + (-78.12 \times 1.0007)^2}}$$

$$= -0.4925$$

$$\alpha_P = -\frac{\frac{\partial g}{\partial P}\big|_{\mu_x} \sigma_P}{\sqrt{(\frac{\partial g(E_2,I_2,P_2)}{\partial E}\sigma_E)^2 + (\frac{\partial g(E_2,I_2,P_2)}{\partial I}\sigma_I)^2 + (\frac{\partial g(E_2,I_2,P_2)}{\partial P}\sigma_P)^2}}$$

$$= -\frac{-78.12 \times 1.0007}{\sqrt{(6.726 \times 10^{-5} \times 0.5 \times 10^7)^2 + (9.7687 \times 10^6 \times 0.2 \times 10^{-4})^2 + (-78.12 \times 1.0007)^2}}$$

$$= 0.1971$$

(d) Compute the coordinates of the new design point using Equation 4.28:

$$E_3 = \mu_E + \beta_2 \sigma_E \alpha_E$$
$$= 2 \times 10^7 + 3.4449 \times 0.5 \times 10^7 \times (-0.8477) = 5.3985 \times 10^6$$

$$I_3 = \mu_I + \beta_2 \sigma_I \alpha_I$$
$$= 10^{-4} + 3.4449 \times 0.2 \times 10^{-4} \times (-0.4925) = 0.6607 \times 10^{-4}$$

$$P_3 = \mu_{P'} + \beta_2 \sigma_{P'} \alpha_P$$
$$= 3.8197 + 3.4449 \times 1.0007 \times 0.1971 = 4.4990$$

$$u_{E,3} = \frac{E_3 - \mu_E}{\sigma_E} = \frac{5.3985 \times 10^6 - 2 \times 10^7}{0.5 \times 10^7} = -2.9203$$

$$u_{I,3} = \frac{I_3 - \mu_I}{\sigma_I} = \frac{1.3393 \times 10^{-4} - 10^{-4}}{0.2 \times 10^{-4}} = -1.6966$$

$$u_{P,3} = \frac{P_3 - \mu_{P'}}{\sigma_{P'}} = \frac{4.4990 - 3.8197}{1.0007} = 0.6788$$

(e) Check β convergence:

$$\varepsilon = \frac{|\beta_2 - \beta_1|}{\beta_1} = \frac{3.4449 - 2.6383}{2.6383} = 0.3057$$

Since $\varepsilon > \varepsilon_r (0.001)$, continue the process.

Table 4.3. Iteration Results in the HL-RF Method (Example 4.5)

Iteration No.	1	2	3	4	5	6
$f_P(P_k)$	0.4107	0.3808	0.2824	0.2316	0.2535	0.2732
$F_P(P_k)$	0.5704	0.6189	0.7438	0.7973	0.7748	0.7538
$\sigma_{P'}$	0.9561	1.0007	1.13998	1.2184	1.1835	1.1535
$\mu_{P'}$	3.8304	3.8197	3.7524	3.6939	3.7217	3.7433
$g(E_k, I_k, P_k)$	1687.52	334.9737	5.2055	-12.8426	-1.6961	-0.2352
$\frac{\partial g(X_k)}{\partial E}$	1.0×10^{-4}	6.726×10^{-5}	6.6069×10^{-5}	7.9676×10^{-5}	8.5848×10^{-5}	8.7569×10^{-5}
$\frac{\partial g(X_k)}{\partial I}$	2×10^{7}	9.7687×10^{6}	5.3985×10^{6}	4.4547×10^{6}	4.1799×10^{6}	4.0432×10^{6}
$\frac{\partial g(X_k)}{\partial P}$	-78.12	-78.12	-78.12	-78.12	-78.12	-78.12
β	2.6383	3.4449	3.3766	3.3292	3.3232	3.3222
α_E	-0.7756	-0.8477	-0.9208	-0.9504	-0.9603	-0.9638
α_I	-0.6205	-0.4925	-0.3009	-0.2125	-0.1870	-0.1780
α_P	0.1159	0.1971	0.2482	0.2271	0.2068	0.1984
E_k	9.7687×10^{6}	5.3985×10^{6}	4.4547×10^{6}	4.1799×10^{6}	4.0432×10^{6}	3.9897×10^{6}
I_k	0.6726×10^{-4}	1.3393×10^{-4}	0.7968×10^{-4}	0.8585×10^{-4}	0.8757×10^{-4}	0.8817×10^{-4}
P_k	4.1227	4.4990	4.7079	4.6151	4.5352	4.5035
$u_{E,k}$	-2.0463	-2.9203	-3.1091	-3.1640	-3.1914	-3.2021
$u_{I,k}$	-1.6370	-1.6966	-1.0162	-0.7076	-0.6215	-0.5914
$u_{P,k}$	0.3057	0.6788	0.8381	0.7560	0.6874	0.6590
ε	-	0.3057	0.0198	0.094	0.10	0.0003

The same procedures can be repeated until the stopping criterion ($\varepsilon < \varepsilon_r$) is satisfied. The iteration results are summarized in Table 4.3. The safety-index β is 3.3222. Since the limit-state function value at MPP, X^*, is close to zero compared to the starting value, this safety-index can be considered as the shortest distance from the origin to the limit-state surface.

4.1.6 FORM with Adaptive Approximations

In the previous algorithm, the limit-state function, $g(U)$, was approximated by the first-order Taylor expansion at the MPP. For nonlinear problems, this approach is only an approximation, and several iterations are usually required. How fast the algorithm converges depends on how well the linearized limit-state function approximates the nonlinear function $g(U)$. The limit-state function could be approximated by other functions, such as the *Two-point Adaptive Nonlinear*

Approximations (TANA), including TANA and TANA2 (the origin of function approximations and detailed developments are presented in Appendix A). This new class of approximations is constructed by using the Taylor series expansion in terms of adaptive intervening variables. The nonlinearity of the adaptive approximations is automatically changed by using the known information generated during the iteration process. TANA2 also has a correction term for second-order terms.

To compute the approximate U-space limit-state surface $\tilde{g}(U)$ using TANA, we must first obtain the adaptive approximate limit-state surface in X-space. Two possible methods, TANA and TANA2, for performing this step are given below.

TANA:

$$\tilde{g}(X) = g(X_k) + \frac{1}{r}\sum_{i=1}^{n} x_{i,k}^{1-r} \frac{\partial g(X_k)}{\partial x_i}(x_i^r - x_{i,k}^r) \tag{4.45}$$

where $x_{i,k}$ is the ith element in the vector X_k of the kth point/iteration. The comma notation does not signify differentiation.

The nonlinear index r in Equation 4.45 can be determined from

$$g(X_{k-1}) - \{g(X_k) + \frac{1}{r}\sum_{i=1}^{n} x_{i,k}^{1-r} \frac{\partial g(X_k)}{\partial x_i}(x_{i,k-1}^r - x_{i,k}^r)\} = 0 \tag{4.46}$$

TANA2:

$$\tilde{g}(X) = g(X_k) + \sum_{i=1}^{2} \frac{\partial g(X_k)}{\partial x_i} \frac{(x_{i,k})^{1-p_i}}{p_i}(x_i^{p_i} - (x_{i,k})^{p_i}) + \frac{1}{2}\varepsilon_2 \sum_{i=1}^{2}(x_i^{p_i} - (x_{i,k})^{p_i})^2 \tag{4.47}$$

p_i and ε in Equation 4.46 can be determined from

$$\frac{\partial g(X_{k-1})}{\partial x_i} = (\frac{x_{i,k-1}}{x_{i,k}})^{p_i-1} \frac{\partial g(X_k)}{\partial gx_i} + \varepsilon_2(x_{i,k-1}^{p_i} - x_{i,k}^{p_i})x_{i,k-1}^{p_i-1}p_i \tag{4.48}$$

$$g(X_{k-1}) = g(X_k) + \sum_{i=1}^{n} \frac{\partial g(X_k)}{\partial x_i} \frac{x_{i,k}^{1-p_i}}{p_i}(x_i^{p_i} - x_{i,k}^{p_i}) + \frac{1}{2}\varepsilon_2 \sum_{i=1}^{n}(x_{i,k-1}^{p_i} - x_{i,k}^{p_i})^2 \tag{4.49}$$

$$(i = 1, 2, ..., n)$$

The next step is to map $\tilde{g}(X)$ into $\tilde{g}(U)$ by using the standard normal or equivalent normal transformations:

$$\tilde{g}(X) = \tilde{g}(\sigma_{x_1'}u_1 + \mu_{x_1'}, \sigma_{x_2'}u_2 + \mu_{x_2'}, ..., \sigma_{x_n'}u_n + \mu_{x_n'}) \tag{4.50}$$

The nonlinear index, r, is numerically calculated by minimizing the difference between the exact and the approximate limit-state functions at the previous point X_{k-1}. In theory, r can be any positive or negative real number (not equal to 0). In practice, r can be restricted from, say, -5 to 5, for the X-space iterations to avoid numerical difficulties associated with higher order polynomials [20]. r can be solved using any nonlinear equation solver, but here a simple iterative scheme is used. The iteration searching for r starts from $r = 1$. When r is increased or decreased a step length (0, 1), the difference ε between the exact and approximate function is calculated. If ε is smaller than the initial error (e.g., corresponding to $r = 1$), the above iteration is repeated until the allowable error, $\varepsilon = 0.001$, or limitation of r is reached and the nonlinear index r is determined. Otherwise, r is decreased by one half and the above iteration process is repeated until the final r is obtained. This search is computationally inexpensive because Equation 4.46 is available in a closed form equation and is very easy to implement.

Usually, the adaptive safety-index algorithm is better than the HL-RF method, because the nonlinear index r is determined by comparing linear approximations (starting from 1) and minimizing the difference between exact and approximate limit-state functions. In the process of searching for r, the nonlinear index will automatically become 1 if other values of r cannot provide any improvement over the linear approximation.

The main steps of this algorithm are summarized as follows:

1) In the first iteration, compute the mean and standard deviation of the equivalent normal distribution at the mean value point for non-Gaussian distribution variables. Construct a linear approximation of Equation 4.23 by using the first-order Taylor series expansion at an initial point (if the initial point is selected as the mean value point, μ, the linear approximation is expanded at μ) and compute the limit-state function value and gradients at the initial point.

2) Compute the initial safety-index β_1 using the HL-RF method and its direction cosine, α_i (if the initial point is the mean value point, the mean-value method is used.).

3) Compute the new design point X_k using Equation 4.28.

4) Compute the mean and standard distribution of the equivalent normal distribution at X_k for non-normal distribution variables. Calculate the limit-state function value gradients at the new design point, X_k.

5) Determine the nonlinear index r by solving Equation 4.46 for TANA; or determine the nonlinear index p_k by solving Equation 4.48 and Equation 4.49 for TANA2; based on the information of the current and previous points (when k is equal to 2, previous design point is the mean value $\overline{X}$).

6) Obtain the adaptive nonlinear approximation, Equation 4.45 for TANA or Equation 4.47 for TANA2.

7) Transform the X-space approximate limit-state function into the U-space function using Equation 4.50.

8) Find the most probable failure point X_{k+1} of the approximate safety-index model given in Equation 4.20 using the HL-RF method or by solving the optimization problem using any commercial software. Compute the safety-index β_{k+1}.

9) Check the convergence: $\varepsilon = \dfrac{|\beta_{k+1} - \beta_k|}{\beta_k}$

10) Stop the process if ε satisfies the required convergence tolerance limit (e.g., 0.001); otherwise, continue.

11) Compute the exact limit-state function value and approximate gradients at X_{k+1} and estimate the approximate safety-index $\tilde{\beta}_{k+1}$ using the HL-RF method.

12) Approximate β convergence check, $\varepsilon = \dfrac{|\tilde{\beta}_{k+1} - \beta_k|}{\beta_k}$

13) Continue the process if ε satisfies the required convergence tolerance (0.001); otherwise, stop.

14) Compute the exact gradients of the limit-state function at X_{k+1} and go to step 5); repeat the process until β converges.

In Step 8, the safety-index β of the approximate model given in Equation 4.20 can be obtained easily by computing the explicit function $\tilde{g}(U)$, in which any optimization scheme or iteration algorithm can be used. Computation of the exact performance function $g(X)$ is not required; therefore, computer time is greatly reduced for problems involving complex and implicit performance functions, particularly with finite element models for structural response simulation.

Example 4.6

For Example 4.3b, solve the safety-index β using the TANA algorithm (Equation 4.45).

Solution:

(1) Iteration 1:

(a) Set the mean value point as an initial design point and the required β convergence tolerance as $\varepsilon_r = 0.001$. Compute the limit-state function value and gradients at the mean value point.

$$g(X^*) = g(\mu_{x_1}, \mu_{x_2}) = \mu_{x_1}^3 + \mu_{x_2}^3 - 18$$
$$= 10.0^3 + 9.9^3 - 18 = 1952.299$$

$$\dfrac{\partial g}{\partial x_1}\Big|_\mu = 3\mu_{x_1}^2 = 3 \times 10^2 = 300, \quad \dfrac{\partial g}{\partial x_2}\Big|_\mu = 3\mu_{x_2}^2 = 3 \times 9.9^2 = 294.03$$

(b) Compute the initial β using the mean-value method and its direction cosine α_i

$$\beta_1 = \frac{\mu_{\tilde{g}}}{\sigma_{\tilde{g}}} = \frac{g(X^*)}{\sqrt{(\frac{\partial g(\mu_{x_1},\mu_{x_2})}{\partial x_1}\sigma_{x_1})^2 + (\frac{\partial g(\mu_{x_1},\mu_{x_2})}{\partial x_2}\sigma_{x_2})^2}}$$

$$= \frac{1952.299}{\sqrt{(300\times 5.0)^2 + (294.03\times 5.0)^2}} = 0.9295$$

$$\alpha_1 = -\frac{\frac{\partial g}{\partial x_1}\big|_\mu \sigma_{x_1}}{\sqrt{(\frac{\partial g(\mu_{x_1},\mu_{x_2})}{\partial x_1}\sigma_{x_1})^2 + (\frac{\partial g(\mu_{x_1},\mu_{x_2})}{\partial x_2}\sigma_{x_2})^2}}$$

$$= -\frac{300\times 5.0}{\sqrt{(300\times 5.0)^2 + (294.03\times 5.0)^2}} = -0.7142$$

$$\alpha_2 = -\frac{\frac{\partial g}{\partial x_2}\big|_\mu \sigma_{x_2}}{\sqrt{(\frac{\partial g(\mu_{x_1},\mu_{x_2})}{\partial x_1}\sigma_{x_1})^2 + (\frac{\partial g(\mu_{x_1},\mu_{x_2})}{\partial x_2}\sigma_{x_2})^2}}$$

$$= -\frac{294.03\times 5.0}{\sqrt{(300\times 5.0)^2 + (294.03\times 5.0)^2}} = -0.6999$$

(c) Compute a new design point X^* using Equation 4.28

$$x_1^* = \mu_{x_1} + \beta_1 \sigma_{x_1} \alpha_1 = 10.0 + 0.9295 \times 5.0 \times (-0.7142) = 6.6808$$

$$x_2^* = \mu_{x_2} + \beta_1 \sigma_{x_2} \alpha_2 = 9.9 + 0.9295 \times 5.0 \times (-0.6999) = 6.6468$$

$$u_1^* = \frac{x_1^* - \mu_{x_1}}{\sigma_{x_1}} = \frac{6.6808 - 10.0}{5.0} = -0.6638$$

$$u_2^* = \frac{x_2^* - \mu_{x_2}}{\sigma_{x_2}} = \frac{6.6468 - 9.9}{5.0} = -0.6506$$

(2) Iteration 2:

(a) Compute the limit-state function value and gradients at X^*:

$$g(X^*) = g(x_1^*, x_2^*) = x_1^{*3} + x_2^{*3} - 18$$
$$= 6.6808^3 + 6.6468^3 - 18 = 573.8398$$

$$\frac{\partial g}{\partial x_1}|_\mu = 3x_1^{*2} = 3 \times 6.6808^2 = 133.8982$$

$$\frac{\partial g}{\partial x_2}|_\mu = 3x_2^{*2} = 3 \times 6.6468^2 = 132.5409$$

(b) Compute the nonlinearity index r based on the function values and gradients of the two points, $\mu(10.0, 9.9)$ and $X^*(6.6808, 6.6468)$, using Equation 4.46:

$$g(X_{k-1}) - \{g(X_k) + \frac{1}{r}\sum_{i=1}^{2} x_{i,k}^{1-r} \frac{\partial g(X_k)}{\partial x_i}(x_{i,k-1}^r - x_{i,k}^r)\}$$

$$= 1952.299 - \{573.8398 + \frac{1}{r}[6.6808^{1-r} \times 133.8982 \times (10^r - 6.6808^r) +$$

$$6.6468^{1-r} \times 132.5409 \times (9.9^r - 6.6468^r)] \leq 0.001$$

where $X_{k-1} = \mu(10.0, 9.9)$ and $X_k = X^*(6.6808, 6.6468)$

Using the adaptive search procedure mentioned before, r can be solved as $r = 3.0$.

(c) Construct the two-point adaptive nonlinear approximation (TANA) using Equation 4.45:

$$\tilde{g}(X) = g(X_k) + \frac{1}{r}\sum_{i=1}^{2} x_{i,k}^{1-r} \frac{\partial g(X_k)}{\partial x_i}(x_i^r - x_{i,k}^r)$$

$$= 573.8398 + \frac{1}{3}[6.6808^{-2} \times 133.8982 \times (x_1^3 - 6.6808^3) +$$

$$6.6468^{-2} \times 132.5409 \times (x_2^3 - 6.6468^3)] = x_1^3 + x_2^3 - 18.0$$

The approximate function is same as the exact problem given, because it is separable function and TANA predicted the nonlinearity index precisely.

(d) Transfer the above X-space approximate limit-state function into the U-space function using Equation 4.50:

$$\tilde{g}(U) = \tilde{g}(\sigma_{x_1}\mu_1 + \mu_{x_1}, \sigma_{x_2}\mu_2 + \mu_{x_2})$$
$$= (5u_1 + 10)^3 + (5u_2 + 9.9)^3 - 18$$

(e) Find the most probable failure point X^* of the approximate safety-index model given in Equation 4.20 using an optimization algorithm.

After four iterations, the MPP point is found as

$$x_1^* = 2.0718, \ x_2^* = 2.088, \ u_1^* = -1.5856, \ u_2^* = -1.5623$$

(f) Compute the safety-index β_2:

$$\beta_2 = \sqrt{u_1^{*2} + u_2^{*2}} = \sqrt{(-1.5856)^2 + (-1.5623)^2} = 2.2260$$

(g) Convergence check:

$$\varepsilon = \frac{|\beta_2 - \beta_1|}{\beta_1} = \frac{|2.2260 - 0.9295|}{0.9295} = 1.3948$$

Since $\varepsilon > \varepsilon_r (0.001)$, continue the process.

(3) Iteration 3:

(a) Compute the limit-state function value and gradients at X^*:

$$\begin{aligned} g(X^*) = g(x_1^*, x_2^*) &= x_1^{*3} + x_2^{*3} - 18 \\ &= 2.0718^3 + 2.0883^3 - 18 \\ &= -0.1276 \times 10^{-5} \end{aligned}$$

(b) Compute approximate gradients using the approximate limit-state function:

$$\frac{\partial \tilde{g}}{\partial x_1}\bigg|_\mu = 3x_{1_i}^{*2} = 3 \times 2.0718^2 = 12.8769$$

$$\frac{\partial \tilde{g}}{\partial x_2}\bigg|_\mu = 3x_2^{*2} = 3 \times 2.0883^2 = 13.0832$$

(c) Compute approximate safety-index $\tilde{\beta}$ using the HL-RF method (Equation 4.25) and the direction cosine α_i:

$$\begin{aligned} \tilde{\beta}_3 &= \frac{g(X^*) - \frac{\partial \tilde{g}(X^*)}{\partial x_1} \sigma_{x_1} u_{x_1}^* - \frac{\partial \tilde{g}(X^*)}{\partial x_2} \sigma_{x_2} u_{x_2}^*}{\sqrt{(\frac{\partial \tilde{g}(X^*)}{\partial x_1} \sigma_{x_1})^2 + (\frac{\partial \tilde{g}(X^*)}{\partial x_2} \sigma_{x_2})^2}} \\ &= \frac{-0.1276 \times 10^{-5} - 12.8769 \times 5 \times -1.5856 - 13.0832 \times 5 \times -1.5623}{\sqrt{(12.8769 \times 5)^2 + (13.0832 \times 5)^2}} \\ &= 2.2258 \end{aligned}$$

(d) Approximate convergence check

$$\varepsilon = \frac{|\tilde{\beta}_3 - \beta_2|}{\beta_2} = \frac{|2.2258 - 2.2260|}{2.2260} = 0.00009$$

Since $\varepsilon < \varepsilon_r (0.001)$, stop the process. The final safety-index is 2.2258. Compared with the result of Example 4.3b ($\beta = 1.1657$), the safety-index algorithm using TANA is much more efficient for this example. It needs only 3 g-function and 2 gradient calculations to reach the convergent point. Since the g-function value is very small, the final MPP is on the limit-state surface.

Example 4.7

The performance function is

$$g(x_1, x_2) = x_1 x_2 - 1400$$

in which x_1 and x_2 are the random variables with lognormal distributions. The mean values and standard deviations of two variables are $\mu_{x_1} = 40.0$, $\mu_{x_2} = 50.0$, $\sigma_{x_1} = 5.0$, and $\sigma_{x_2} = 2.5$. Solve the safety-index β by using the TANA2 algorithm (Equation 4.47).

Solution:

(1) Compute the mean values and standard deviations of normally distributed variables y_1 and y_2 ($y_1 = \ln x_1$, $y_2 = \ln x_2$) using Equations 2.44 and 2.45:

$$\sigma_{y_1} = \sqrt{\ln\left[\left(\frac{\sigma_{x_1}}{\mu_{x_1}}\right)^2 + 1\right]} = \sqrt{\ln\left[\left(\frac{5.0}{40.0}\right)^2 + 1\right]} = 0.1245$$

$$\sigma_{y_2} = \sqrt{\ln\left[\left(\frac{\sigma_{x_2}}{\mu_{x_2}}\right)^2 + 1\right]} = \sqrt{\ln\left[\left(\frac{2.5}{50.0}\right)^2 + 1\right]} = 4.9969 \times 10^{-2}$$

$$\mu_{y_1} = \ln \mu_{x_1} - \frac{1}{2}\sigma_{y_1}^2 = \ln 40 - \frac{1}{2} \times 0.1245^2 = 3.6811$$

$$\mu_{y_2} = \ln \mu_{x_2} - \frac{1}{2}\sigma_{y_2}^2 = \ln 50 - \frac{1}{2} \times (4.9969 \times 10^{-2})^2 = 3.9108$$

(2) Iteration 1:

118 Reliability-based Structural Design

(a) Compute the mean value and standard deviations of the equivalent normal distributions for x_1 and x_2:

First, assuming the design point $X^* = \{x_1^*, x_2^*\}^T$ as the mean value point, the coordinates of the initial design point are

$$x_1^* = \mu_{x_1} = 2 \times 40.0, \quad x_2^* = \mu_{x_2} = 50.0$$

The density function values at x_1^* and x_2^* are

$$f_{x_1}(x_1^*) = \frac{1}{\sqrt{2\pi}\, x_1^* \sigma_{y_1}} \exp[-\frac{1}{2}(\frac{\ln x_1^* - \mu_{y_1}}{\sigma_{y_1}})^2]$$

$$= \frac{1}{\sqrt{2\pi} \times 40 \times 0.1245} \exp[-\frac{1}{2}(\frac{\ln 40 - 3.6811}{0.1245})^2] = 7.9944 \times 10^{-2}$$

$$f_{x_2}(x_2^*) = \frac{1}{\sqrt{2\pi}\, x_2^* \sigma_{y_2}} \exp[-\frac{1}{2}(\frac{\ln x_2^* - \mu_{y_2}}{\sigma_{y_2}})^2]$$

$$= \frac{1}{\sqrt{2\pi} \times 50 \times 4.9969 \times 10^{-2}} \exp[-\frac{1}{2}(\frac{\ln 50 - 3.9108}{4.9969 \times 10^{-2}})^2] = 0.1596$$

$$\phi(\Phi^{-1}[F_{x_1}(x_1^*)]) = \frac{1}{\sqrt{2\pi}} \exp[-\frac{1}{2}(\frac{\ln x_1^* - \mu_{y_1}}{\sigma_{y_1}})^2]$$

$$= \frac{1}{\sqrt{2\pi}} \exp[-\frac{1}{2}(\frac{\ln 40 - 3.6811}{0.1245})^2] = 0.3982$$

$$\phi(\Phi^{-1}[F_{x_2}(x_2^*)]) = \frac{1}{\sqrt{2\pi}} \exp[-\frac{1}{2}(\frac{\ln x_2^* - \mu_{y_2}}{\sigma_{y_2}})^2]$$

$$= \frac{1}{\sqrt{2\pi}} \exp[-\frac{1}{2}(\frac{\ln 50 - 3.9108}{4.9969 \times 10^{-2}})^2] = 0.3988$$

Therefore, the standard deviation and mean value of the equivalent normal variable at P^* using Equations 4.44a and 4.44b are

$$\sigma_{x_1'} = \frac{\phi(\Phi^{-1}[F_{x_1}(x_1^*)])}{f_{x_1}(x_1^*)} = \frac{0.3982}{7.9944 \times 10^{-2}} = 4.9806$$

$$\sigma_{x_2'} = \frac{\phi(\Phi^{-1}[F_{x_2}(x_2^*)])}{f_{x_2}(x_2^*)} = \frac{0.3988}{0.1596} = 2.4984$$

$$\mu_{x_1'} = x_1^* - \Phi^{-1}[F_{x_1}(x_1^*)]\sigma_{x_1'} = 40 - 6.2258 \times 10^{-2} \times 4.9806 = 39.6899$$

$$\mu_{x_2'} = x_2^* - \Phi^{-1}[F_{x_2}(x_2^*)]\sigma_{x_2'} = 50 - 2.4984 \times 10^{-2} \times 2.4984 = 49.9376$$

(b) Set the mean value point as an initial design point and the required β convergence tolerance as $\varepsilon_r = 0.001$. Compute the limit-state function value and the gradients at the mean value point:

$$g(X^*) = g(\mu_{x_1}, \mu_{x_2}) = \mu_{x_1}\mu_{x_2} - 1400 = 40 \times 50 - 1400 = 600$$

$$\left.\frac{\partial g}{\partial x_1}\right|_\mu = \mu_{x_2} = 50, \quad \left.\frac{\partial g}{\partial x_2}\right|_\mu = \mu_{x_1} = 40$$

(c) Compute the initial β using the mean-value method and its direction cosine, α_i:

$$\beta_1 = \frac{\mu_{\tilde{g}}}{\sigma_{\tilde{g}}} = \frac{g(X^*)}{\sqrt{(\frac{\partial g(\mu_{x_1},\mu_{x_2})}{\partial x_1}\sigma_{x_1})^2 + (\frac{\partial g(\mu_{x_1},\mu_{x_2})}{\partial x_2}\sigma_{x_2})^2}}$$

$$= \frac{600}{\sqrt{(50 \times 4.9806)^2 + (40 \times 2.4984)^2}} = 2.1689$$

$$\alpha_1 = -\frac{\left.\frac{\partial g}{\partial x_1}\right|_\mu \sigma_{x_1}}{\sqrt{(\frac{\partial g(\mu_{x_1},\mu_{x_2})}{\partial x_1}\sigma_{x_1})^2 + (\frac{\partial g(\mu_{x_1},\mu_{x_2})}{\partial x_2}\sigma_{x_2})^2}}$$

$$= -\frac{50 \times 4.9806}{\sqrt{(50 \times 4.9806)^2 + (40 \times 2.4984)^2}} = -0.9281$$

$$\alpha_2 = -\frac{\left.\frac{\partial g}{\partial x_2}\right|_\mu \sigma_{x_2}}{\sqrt{(\frac{\partial g(\mu_{x_1},\mu_{x_2})}{\partial x_1}\sigma_{x_1})^2 + (\frac{\partial g(\mu_{x_1},\mu_{x_2})}{\partial x_2}\sigma_{x_2})^2}}$$

$$= -\frac{40 \times 2.4984}{\sqrt{(50 \times 4.9806)^2 + (40 \times 2.4984)^2}} = -0.3724$$

(d) Compute a new design point X^* from Equation 4.28:

$$x_1^* = \mu_{x_1} + \beta_1 \sigma_{x_1} \alpha_1 = 39.6899 + 2.1689 \times 4.9806 \times (-0.9281) = 29.6645$$

$$x_2^* = \mu_{x_2} + \beta_1 \sigma_{x_2} \alpha_2 = 49.9376 + 2.1689 \times 2.4984 \times (-0.3724) = 47.9194$$

$$u_1^* = \frac{x_1^* - \mu_{x_1}}{\sigma_{x_1}} = -2.0129, \quad u_2^* = \frac{x_2^* - \mu_{x_2}}{\sigma_{x_2}} = -0.8078$$

(3) Iteration 2:

(a) Compute the mean values and standard deviations of the equivalent normal distributions for x_1^* and x_2^*:

The density function values at x_1^* and x_2^* are

$$f_{x_1}(x_1^*) = \frac{1}{\sqrt{2\pi}\, x_1^* \sigma_{y_1}} \exp\left[-\frac{1}{2}\left(\frac{\ln x_1^* - \mu_{y_1}}{\sigma_{y_1}}\right)^2\right]$$

$$= \frac{1}{\sqrt{2\pi} \times 29.6645 \times 0.1245} \exp\left[-\frac{1}{2}\left(\frac{\ln 29.6645 - 3.6811}{0.1245}\right)^2\right] = 7.0144 \times 10^{-3}$$

$$f_{x_2}(x_2^*) = \frac{1}{\sqrt{2\pi}\, x_2^* \sigma_{y_2}} \exp\left[-\frac{1}{2}\left(\frac{\ln x_2^* - \mu_{y_2}}{\sigma_{y_2}}\right)^2\right]$$

$$= \frac{1}{\sqrt{2\pi} \times 47.9194 \times 4.9969 \times 10^{-2}} \exp\left[-\frac{1}{2}\left(\frac{\ln 47.9194 - 3.9108}{4.9969 \times 10^{-2}}\right)^2\right] = 0.1185$$

$$\phi(\Phi^{-1}[F_{x_1}(x_1^*)]) = \frac{1}{\sqrt{2\pi}} \exp\left[-\frac{1}{2}\left(\frac{\ln x_1^* - \mu_{y_1}}{\sigma_{y_1}}\right)^2\right]$$

$$= \frac{1}{\sqrt{2\pi}} \exp\left[-\frac{1}{2}\left(\frac{\ln 29.6645 - 3.6811}{0.1245}\right)^2\right] = 2.5909 \times 10^{-2}$$

$$\phi(\Phi^{-1}[F_{x_2}(x_2^*)]) = \frac{1}{\sqrt{2\pi}} \exp\left[-\frac{1}{2}\left(\frac{\ln x_2^* - \mu_{y_2}}{\sigma_{y_2}}\right)^2\right]$$

$$= \frac{1}{\sqrt{2\pi}} \exp\left[-\frac{1}{2}\left(\frac{\ln 47.9194 - 3.9108}{4.9969 \times 10^{-2}}\right)^2\right] = 0.2837$$

Therefore, the standard deviation and mean value of the equivalent normal variable at P^* using Equations 4.44a and 4.44b are

$$\sigma_{x_1'} = \frac{\phi(\Phi^{-1}[F_{x_1}(x_1^*)])}{f_{x_1}(x_1^*)} = \frac{2.5909 \times 10^{-2}}{7.0144 \times 10^{-3}} = 3.6937$$

$$\sigma_{x_2'} = \frac{\phi(\Phi^{-1}[F_{x_2}(x_2^*)])}{f_{x_2}(x_2^*)} = \frac{0.2837}{0.1185} = 2.3945$$

$$\mu_{x_1'} = x_1^* - \Phi^{-1}[F_{x_1}(x_1^*)]\sigma_{x_1'} = 29.6645 - (-2.3384) \times 3.6937 = 38.3021$$

$$\mu_{x_2'} = x_2^* - \Phi^{-1}[F_{x_2}(x_2^*)]\sigma_{x_2'} = 47.9194 - (-0.8256) \times 2.3945 = 49.8963$$

(b) Compute the limit-state function value and gradients at X^*:

$$g(X^*) = g(x_1^*, x_2^*) = x_1^* x_2^* - 1400 = 29.6645 \times 47.9194 - 1400 = 21.5041$$

$$\left.\frac{\partial g}{\partial x_1}\right|_{X^*} = x_2^* = 47.9194, \quad \left.\frac{\partial g}{\partial x_2}\right|_{X^*} = x_1^* = 29.6645$$

(c) Compute the nonlinearity indices p_1 and p_2 based on the function values and gradients of the two points μ (40,50) and X^* (29.6645, 47.9194) using Equations 4.48 and 4.49:

$$\frac{\partial g(\mu)}{\partial x_1} = (\frac{\mu_1}{x_1^*})^{p_1-1} \frac{\partial g(X^*)}{\partial x_1} + \varepsilon_2 \, (\mu_1^{p_1} - (x_1^*)^{p_1}) \mu_1^{p_1-1} p_1$$

$$50 = (\frac{40}{29.6645})^{p_1-1} 47.9194 + \varepsilon_2 \, (40^{p_1} - (47.9194)^{p_1}) 40^{p_1-1} p_1$$

$$\frac{\partial g(\mu)}{\partial x_2} = (\frac{\mu_2}{x_2^*})^{p_2-1} \frac{\partial g(X^*)}{\partial x_2} + \varepsilon_2 \, (\mu_2^{p_2} - (x_2^*)^{p_2}) \mu_2^{p_2-1} p_2$$

$$40 = (\frac{50}{47.9194})^{p_2-1} 29.6645 + \varepsilon_2 (50^{p_2} - (47.9194)^{p_2}) 50^{p_2-1} p_2$$

$$g(\mu) = g(X^*) + \sum_{i=1}^{2} \frac{\partial g(X^*)}{\partial x_i} \frac{(x_i^*)^{1-p_i}}{p_i} (\mu_i^{p_i} - (x_i^*)^{p_i}) + 0.5\varepsilon_2 \sum_{i=1}^{2} (\mu_i^{p_i} - (x_i^*)^{p_i})^2$$

$$600 = 21.5041 + 47.9194 \times \frac{29.6645^{1-p_1}}{p_1} (40^{p_1} - 29.6645^{p_1})$$

$$+ 29.6645 \times \frac{47.9194^{1-p_2}}{p_2} (50^{p_1} - 47.9194^{p_2})$$

$$+ 0.5\varepsilon_2 [(\mu_1^{p_1} - (x_1^*)^{p_1})^2 + (\mu_2^{p_2} - (x_2^*)^{p_2})^2]$$

Based on the above three equations, p_1, p_2 and ε_2 are solved using the adaptive search procedure.

$$p_1 = 1.0375, \quad p_2 = 1.4125, \quad \varepsilon = 0.1$$

(d) Construct TANA2 model using Equation 4.47:

$$\widetilde{g}(X) = g(X^*) + \sum_{i=1}^{2} \frac{\partial g(X^*)}{\partial x_i} \frac{(x_i^*)^{1-p_i}}{p_i} (x_i^{p_i} - (x_i^*)^{p_i}) + 0.5\varepsilon_2 \sum_{i=1}^{2} (x_i^{p_i} - (x_i^*)^{p_i})^2$$

$$= 21.5041 + 47.9194 \times \frac{29.6645^{1-1.0375}}{1.0375} (x_1^{1.0375} - 29.6645^{1.0375})$$

$$+ 29.6645 \times \frac{47.9194^{1-1.4125}}{1.4125} (x_2^{1.4125} - 47.9194^{1.4125})$$

$$+\frac{0.1}{2}[(x_1^{1.0375} - 29.6645^{1.0375})^2 + (x_2^{1.4125} - 47.9194^{1.4125})^2]$$

$$= 21.5041 + 40.6738(x_1^{1.0375} - 29.6645^{1.0375}) + 4.2564(x_2^{1.4125} - 47.9194^{1.4125})$$
$$+\frac{0.1}{2}[(x_1^{1.0375} - 29.6645^{1.0375})^2 + (x_2^{1.4125} - 47.9194^{1.4125})^2]$$

(e) Transfer the above X-space approximate limit-state function into the U-space function using Equation 4.50:

$$\tilde{g}(U) = \tilde{g}(\sigma_{x_1} u_1 + \mu_{x_1}, \sigma_{x_2} u_2 + \mu_{x_2})$$
$$= 21.5041 + 40.6738[(3.6937 \mu_1 + 38.3021)^{1.0375} - 29.6645^{1.0375}]$$
$$+ 4.2564[(2.3945 \mu_2 + 49.8963)^{1.4125} - 47.9194^{1.4125}]$$
$$+\frac{0.1}{2}[((3.6937 \mu_1 + 38.3021)^{1.0375} - 29.6645^{1.0375})^2$$
$$+ ((2.3945 \mu_2 + 49.8963)^{1.4125} - 47.9194^{1.4125})^2]$$

(f) Find the most probable failure point X^* of the approximate safety-index model given in Equation 4.20 using an optimization algorithm.

After two iterations, the MPP point is found as

$$x_1^* = 29.3517, \; x_2^* = 47.6961, \; u_1^* = -2.4236, \; u_2^* = -0.9191$$

At each iteration, the mean value and standard deviation of the equivalent normal distributions at the new design point X^* must be calculated.

(g) Compute the safety-index β_2:

$$\beta_2 = \sqrt{(u_1^*)^2 + (u_2^*)^2} = \sqrt{(-2.4236)^2 + (-0.9191)^2} = 2.5920$$

(h) Convergence Check:

$$\varepsilon = \frac{|\beta_2 - \beta_1|}{\beta_1} = \frac{2.5920 - 2.1689}{2.1689} = 0.1951$$

Since $\varepsilon > \varepsilon_r (0.001)$, continue the process

(4) Iteration 3:

(a) Compute the mean values and standard deviations of the equivalent normal distributions for x_1^* and x_2^*:

The density function values at x_1^* and x_2^* are

$$f_{x_1}(x_1^*) = \frac{1}{\sqrt{2\pi}\, x_1^* \sigma_{y_1}} \exp\left[-\frac{1}{2}\left(\frac{\ln x_1^* - \mu_{y_1}}{\sigma_{y_1}}\right)^2\right]$$

$$= \frac{1}{\sqrt{2\pi} \times 29.3517 \times 0.1245} \exp\left[-\frac{1}{2}\left(\frac{\ln 29.3517 - 3.6811}{0.1245}\right)^2\right]$$

$$= 5.7886 \times 10^{-3}$$

$$f_{x_2}(x_2^*) = \frac{1}{\sqrt{2\pi}\, x_2^* \sigma_{y_2}} \exp\left[-\frac{1}{2}\left(\frac{\ln x_2^* - \mu_{y_2}}{\sigma_{y_2}}\right)^2\right]$$

$$= \frac{1}{\sqrt{2\pi} \times 47.6961 \times 4.9969 \times 10^{-2}} \exp\left[-\frac{1}{2}\left(\frac{\ln 47.6961 - 3.9108}{4.9969 \times 10^{-2}}\right)^2\right]$$

$$= 0.1097$$

$$\phi(\Phi^{-1}[F_{x_1}(x_1^*)]) = \frac{1}{\sqrt{2\pi}} \exp\left[-\frac{1}{2}\left(\frac{\ln x_1^* - \mu_{y_1}}{\sigma_{y_1}}\right)^2\right]$$

$$= \frac{1}{\sqrt{2\pi}} \exp\left[-\frac{1}{2}\left(\frac{\ln 29.3517 - 3.6811}{0.1245}\right)^2\right] = 2.1156 \times 10^{-2}$$

$$\phi(\Phi^{-1}[F_{x_2}(x_2^*)]) = \frac{1}{\sqrt{2\pi}} \exp\left[-\frac{1}{2}\left(\frac{\ln x_2^* - \mu_{y_2}}{\sigma_{y_2}}\right)^2\right]$$

$$= \frac{1}{\sqrt{2\pi}} \exp\left[-\frac{1}{2}\left(\frac{\ln 47.6961 - 3.9108}{4.9969 \times 10^{-2}}\right)^2\right] = 0.2615$$

Therefore, the standard deviation and mean value of the equivalent normal variable at P^* using Equations 4.44a and 4.44b are

$$\sigma_{x_1'} = \frac{\phi(\Phi^{-1}[F_{x_1}(x_1^*)])}{f_{x_1}(x_1^*)} = \frac{2.1156 \times 10^{-2}}{5.7886 \times 10^{-3}} = 3.6548$$

$$\sigma_{x_2'} = \frac{\phi(\Phi^{-1}[F_{x_2}(x_2^*)])}{f_{x_2}(x_2^*)} = \frac{0.2615}{0.1097} = 2.3833$$

$$\mu_{x_1'} = x_1^* - \Phi^{-1}[F_{x_1}(x_1^*)]\sigma_{x_1'} = 29.3517 - (-2.4236) \times 3.6548 = 38.2094$$

$$\mu_{x_2'} = x_2^* - \Phi^{-1}[F_{x_2}(x_2^*)]\sigma_{x_2'} = 47.6961 - (-0.9191) \times 2.3833 = 49.8865$$

(b) Compute the limit-state function value at X^*:

$$g(X^*) = g(x_1^*, x_2^*) = x_1^* x_2^* - 1400 = 29.3517 \times 47.6961 - 1400 = -0.0359$$

(c) Compute approximate gradients using the approximate limit-state function:

$$\left.\frac{\partial \widetilde{g}}{\partial x_1}\right|_\mu = 47.8569, \quad \left.\frac{\partial \widetilde{g}}{\partial x_2}\right|_\mu = 28.5261$$

(d) Compute the approximate safety-index $\widetilde{\beta}$ using the HL-RF method (Equation 4.25) and the direction cosine, α_i:

$$\widetilde{\beta}_3 = \frac{g(X^*) - \frac{\partial \widetilde{g}(X^*)}{\partial x_1}\sigma'_{x_1} u^*_{x_1} - \frac{\partial \widetilde{g}(X^*)}{\partial x_2}\sigma'_{x_2} u^*_{x_2}}{\sqrt{(\frac{\partial \widetilde{g}(X^*)}{\partial x_1}\sigma'_{x_1})^2 + (\frac{\partial \widetilde{g}(X^*)}{\partial x_2}\sigma'_{x_2})^2}}$$

$$= \frac{-0.0359 - 47.8569 \times 3.6548 \times (-2.4236) - 28.5261 \times 2.3833 \times (-0.9191)}{\sqrt{(47.8569 \times 3.6548)^2 + (28.5261 \times 2.3833)^2}}$$

$$= 2.5917$$

(e) Approximate convergence check:

$$\varepsilon = \frac{|\widetilde{\beta}_3 - \widetilde{\beta}_2|}{\widetilde{\beta}_2} = \frac{|2.5917 - 2.5920|}{2.5920} = 0.0001$$

Since $\varepsilon < \varepsilon_r (0.001)$, stop the process. The final safety-index is 2.5917.

4.2 Second-order Reliability Method (SORM)

FORM usually works well when the limit-state surface has only one minimal distance point and the function is nearly linear in the neighborhood of the design point. However, if the failure surface has large curvatures (high nonlinearity), the failure probability estimated by FORM using the safety-index β may give unreasonable and inaccurate results [13]. To resolve this problem, the second-order Taylor series (or other polynomials) is considered. Various nonlinear approximate methods have been proposed in the literature.

Breitung [3], Tvedt [18], [19], Hohenbichler and Rackwitz [9], Koyluoglu and Nielsen [11], and Cai and Elishakoff [4] have developed SORM using the second-order approximation to replace the original surfaces. Wang and Grandhi [20] and Der Kiureghian, et al. [5] calculated second-order failure probabilities using approximate curvatures to avoid exact second-order derivatives calculations of the limit-state surface.

First, in Section 4.2.1, we present the fundamentals of the second-order approximation of the response surface with orthogonal transformation. Then, Breitung's and Tvedt's formulations are introduced in Sections 4.2.2 and 4.2.3, respectively. Wang and Grandhi's SORM with approximate curvatures calculations is given in Section 4.2.4.

4.2.1 First- and Second-order Approximation of Limit-state Function

Orthogonal Transformations

To facilitate the integration of Equation 3.10 the rotated new standard normal Y-space instead of U-space can be considered in most failure probability calculations. To conduct the rotation from the standard normal U-space to the Y-space, an orthogonal matrix H needs to be generated in which the nth row of H is the unit normal of the limit-state function at the MPP, i.e., $-\nabla g(U^*)/|\nabla g(U^*)|$. To generate H, first, an initial matrix is selected as follows:

$$\begin{pmatrix} \dfrac{-\partial g(U^*)/\partial U_1}{|\nabla g(U^*)|} & \dfrac{-\partial g(U^*)/\partial U_2}{|\nabla g(U^*)|} & \cdots & \dfrac{-\partial g(U^*)/\partial U_n}{|\nabla g(U^*)|} \\ 0 & 1 & \cdots & 0 \\ 0 & 0 & \cdots & 0 \\ \cdots & \cdots & \cdots & \cdots \\ 0 & 0 & \cdots & 1 \end{pmatrix} \quad (4.51)$$

where the last n-1 rows consist of zeros and unity on the diagonal. The Gram-Schmidt algorithm [17] is used to orthogonalize the above matrix to obtain an orthogonal matrix. First, let $f_1, f_2, ..., f_n$ denote the first, second, ..., n^{th} row vector of the above matrix, respectively:

$$f_1 = \{\dfrac{-\partial g(U^*)/\partial U_1}{|\nabla g(U^*)|}, \dfrac{-g(U^*)/\partial U_2}{|\nabla g(U^*)|}, ..., \dfrac{-g(U^*)/\partial U_n}{|\nabla g(U^*)|}\}^T,$$

$$f_2 = \{0,1,0,...,0\}^T, \quad ..., \quad f_n = \{0,0,0,...,1\}^T$$

Set $D_1 = (f_1, f_1)^{\frac{1}{2}}$, $e_{11} = \dfrac{1}{D_1}$, $\gamma_1 = e_{11} f_1$,

$D_2 = [(f_2, f_2) - |(f_2, \gamma_1)|^2]^{\frac{1}{2}}$, $e_{12} = -\dfrac{(f_2, \gamma_1)}{D_2}$,

$e_{22} = \dfrac{1}{D_2}$, $\gamma_2 = e_{12}\gamma_1 + e_{22}f_2$

and in general,

$$D_k = [(f_k,f_k) - |(f_k,\gamma_1)|^2 - |(f_k,\gamma_2)|^2 - ..., |(f_k,\gamma_{k-1})|^2]^{\frac{1}{2}}$$

$$e_{1k} = -\frac{(f_k,\gamma_1)}{D_k}, \quad e_{2k} = -\frac{(f_k,\gamma_2)}{D_k}, \quad ..., \quad e_{k-1,k} = -\frac{(f_k,\gamma_{k-1})}{D_k},$$

$$e_{kk} = \frac{1}{D_k}, \quad \gamma_k = e_{1k}\gamma_1 + e_{2k}\gamma_2 + ..., + e_{k-1,k}\gamma_{k-1} + e_{kk}f_k$$

where (f,f) and (f,γ) represent the scalar product (dot product) of two vectors. It can be verified that the generated vectors $\gamma_1, \gamma_2, ..., \gamma_n$ are orthogonalized. The generated orthogonal matrix H_0 is

$$H_0^T = \{\gamma_1^T, \gamma_2^T, ..., \gamma_n^T\} \tag{4.52}$$

In fact, in the orthogonal matrix of Equation 4.52, the first row is $-\nabla g(U^*)/|\nabla g(U^*)|$ due to $D_1 = 1$. To satisfy that the nth row of H is $-\nabla g(U^*)/|\nabla g U^*)|$, the first row of the orthogonal matrix is moved to the last row. This rearranged matrix is also an orthogonalized matrix and satisfies that the nth row of H equals $-\nabla g(U^*)/|\nabla g(U^*)|$, so it is defined as the H matrix and is given as

$$H^T = \{\gamma_2^T, \gamma_3^T, ..., \gamma_n^T, \gamma_1^T\} \tag{4.53}$$

First-order Approximation

Assuming the most probable failure point (MPP) in U-space to be $U^* = \{u_1^*, u_2^*, ..., u_n^*\}^T$, the linear approximation of the response surface g(U)=0 is given by the first-order Taylor Series expansion at the MPP:

$$\tilde{g}(U) \approx g(U^*) + \nabla g(U^*)(U - U^*) = 0 \tag{4.54}$$

In this equation, $g(U^*)$ equals 0 because U^* point is on the response surface. Dividing by $|\nabla g(U^*)|$, Equation 4.54 is rewritten as

$$\tilde{g}(U) \approx \frac{\nabla g(U^*)}{|\nabla g(U^*)|}(U - U^*) \tag{4.55}$$

From Equation 4.25, we have

$$\frac{\nabla g(U^*)U^*}{|\nabla g(U^*)|} = -\beta \tag{4.56}$$

Substituting this equation into Equation 4.55, we obtain

$$\tilde{g}(U) \approx \frac{\nabla g(U^*)}{|\nabla g(U^*)|}U + \beta = 0 \tag{4.57}$$

By a rotation of U into a new set of mutually independent standard normal random variables Y using the orthogonal matrix H given in Equation 4.53,

$$Y = HU \tag{4.58}$$

and the approximate response surface given in Equation 4.57 becomes

$$\tilde{g}(U) \approx -y_n + \beta = 0 \tag{4.59a}$$
$$\text{or,} \quad y_n = \beta \tag{4.59b}$$

Equation 4.59b is the first-order approximation of the response surface in the rotated standard normal space (denoted as Y-space), as shown in Figure 4.6. If the limit-state functions of the practical problems are linear or close to linear, this approximation closely or exactly represents the response surface. Otherwise, the truncation errors from the first-order Taylor approximation might be large and more accurate approximations need to be employed.

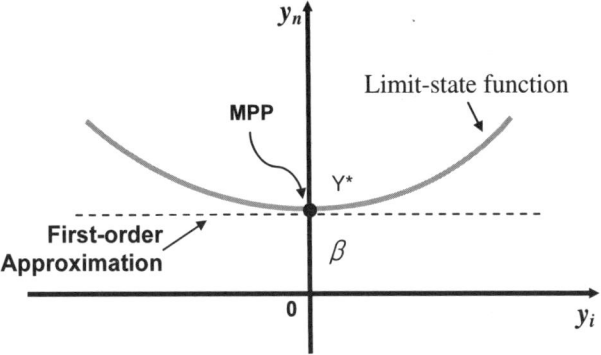

Figure 4.6. First-order Approximation of the Response Surface in Y-space

Second-order Approximation

The second-order approximation of the response surface $g(U) = 0$ is given by the second-order Taylor series expansion at the MPP:

$$\tilde{g}(U) \approx g(U^*) + \nabla g(U^*)^T (U - U^*) + \frac{1}{2}(U - U^*)^T \nabla^2 g(U^*)(U - U^*) \quad (4.60)$$

where $\nabla^2 g(U^*)$ represents the symmetric matrix of the second derivative of the limit-state function:

$$\nabla^2 g(U^*)_{ij} = \frac{\partial^2 g(U^*)}{\partial u_i \partial u_j} \quad (4.61)$$

Dividing by $|\nabla g(U^*)|$ and considering $g(U^*) = 0$, we obtain

$$\tilde{g}(U) \approx \alpha^T (U - U^*) + \frac{1}{2}(U - U^*)^T B(U - U^*) \quad (4.62a)$$

where

$$\alpha = \frac{\nabla g(U^*)}{|\nabla g(U^*)|}, \text{ and } B = \frac{\nabla^2 g(U^*)}{|\nabla g(U^*)|} \quad (4.62b)$$

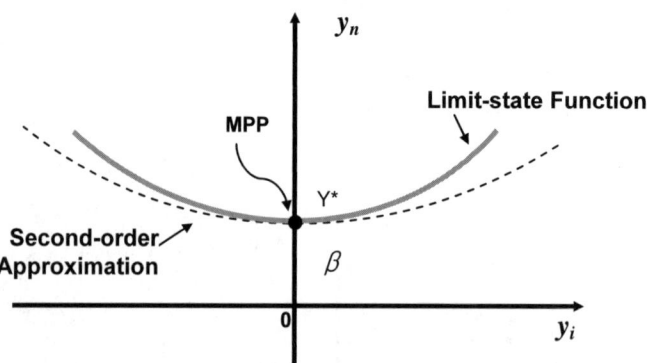

Figure 4.7. Second-order Approximation of the Response Surface in Y-space

Physically, the following transformations are the coordinate rotations to make the y_n axis coincide with the β vector, as shown in Figure 4.7. Substituting

Equation 4.58 into Equation 4.62 and replacing the first term by Equation 4.59a, the U-space approximate response surface is rotated as

$$\widetilde{g}(Y) \approx -y_n + \beta + \frac{1}{2}(H^{-1}Y - H^{-1}Y^*)B(H^{-1}Y - H^{-1}Y^*) \tag{4.63}$$

where Y^* is the Y-space MPP ($Y^* = \{0,0,..., \beta\}^T$) corresponding to the U-space MPP U^*. In Y-space, the y_n axis is in coincidence with the β vector.

Since the H matrix is an orthogonal matrix,

$$H^{-1} = H^T \tag{4.64}$$

Substituting this equation into Equation 4.63, we have

$$\widetilde{g}(Y) \approx -y_n + \beta + \frac{1}{2}(Y - Y^*)^T HBH^T (Y - Y^*) \tag{4.65}$$

where $(Y - Y^*)^T = (y_1, y_2,..., y_n - \beta)^T \tag{4.66}$

By a series of orthogonal transformations, $H_1, H_2,..., H_m$, for the first $n-1$ variables, $\overline{Y} = \{y_1, y_2,..., y_{n-1}\}^T$, i.e.,

$$\overline{Y}' = H_1 H_2 ... H_m \overline{Y} \tag{4.67}$$

Quantities associated with $n-1$ variables are denoted with a bar. Finally, the first $(n-1)\times(n-1)$ order matrix of HBH^T becomes a diagonal matrix:

$$\overline{H}\,\overline{B}\,\overline{H}^T = \begin{pmatrix} k_1 & 0 & ... & 0 \\ 0 & k_2 & ... & 0 \\ 0 & 0 & ... & 0 \\ ... & ... & ... & ... \\ 0 & 0 & ... & k_{n-1} \end{pmatrix} \tag{4.68}$$

and Equation 4.65 becomes

$$y_n = \beta + \frac{1}{2}\sum_{i=1}^{n-1} k_i y_i'^2 \tag{4.69}$$

In fact, the above procedure for finding the diagonal matrix can be treated as an eigenvalue problem. So, k are given by

$$k_{ij} = (\overline{H}\,\overline{B}\,\overline{H}^T)_{ij}, \, (i,j = 1,2,...,n-1) \tag{4.70}$$

where k_{ij} represents the curvature of the response surface at the MPP.

Equation 4.69 is the second-order approximation of the response surface in the rotated standard normal space. The major computational cost is in computing the second derivatives B of the limit-state function at the MPP. The exact second-order derivatives of $g(U)$ require additional $n(n+1)/2$ limit-state function simulations for a finite difference scheme. For problems having a large number of random variables, this calculation is extremely computer intensive. From this procedure, it is clear that one has to increase computational efficiency in calculating the curvature matrix and second-order function derivatives. Then, it enables an accelerated and cost-effective procedure to perform the second-order probability analysis, particularly when finite element-based structural analysis tools are used.

4.2.2 Breitung's Formulation

To explain the Breitung formulation, first, a Laplace method for the asymptotic approximation of multidimensional integrals is needed. Define

$$I(\beta) = \int_{g(Y)<0} \exp(\frac{-\beta^2|Y|^2}{2}) dY \tag{4.71}$$

where $I(\beta)$ is an integral over a fixed domain whose integrand is an exponential function depending linearly on the parameter β^2. An extensive study of the asymptotic behavior for β^2 is described in [2]. Making use of results given in this reference, the asymptotic form of $I(\beta)$ is (details see Appendix B)

$$I(\beta) \sim (2\pi)^{(n-1)/2} \exp(-\frac{\beta^2}{2}) \beta^{-(n+1)} |J|^{-1/2}, \, \beta \to \infty \tag{4.72}$$

where J is a quantity independent of β, depending only on the first and second derivatives of the failure surface at the MPP, which is defined in Equation B.3 in Appendix B:

In the case of independent standard normal random variables, the joint probability density function (PDF) is given by

$$P_f = (2\pi)^{-n/2} \int_{g(U)<0} \exp(-\frac{|U|^2}{2}) dU \tag{4.73}$$

Substituting $(x_1, x_2, ..., x_n) \to (y_1, y_2, ..., y_n)$ with $y_i = \beta^{-1} u_i$

$$P_f = (2\pi)^{-n/2} \beta^n \int_{g(Y)<0} \exp(-\beta^2 \frac{|Y|^2}{2}) dY \tag{4.74}$$

Substituting Equation 4.72 into this equation, we obtain

$$P_f \sim (2\pi)^{-1/2} \beta^{-1} \exp(-\frac{\beta^2}{2}) |J|^{-1/2}, \quad \beta \to \infty \tag{4.75}$$

Since the failure surface is approximated by the quadratic Taylor series expansion at the MPP, $|J|$ can be computed based on Equation B.9 given in Appendix B,

$$|J| = \sum_{i=1}^{p} |J_i| = \sum_{i=1}^{n} \prod_{j=1}^{n-1} (1 + k_{ij}\beta) \tag{4.76}$$

where p is the number of points on $g(U)=0$ with the shortest distance β from the origin to the failure surface, and k_{ij} is the main curvature of the failure surface at the MPP. If there is only one MPP on the surface, by substituting this equation into Equation 4.75 and considering Mill's ratio [2]

$$\Phi(-Y) \sim (2\pi)^{-1/2} Y^{-1} \exp(-Y^2/2) \tag{4.77}$$

P_f can be computed as

$$P_f \approx \Phi(-\beta) \prod_{j=1}^{n-1} (1 + k_j \beta)^{-1/2} \tag{4.78}$$

Since Equation 4.78 is an analytical equation, it is easy to implement the Breitung algorithm. The main steps of Breitung's formulation include:

1) Conduct the safety-index search and locate the MPP, U^*
2) Compute the second-order derivatives of the limit-state surface at U^* and form the B matrix given in Equation 4.62b
3) Calculate the orthogonal matrix H
4) Compute the main curvatures k_j of the failure surface at the MPP using Equation 4.70
5) Compute the failure probability P_f using Equation 4.78

Example 4.8

This example has a highly nonlinear performance function,
$$g(x_1, x_2) = x_1^4 + 2x_2^4 - 20$$
in which x_1 and x_2 are the random variables with normal distributions (mean $\mu_{x_1} = \mu_{x_2} = 10.0$, standard deviations $\sigma_{x_1} = \sigma_{x_2} = 5.0$). Solve the safety-index β or the probability of failure P_f by using FORM, MCS, and Breitung's method.

Solution:

(a) The safety-index β was obtained using FORM: $\beta = 2.3654$. Using Equation 3.9, the P_f is computed as

$$P_f = \Phi(-\beta) = \Phi(-2.3654) = 0.009004$$

(b) The P_f using the Monte Carlo method (sample size=100,000) is 0.001950. Compared with this result, FORM is inaccurate for this highly nonlinear problem. Therefore, more accurate approximate methods are needed.

(c) Compute the failure probability P_f using the Breitung method (Equation 4.78)

1) Compute the safety-index and MPP, U^*: The safety-index β is calculated as $\beta = 2.3654$ using FORM. The MPP is located at U^* (-1.6368, -1.7077) (in X-space, X^* (1.8162, 1.4613)).

2) Compute the second-order derivatives of the limit-state surface at U^* and form the B matrix given in Equation 4.62b:

$$\frac{\partial g}{\partial u_1} = \frac{\partial g}{\partial x_1} \sigma_1 = 4x_1^3 \sigma_1 = 4 \times 1.8162^2 \times 5 = 119.8148$$

$$\frac{\partial g}{\partial u_2} = \frac{\partial g}{\partial x_2} \sigma_2 = 4x_2^3 \sigma_2 = 8 \times 1.4613^2 \times 5 = 124.8218$$

$$\frac{\partial^2 g}{\partial x_1^2} = 12x_1^2 = 12 \times 1.8162^2 = 39.583$$

$$\frac{\partial^2 g}{\partial x_2^2} = 24x_2^2 = 24 \times 1.4613^2 = 51.2495$$

$$\frac{\partial^2 g}{\partial u_1^2} = \frac{\partial^2 g}{\partial x_1^2} \sigma_1^2 = 39.583 \times 5^2 = 989.5592$$

$$\frac{\partial^2 g}{\partial u_2^2} = \frac{\partial^2 g}{\partial x_2^2}\sigma_2^2 = 51.2495 \times 5^2 = 1281.2632$$

$$|\nabla g(U^*)| = \sqrt{(\frac{\partial g}{\partial u_1})^2 + (\frac{\partial g}{\partial u_2})^2}$$

$$= \sqrt{(119.8148)^2 + (124.8218)^2} = 173.205$$

$$B = \frac{\nabla^2 g(U^*)}{|\nabla g(U^*)|} = \frac{1}{|\nabla g(U^*)|}\begin{pmatrix} 989.5592 & 0 \\ 0 & 1281.2632 \end{pmatrix}$$

3) Calculate the orthogonal matrix H:

$$H = \begin{pmatrix} \gamma_2^T \\ \gamma_1^T \end{pmatrix} = \begin{pmatrix} -0.7214 & 0.6925 \\ -0.6924 & -0.7214 \end{pmatrix}$$

4) Compute the main curvatures k_j by solving the eigenvalues of HBH^T:

$$HBH^T = \begin{pmatrix} 6.5278 & -0.8423 \\ -0.8423 & 6.5968 \end{pmatrix}$$

From Equation 4.70, $k_1 = 6.5278$, so the main curvature of the failure surface at the MPP is 6.5278.

5) Compute P_f using the Breitung formula:

$$P_f = \Phi(-\beta)\prod_{j=1}^{n-1}(1+k_j\beta)^{-1/2}$$

$$= \Phi(-2.3654)(1+6.5278\times 2.3654)^{-1/2}$$

$$= 9.0040\times 10^{-3} \times 0.2466 = 0.00222059$$

6) Compared with FORM's result ($P_f = 0.009004$), Breitung's method gives a value closer to the Monte Carlo result ($P_f = 0.001950$). It generally provides better results than FORM due to the second-order approximation. However, this method is not valid for $\beta_{kj} \leq -1$ and does not work well for negative curvatures.

4.2.3 Tvedt's Formulation

Based on the second-order approximation of the failure surface given in Equation 4.69, the approximate failure region Ω is defined as

$$\Omega = \left\{ Y \,\Big|\, y_n - \left(\beta + \frac{1}{2}\sum_{i=1}^{n-1} k_i y_i'^2\right) > 0 \right\} \qquad (4.79)$$

The failure probability given in Equation 3.10 can be computed from a formulation in Y-space:

$$P_f = 1 - \int_{-\infty}^{\infty}\cdots\int_{-\infty}^{\infty}\phi(y_1)\cdots\phi(y_{n-1})\int_{\beta+\frac{1}{2}\sum_{i=1}^{n-1}k_i y_i'^2}^{\infty}\phi(y_n)\,dy_n\,dy_{n-1}\cdots dy_1 \qquad (4.80)$$

Tvedt has derived a three-term approximation for this equation by a power series expansion in terms of $\frac{1}{2}\sum_{i=1}^{n-1} k_i y_i'^2$, ignoring terms of orders higher than two. The resulting approximation for P_f is

$$A_1 = \Phi(-\beta)\prod_{i=1}^{n-1}(1+\beta k_i)^{-1/2} \qquad (4.81a)$$

$$A_2 = [\beta\Phi(-\beta) - \phi(\beta)]\left\{\prod_{i=1}^{n-1}(1+\beta k_i)^{-1/2} - \prod_{i=1}^{n-1}(1+(\beta+1)k_i)^{-1/2}\right\} \qquad (4.81b)$$

$$A_3 = (\beta+1)[\beta\Phi(-\beta) - \phi(\beta)]\left\{\prod_{i=1}^{n-1}(1+\beta k_i)^{-1/2}\right\}$$

$$- \mathrm{Re}\left\{\prod_{i=1}^{n-1}(1+(\beta+1)k_i)^{-1/2}\right\} \qquad (4.81c)$$

$$P_f = A_1 + A_2 + A_3 \qquad (4.81d)$$

The first term, A_1, is the Breitung formula of Equation 4.78. $Re\{.\}$ denotes the real part. This method has been found to give very good approximation in most cases. The asymptotic behavior of the three terms can be compared in the asymptotic sense used in Equation 4.78. It may be shown that the ratio of the second term to the first term is

$$\frac{A_2}{A_1} \sim \frac{1}{2\beta^2}\sum_{j=1}^{n-1}\frac{\beta\, k_j}{1-\beta\, k_j}, \quad \beta \to \infty \qquad (4.82)$$

Similarly, the ratio of the third to the first term is

$$\frac{A_3}{A_1} \sim -\frac{3}{8\beta^2} \sum_{j=1}^{n-1}(\frac{\beta\ k_j}{1-\beta\ k_j})^2 - \frac{1}{2\beta^2} \sum_{j=1}^{n-1}\sum_{m=j+1}^{n-1} \frac{\beta^2 k_j k_m}{(1-\beta\ k_j)(1-\beta\ k_m)}, \beta \to \infty \quad (4.83)$$

Since Equation 4.78 is an analytical equation, it is easy to implement the algorithm. The main steps of the Tvedt's formulations are the same as Breitung's, except Step 5) of the previous section, where the failure probability P_f is calculated using Equation 4.81.

Example 4.9

Compute the P_f using the Tvedt method (Equation 4.81) for Example 4.8.

Solution:

Step 1) ~ 4): The first four steps are the same as Example 4.8:

$$\left|\nabla g(U^*)\right| = \sqrt{(\frac{\partial g}{\partial u_1})^2 + (\frac{\partial g}{\partial u_2})^2} = 173.205$$

$$B = \frac{\nabla^2 g(U^*)}{\left|\nabla g(U^*)\right|} = \frac{1}{\left|\nabla g(U^*)\right|} \begin{pmatrix} 989.5592 & 0 \\ 0 & 1281.2632 \end{pmatrix}$$

$$H = \begin{pmatrix} \gamma_2 \\ \gamma_1 \end{pmatrix} = \begin{pmatrix} -0.7214 & 0.6925 \\ -0.6924 & -0.7214 \end{pmatrix}$$

$$HBH^T = \begin{pmatrix} 6.5278 & -0.8423 \\ -0.8423 & 6.5968 \end{pmatrix}$$

From Equation 4.70, $k_1 = 6.5278$

Step 5) Compute the failure probability P_f using the Tvedt formula (Equation 4.81)

The first term of the Tvedt formula is the same as Breitung's method, so

$A_1 = 0.00222059$

$$A_2 = [\beta\Phi(-\beta) - \phi(\beta)]\{\prod_{i=1}^{n-1}(1+\beta k_i)^{-1/2} - \prod_{i=1}^{n-1}(1+(\beta+1)k_i)^{-1/2}\}$$

$$= [2.3654 \times \Phi(-2.3654) - \phi(2.3654)]$$
$$\times \{(1+2.3654 \times 6.5278)^{-1/2} - (1+(2.3654+1)\times 6.5278)^{-1/2}\}$$
$$= 2.2205 \times 10^{-3}$$

$$A_3 = (\beta+1)[\beta\Phi(-\beta)-\phi(\beta)]\{\prod_{i=1}^{n-1}(1+\beta k_i)^{-\frac{1}{2}} - \text{Re}[\prod_{i=1}^{n-1}(1+(\beta+1)k_i)^{-\frac{1}{2}}]\}$$

$$= (2.3654+1)[2.3654\times\Phi(-2.3654)-\phi(2.3654)]$$
$$\times\{(1+2.3654\times 6.5278)^{-1/2} - \text{Re}[(1+(2.3654+1)\times 6.5278)^{-1/2}]\}$$
$$= -1.3297\times 10^{-4}$$

$$P_f = A_1 + A_2 + A_3 = 0.00222059 + 2.2205\times 10^{-3} - 1.3297\times 10^{-4} = 0.002087$$

Compared to the FORM result ($P_f = 0.009004$) and the Breitung result ($P_f = 0.00222059$), Tvedt's method is closer to the Monte Carlo ($P_f = 0.001950$). Like Breitung's algorithm, Tvedt's method is also invalid for $\beta_{kj} \leq -1$ and does not work well in the case of negative curvatures.

4.2.4 SORM with Adaptive Approximations

Wang and Grandhi [20] suggest an adaptive approximation method for SORM. In this method, Breitung's and Tvedt's formulas are used to perform the failure probability calculations. However, the main curvatures are calculated for the nonlinear approximation developed during the safety-index calculations. The second-order derivatives for the closed-form adaptive model representing the original limit-state can be given as

$$\nabla^2 \tilde{g}(U^*)_{ij} = \frac{\partial^2 \tilde{g}(U^*)}{\partial u_i \partial u_j} \quad (4.84)$$

$$= (r-1)\sum_i^n (u_{i,k} + \frac{\bar{x}_i}{\sigma_i} s)^{(1-r)} \frac{\partial g(Y_k)}{\partial u_i} (u_i + \frac{\bar{x}_i}{\sigma_i} s)^{r-2} \nabla^2 \tilde{g}(U^*)_{ij}$$

$$= 0 \quad (i \neq j)$$

By considering Equation 4.62b and the orthogonal transformation of Equation 4.58, the curvature k_i can be approximately determined from Equation 4.70. Since the nonlinear function given in Equation 4.47 is fairly accurate around the MPP when convergence is realized, the calculation of second-order derivatives using this nonlinear approximation would give improved accuracy in failure probability compared to the first-order methods. Also, this procedure avoids the exact second-order derivative computations of the limit-state function at the MPP.

The main steps of SORM having approximate curvatures are summarized as follows for a complete failure probability analysis:

Methods of Structural Reliability   137

1) Conduct the safety search and locate the MPP, U^*
2) Compute the second-order derivatives of the limit-state surface at U^* using Equation 4.84 and form the B matrix given in Equation 4.62b
3) Calculate the orthogonal matrix H based on the procedure given in Section 4.2.1
4) Compute the approximate curvatures k_j of the failure surface at the MPP using Equation 4.70
5) Compute the failure probability P_f using Breitung's formula of Equation 4.78 or Tvedt's formula of Equation 4.81

A significant reduction in computer effort is derived from the use of approximate functions in Step 2 for second-order derivatives, because exact analysis is avoided. Therefore, this method is particularly suitable for problems having implicit performance functions that require large-scale finite element models for structural analysis.

Example 4.10

Compute the failure probability P_f using Wang and Grandhi's SORM for Example 4.8.

Solution:

Steps 1) ~ 4): The first four steps are the same as Example 4.8:

$$|\nabla g(U^*)| = \sqrt{(\frac{\partial g}{\partial u_1})^2 + (\frac{\partial g}{\partial u_2})^2} = 173.205$$

$$B = \frac{\nabla^2 g(U^*)}{|\nabla g(U^*)|} = \frac{1}{|\nabla g(U^*)|}\begin{pmatrix} 989.5592 & 0 \\ 0 & 1281.2632 \end{pmatrix}$$

$$H = \begin{pmatrix} \gamma_2 \\ \gamma_1 \end{pmatrix} = \begin{pmatrix} -0.7214 & 0.6925 \\ -0.6924 & -0.7214 \end{pmatrix}$$

$$HBH^T = \begin{pmatrix} 6.5278 & -0.8423 \\ -0.8423 & 6.5968 \end{pmatrix}$$

From Equation 4.70, $k_1 = 6.5278$

Step 5) Compute the failure probability P_f using the Tvedt formula (Equation 4.81)

The first term of the Tvedt formula is the same as Breitung's method, so $A_1 = 0.00222059$

$$A_2 = [\beta\Phi(-\beta) - \phi(\beta)]\{\prod_{i=1}^{n-1}(1+\beta k_i)^{-1/2} - \prod_{i=1}^{n-1}(1+(\beta+1)k_i)^{-1/2}\}$$

$$= [2.3654 \times \Phi(-2.3654) - \phi(2.3654)]$$

$$\times \{(1+2.3654 \times 6.5278)^{-1/2} - (1+(2.3654+1) \times 6.5278)^{-1/2}\}$$

$$= 2.2206 \times 10^{-3}$$

$$A_3 = (\beta+1)[\beta\Phi(-\beta) - \phi(\beta)]\{\prod_{i=1}^{n-1}(1+\beta k_i)^{-1/2} - \text{Re}[\prod_{i=1}^{n-1}(1+(\beta+1)k_i)^{-1/2}]\}$$

$$= (2.3654+1)[2.3654 \times \Phi(-2.3654) - \phi(2.3654)]$$

$$\times \{(1+2.3654 \times 6.5278)^{-1/2} - \text{Re}[(1+(2.3654+1) \times 6.5278)^{-1/2}]\}$$

$$= -1.3298 \times 10^{-4}$$

$$P_f = A_1 + A_2 + A_3 = 0.00222059 + 2.2206 \times 10^{-3} - 1.3298 \times 10^{-4} = 0.002088$$

This result is very close to the Tvedt's result ($P_f = 0.002087$), with the exact second-order gradients of the limit-state surface. This means that the approximation given in Equation 4.71 accurately represents the real failure surface in this example. Since this method does not require any exact second-order gradient calculations, it can be used for problems where the second-order gradients are expensive or impossible to calculate.

This approach is suitable where the limit states are computed from finite element analysis or computational fluid dynamics. A higher-order surrogate model representing the nonlinear behavior is used as a closed-form equation in safety-index and P_f calculations. As demonstrated, these surrogate models are iteratively updated until the MPP is converged.

4.3 Engineering Applications

In the following sections, several engineering applications are presented as case studies for the purpose of checking new and existing methods.

4.3.1 Ten-bar Truss

The ten-bar truss structure shown in Figure 4.8 is widely used by the design optimization community. The weight of the ten-bar truss is optimized with stress and displacement constraints. Since the horizontal, vertical, and diagonal members are cut from three different aluminum rods, the cross-sectional areas A_1, A_2 and A_3 are considered design variables. These three variables can be potential sources of uncertainties. Therefore, the cross-sectional areas of the horizontal members A_1, vertical members A_2, and diagonal members A_3, are assumed to be random. The random quantities of these three variables are considered as Gaussian distribution with the mean values of 13, 2, and 9 in^2, respectively. The coefficient of variation

(COV) is 0.1 for all three variables. Young's modulus, material density, length, and load are assumed to be deterministic:

- Force: $P = 100{,}000$ lb
- Length: $L = 360$ in
- Young's modulus: $E = 10^7$ psi (for Aluminum)
- Material density: $\rho = 0.1$ lb/in^3
- Allowable stress for members 3 and 7: $\sigma_{allow} = 20{,}000$ psi
- Allowable tip displacement at Node (2): $d_{allow} = 4.0$ in

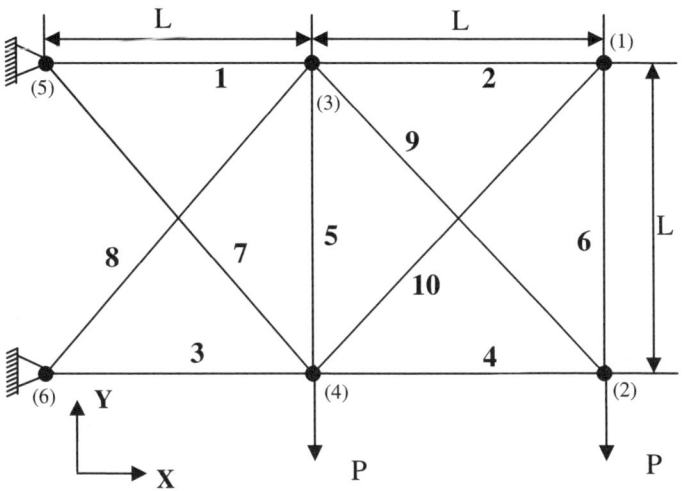

Figure 4.8. Ten-bar Truss Structure

The reliability analysis is applied to check the statistical characteristics of the displacement and stress constraints at the optimum design. To set up the limit-state function, the following closed-form analytical expressions are derived using the finite element analysis procedure. The limit-state function for the tip displacement at Node (2) is given by

$$g_{disp} = \frac{PL}{A_1 A_3 E}\left\{\frac{4\sqrt{2}A_1^3(24A_2^2 + A_3^2) + A_3^3(7A_1^2 + 26A_2^2)}{D_T} + \frac{4A_1 A_2 A_3 (20A_1^2 + 76A_1 A_2 + 10A_3^2) + \sqrt{2}A_3(25A_1 + 29A_2)}{D_T}\right\} - d_{allow} = 0$$

where $D_T = 4A_2^2(8A_1^2 + A_3^2) + 4\sqrt{2}A_1 A_2 A_3(3A_1 + 4A_2) + A_1 A_3^2(A_1 + 6A_2)$

Table 4.4. Probability of Failure for Tip Displacement at Node 2

Limit d_{allow}	MCS(100,000)	FORM	SORM	
			Breitung	Tvedt
3.0000	0.99960	0.99978	0.98328	0.99949
3.5000	0.84420	0.81994	0.81674	0.82106
3.7188	0.52440	0.50000	0.50000	0.50171
4.0000	0.17640	0.15243	0.15303	0.15334
4.5000	0.00720	0.00573	0.00578	0.00579

Table 4.5. Probability of Failure for Stress in Member 3

Limit σ_{allow}	MCS(100,000)	FORM	SORM	
			Breitung	Tvedt
15,000	0.77412	0.77346	0.77331	0.77354
16,124	0.50100	0.50000	0.50000	0.50003
20,000	0.02592	0.02611	0.02612	0.02613
21,000	0.01000	0.01001	0.01002	0.01002
22,000	0.00338	0.00373	0.00373	0.00373

Table 4.6. Probability of Failure for Stress in Member 7

Limit σ_{allow}	MCS(100,000)	FORM	SORM	
			Breitung	Tvedt
17,223	0.49918	0.50000	0.49999	0.49999
20,000	0.07932	0.07975	0.07972	0.07971
21,000	0.03456	0.03418	0.03417	0.03417
22,000	0.01420	0.01388	0.01387	0.01387

The limit-state functions for the stress in member 3 and member 7 are given as:

$$g_3 = \frac{P}{A_1}\left\{2 + \frac{A_1 A_2 A_3 (2\sqrt{2} A_1 + A_3)}{D_T}\right\} - \sigma_{allow} = 0 \text{ and}$$

$$g_7 = \frac{P}{A_3}\left\{2 + \frac{\sqrt{2} A_1 A_2 A_3 (2\sqrt{2} A_1 + A_3)}{D_T}\right\} - \sigma_{allow} = 0$$

With various values of the allowable constraints (σ_{allow} and d_{allow}), the corresponding probability of failure for each limit-state function was obtained by using FORM, SORM, and MCS. In SORM, Breitung was a second-order method and Tvedt used two terms in the expansion. The results are summarized in Table 4.4, Table 4.5, and Table 4.6. Additionally, the CDFs of each limit-state function are plotted in Figure 4.9 and Figure 4.10.

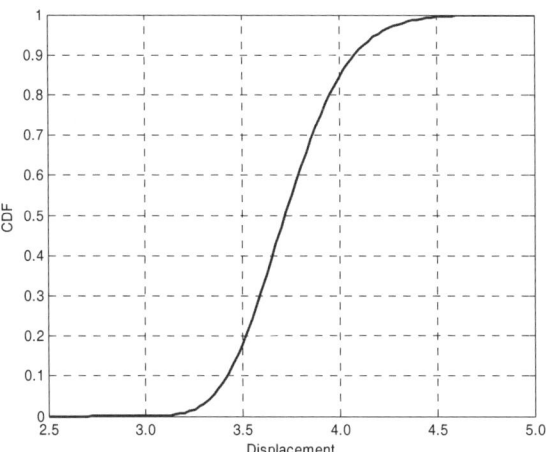

Figure 4.9. CDF Plot of Tip Displacement at Node 2

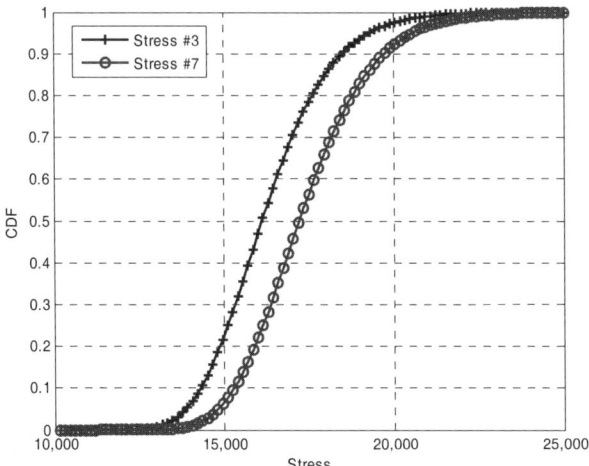

Figure 4.10. CDF Plot of Stresses in Members 3 and 7

4.3.2 Fatigue Crack Growth

This is an example of fatigue crack growth probabilistic analysis of a finite width rectangular plate with an edge crack subject to constant amplitude. The basis of this problem follows the classic *Paris equation* [14], which reflects marginal crack growth by cycle to the cyclic stress at the crack:

$$\frac{da}{dN} = c\Delta K^m \tag{4.85}$$

where $m = 3.32$, a is the crack length, N is the cycle count, c is the *Paris constant*, and ΔK is the change in stress intensity at the crack. The stress intensity is defined by:

$$K = \sigma\sqrt{\pi a} \tag{4.86}$$

where σ is the stress at the crack. The change of the stress intensity is given by:

$$\Delta K = \Delta\sigma\sqrt{\pi a} \tag{4.87}$$

where $\Delta\sigma$ is the change of the stress at the crack.

To find the number of cycles to failure N_f, Equation 4.85 is integrated with respect to a from a initial crack size, a_i, to a final crack size, a_f:

$$N_f = \int_{a_i}^{a_f} \frac{da}{c(\Delta\sigma\sqrt{\pi a})^m} \tag{4.88}$$

because

$$dN = \frac{da}{c\Delta K^m} = \frac{da}{c(\Delta\sigma\sqrt{\pi a})^m} \tag{4.89}$$

The final crack size is determined from

$$a_f = \frac{1}{\pi}\left(\frac{K_{IC}}{1.1215\Delta\sigma}\right)^2 \tag{4.90}$$

where K_{IC} is the fracture toughness. The failure occurs when $K_I > K_{IC}$. Here, the stress intensity factor is given by $K_I = 1.1215\sigma\sqrt{\pi a}$.

After integration and substitution of the final crack size Equation 4.90 into Equation 4.88, the number of cycles to failure N_f is obtained as

$$N_f = \frac{a_f^{1-m/2} - a_i^{1-m/2}}{c(1.1215\Delta\sigma)^m \pi^{m/2}(1-m/2)} \tag{4.91}$$

Table 4.7. Random Variable Properties for Crack Growth Problem

Random Variable	Distribution	Mean	Standard Deviation
Load (ksi), $\Delta\sigma$	Lognormal	100	10
Initial Crack Size (in), a_i	Lognormal	0.01	0.005
Paris constant, c	Lognormal	1.2×10^{-10}	1.2×10^{-11}
Fracture Toughness(ksi/$\sqrt{in}$), K_{IC}	Normal	60	6

After production, every disk has some small scratches, nicks, or gouges that serve as initial cracks of varying sizes. The crack's radial position changes the stress at the crack from the inertial load of the spinning disk. Material variability is also taken into account by the random fracture toughness K_{IC} and the random Paris constant c in the crack growth equation. The random variables and their distributions were validated by previous operational experience [14]. The corresponding random properties are given in Table 4.7.

Table 4.8. Probability of Failure for Crack Growth Problem

Limit N_{allow}	MCS(100,000)	FORM	SORM	
			Breitung	Tvedt
3,000	0.20670	0.15774	0.14473	0.13987
4,000	0.37280	0.30410	0.28996	0.27646
5,000	0.52760	0.45354	0.44831	0.42068
5,334	0.55450	0.50001	0.50007	0.46648
6,000	0.63860	0.58516	0.59857	0.55125
7,000	0.71980	0.69900	0.73135	0.65903

The probability of failure that N_f is less than the limit, N_{allow}, is computed. The results of this crack growth problem are summarized in Table 4.8, and the CDF plot is shown in Figure 4.11.

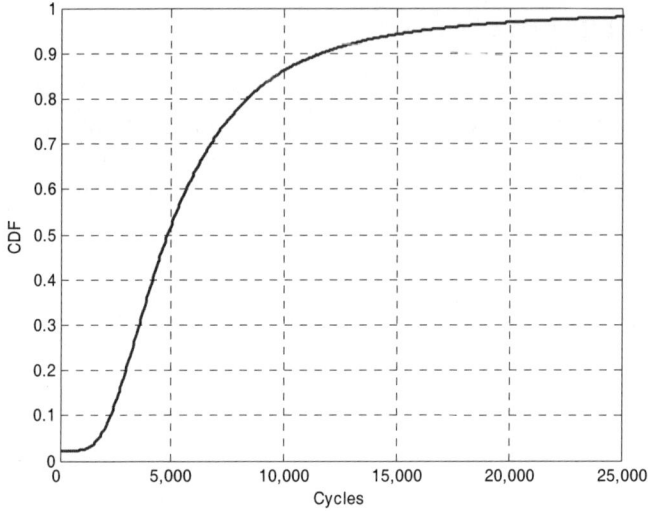

Figure 4.11. Cumulative Distribution Function of Cycles to Failure

4.3.3 Disk Burst Margin

In the disk burst margin problem [15], a margin is expressed as the square root of the ratio of strength to the applied load:

$$M_b = \sqrt{\frac{F\sigma_u}{\sigma_a}} = \sqrt{\frac{F\sigma_u}{\dfrac{\rho}{3(385.82)}\left(\dfrac{2\pi\omega}{60}\right)^2 \dfrac{r_o^3 - r_i^3}{r_o - r_i}}} \qquad (4.92)$$

The ultimate tensile strength of the material (σ_u), is adjusted by a material utilization factor (F) that accounts for uncertainties and unknown material properties. The denominator represents the average tangential stress in the disk. The density of the material (ρ), the disk or rotor speed (ω), and the thickness of the disk [outer radius (r_o) - inner radius (r_i)] are used to determine the average tangential stress (σ_a). A margin of safety is defined as the burst margin of less than one indicates the failure condition or a burst disk.

A six-variable (Table 4.9) limit-state is considered for the reliability analysis. The probability of failure that M_b is less than the limit, M_{allow}, is computed. The results of FORM, the importance sampling method, and MCS are summarized in Table 4.10, and the CDF plot is shown in Figure 4.12.

Table 4.9. Random Variable Properties for Disk Burst Margin

	Distribution Type	Mean	Standard Deviation
σ_u	Normal	220,000	5,000
ω	Normal	21,000	1,000
r_o	Normal	24	0.5
r_i	Normal	8	0.3
ρ	Uniform	[0.28, 0.30]	
F	Weibull	(25.508, 0.958)	

Table 4.10. Probability of Failure

Limit	MCS (100,000)	Importance Sampling (10,000)	FORM
0.440	0.19132	0.18019	0.18025
0.460	0.45669	0.46942	0.43838
0.463	0.50157	0.52000	0.51756
0.480	0.73331	0.70000	0.71575
0.500	0.90556	0.87000	0.89616

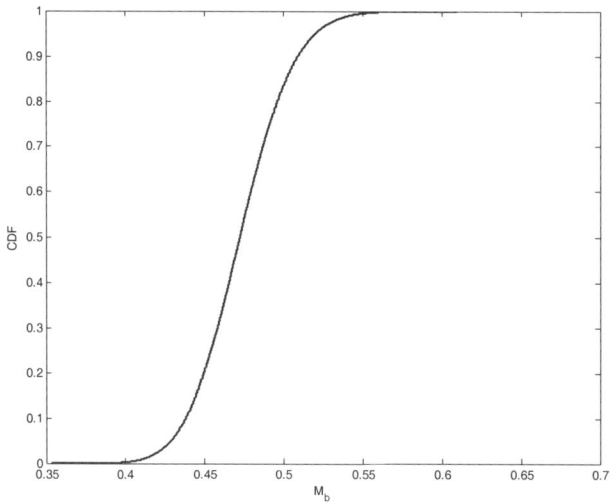

Figure 4.12. Cumulative Distribution Function of Disk Burst Margin

4.3.4 Two-member Frame

Consider the design of a two-member frame subjected to out-of-plane loads as shown in Figure 4.13. Such frames are encountered in numerous automotive, aerospace, mechanical, and structural engineering applications. The problem of minimizing the volume of the frame subject to stress and size limitations is presented by Arora [1]. Since the optimum structure is symmetric, the two members of the frame are identical. Also, hollow rectangular sections are used as members with three design variables: width d (in), height h (in), and wall thickness t (in). The volume of the frame structure is considered as the cost function:

$$f(x) = 2L(2dt + 2ht - 4t^2) \tag{4.93}$$

The members are subjected to both bending and torsional stresses, and the combined stress constraint is imposed at points 1 and 2. Let σ and τ be the maximum bending and shear stress in the member, respectively. The failure criterion for the member is based on the von Mises yield criterion. The effective stress σ_e is given as $\sqrt{\sigma^2 + 3\tau^2}$, and the stress constraint is written in a normalized form as

$$\frac{1}{\sigma_a^2}(\sigma^2 + 3\tau^2) - 1.0 \leq 0 \tag{4.94}$$

where σ_a is the allowable design stress. ($\sigma_a = 40,000$ psi)

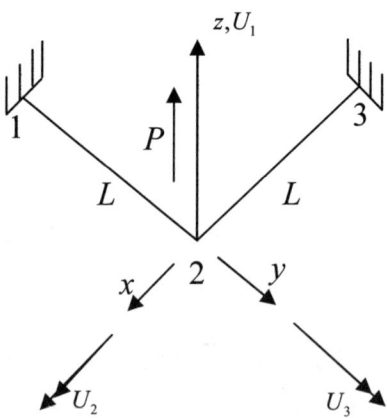

Figure 4.13. Two-member Frame

The stresses are calculated from the member-end moments and torques. The three generalized nodal displacements (deflections and rotations) of the frame structure

(Figure 4.13) are defined as U_1 (vertical displacement) at node 2, U_2 (rotation about line 3-2), and U_3 (rotation about line 1-2). With this configuration, the equilibrium equation of the finite element analysis can determine the displacements U_1, U_2, and U_3:

$$\frac{EI}{L^3}\begin{bmatrix} 24 & -6L & 6L \\ -6L & \left(4L^2+\frac{GJ}{EI}L^2\right) & 0 \\ 6L & 0 & \left(4L^2+\frac{GJ}{EI}L^2\right) \end{bmatrix}\begin{bmatrix} U_1 \\ U_2 \\ U_3 \end{bmatrix} = \begin{bmatrix} P \\ 0 \\ 0 \end{bmatrix} \quad (4.95)$$

where
- E = Young's modulus, (3.0×10^7) psi,
- L = Member length, 100 in,
- G = Shear modulus, (1.154×10^7) psi,
- P = Load at node 2, -10,000 lb,
- I = Moment of inertia = $\frac{1}{12}\{dh^3 - (d-2t)(h-2t)^3\}$ in^4,
- J = Polar moment of inertia = $\frac{2t(d-t)^2(h-t)^2}{(d+h-2t)}$, in^4 and
- A = Area of torsional shear stress = $(d-t)(h-t)$ in^2.

By solving the system equation (Equation 4.95), the closed-form expressions for U_1, U_2, and U_3 are obtained as:

$$U_1 = \frac{PL^3(4EI+GJ)}{24EI(EI+GJ)}, \quad U_2 = \frac{PL^2}{4(EI+GJ)}, \quad U_3 = \frac{-PL^2}{4(EI+GJ)}$$

The torque and bending moments at points 1 and 2 for member 1-2 (symmetry) are calculated as

$$T = -\frac{GJ}{L^3}U_3 \text{ lb-in} \quad (4.96)$$

$$M_1 = \frac{2EI}{L^2}(-3U_1 + U_2L) \text{ lb-in (moment at end 1)} \quad (4.97)$$

$$M_2 = \frac{2EI}{L^2}(-3U_1 + 2U_2L) \text{ lb-in (moment at end 2)} \quad (4.98)$$

Using these moments, the torsional shear and bending stresses are calculated as

$$\tau = \frac{T}{2At} \text{ psi} \tag{4.99}$$

$$\sigma_1 = \frac{1}{2I} M_1 h \text{ psi (bending stress at end 1)} \tag{4.100}$$

$$\sigma_2 = \frac{1}{2I} M_2 h \text{ psi (bending stress at end 2)} \tag{4.101}$$

After conducting the optimization procedure for minimizing the volume of the frame, the optimum design of the cross-sectional variables are obtained as $d = 8$, $h = 10$, and $t = 0.1$. With this optimum, the other calculations can be readily conducted:

$$I = \frac{1}{12}\{(8)(10)^3 - (7.8)(9.8)^3\} = 54.8919 \text{ in}^4$$

$$J = \frac{2(0.1)(7.9)^2 (9.9)^2}{(17.8)} = 68.7281 \text{ in}^4 \quad A = (7.9)(9.9) = 78.21 \text{ in}^2$$

$$GJ = (1.154 \times 10^7)(68.7281) = 7.9312 \times 10^8$$

$$EI = (3.0 \times 10^7)(54.8919) = 1.6468 \times 10^9$$

$$\left(4L^2 + \frac{GJ}{EI}L^2\right) = \left(4 + \frac{7.9312 \times 10^8}{1.6468 \times 10^9}\right)(100)^2 = 4.4816 \times 10^4$$

Thus, the equilibrium equation, Equation 4.95, is calculated as

$$1646.756 \begin{bmatrix} 24 & -600 & 600 \\ -600 & 44816.2732 & 0 \\ 600 & 0 & 44816.2732 \end{bmatrix} \begin{bmatrix} U_1 \\ U_2 \\ U_3 \end{bmatrix} = \begin{bmatrix} -10000 \\ 0 \\ 0 \end{bmatrix}$$

Solving the preceding equation, the three generalized displacements of node 2 are obtained as $U_1 = -0.7653$, $U_2 = -0.0102$, and $U_3 = 0.0102$. The torque (Equation 4.96) and the bending moments (Equation 4.98) at points 1 and 2 are calculated as

$$T = -\frac{7.9312 \times 10^8}{100}(0.0102) = -8.1266 \times 10^4 \text{ lb-in}$$

$$M_1 = \frac{2(3.0 \times 10^7)(54.8919)}{100^2} \{-3(-0.7653) + (-0.0102)(100)\}$$
$$= 4.1873 \times 10^5 \text{ lb-in}$$

$$M_2 = \frac{2(3.0 \times 10^7)(54.8919)}{100^2} \{-3(-0.7653) + 2(-0.0102)(100)\}$$
$$= 8.1267 \times 10^4 \text{ lb-in}$$

Because of $M_1 > M_2$, σ_1 will be larger than σ_2 as observed from Equations 4.100 and 4.101. Therefore, the $g_1(x)$ constraint (Equation 4.102) is only imposed. The torsional shear stress and bending stress at point 1 are calculated from Equations 4.99 and 4.100 as

$$\tau = \frac{(-8.1266 \times 10^4)}{2(78.21)(0.1)} = -5.195 \times 10^4 \text{ psi}$$

$$\sigma_1 = \frac{1}{2(54.8919)}(4.1873 \times 10^5)(10) = 3.8142 \times 10^5 \text{ psi}$$

To check the statistical characteristics of the stress constraint, Equation 4.94, set the limit-state as:

$$g_1(x) = \frac{1}{\sigma_a^2}(\sigma_1^2 + 3\tau^2) - 1.0 = 0 \quad (4.102)$$

Six quantities of the two-member frame are assumed to be random as shown in Table 4.11. With various levels of the allowable stress, σ_a, the probability of failures of Equation 4.102 are computed and the results are summarized in Table 4.12 and Figure 4.14.

Table 4.11. Random Variables of Two-member Frame

Random Variable	Distribution	Mean	Standard Deviation
Width (d)	Normal	8	0.8
Height (h)	Normal	10	1
Wall Thickness (t)	Normal	0.1	0.01
Young's Modulus (E)	Lognormal	3×10^7	3×10^6
Length (L)	Normal	100	10
Load (P)	Lognormal	10,000	1,000

Table 4.12. Probability of Failure for Two-member Frame

σ_a	MCS(100,000)	FORM	SORM	
			Breitung	Tvedt
20,000	0.99846	0.99869	0.99869	0.99869
30,000	0.88524	0.88524	0.88524	0.88524
38977	0.48218	0.50000	0.50000	0.50000
40,000	0.49532	0.45436	0.45436	0.45436
50,000	0.14324	0.13776	0.13776	0.13776
60,000	0.03252	0.03079	0.03079	0.03079

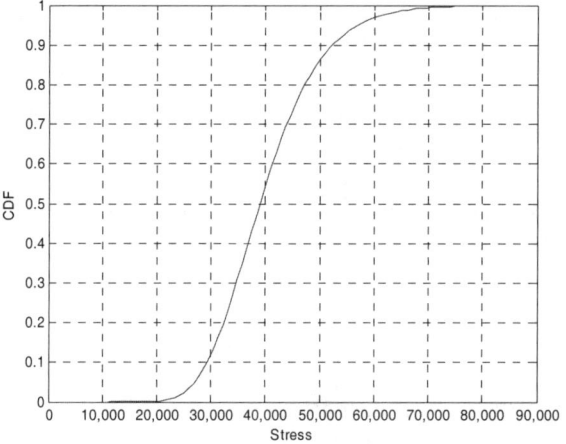

Figure 4.14. CDF Plot of Two-member Frame

Methodologies of FORM and SORM have been presented and implemented for structural reliability analysis of various engineering structures such as truss, frame, crack growth, and disk problems. The key feature of the methods is that the efficient selection of suitable approximations at different stages of reliability analysis makes these tools practical for many large-scale engineering problems. Potential extentions of the methods in the design procedure will be shown in the following chapters.

4.4 References

[1] Arora, J.S., *Introduction to Optimum Design*, Elsevier Academic Press, London, UK, 2004.
[2] Bleistein, N., and Handelsman, R.A., "Asymptotic Expansions of Integrals," Holt, Rinehart and Winston, New York, NY, 1975.

[3] Breitung, K., "Asymptotic Approximations for Multinormal Integrals," *Journal of the Engineering Mechanics Division*, ASCE, Vol. 110, No. 3, Mar., 1984, pp 357-366.

[4] Cai, G.Q., and Elishakoff, I., "Refined Second-order Approximations Analysis," *Structural Safety*, Vol. 14, No. 4, April, 1994, pp. 267-276.

[5] Der Kiureghian, A., Lin, H.Z., and Hwang, S.J., "Second Order Reliability Approximations," *Journal of Engineering Mechanics*, ASCE, Vol. 113, 1987, pp. 1208-1225.

[6] Freudenthal, A. M., Garrelts, J. M., and Shinozuka, M., "The Analysis of Structural Safety," *Journal of the Structural Division*, ASCE, Vol. 92, No. ST1, 1966, pp. 267-325.

[7] Grandhi, R.V., and Wang, L.P., *Structural Reliability Analysis and Optimization: Use of Approximations*, NASA CR-1999-209154, 1999.

[8] Hasofer, A.M., and Lind, N.C., "Exact and Invariant Second-Moment Code Format," *Journal of the Engineering Mechanics Division*, ASCE, 100(EM), 1974, pp.111-121.

[9] Hohenbichler, M., and Rackwitz, R., "Improvement of Second-order Reliability Estimates by Importance Sampling," *Journal of the Engineering Mechanics Division*, ASCE, Vol. 116, 1990, pp. 1183-1197.

[10] Hohenbichler, M., and Rackwitz, R., "Non-normal Dependent Vectors in Structural Safety," *Journal of the Engineering Mechanics Division*, ASCE, Vol. 107, No. EM6, Dec. 1981, pp.1227-1238.

[11] Koyluoglu, H.U., and Nielsen, S.R.K., "New Approximations for SORM Integrals," *Structural Safety*, Vol. 13, No. 4, April, 1994, pp. 235-246.

[12] Madsen, H.O., Krenk, S., and Lind, N.C., *Methods of Structural Safety*, Prentice-Hall, Englewood Cliffs, New Jersey, 1986.

[13] Melchers, R.E., *Structural Reliability Analysis and Prediction*, Ellis Horwood Limited, UK., 1987.

[14] Millwater, H.R., Wu, Y-.T., and Cardinal, J.W., "Probabilistic Structural Analysis of Fatigue and Fracture," AIAA-94-1507-CP, *Proceedings of the 35th AIAA Structures, Structural Dynamics and Materials Conference*, Hilton Head, SC, April, 1994.

[15] Penmetsa, R.C., and Grandhi, R.V., "Adaptation of Fast Fourier Transformations to Estimate Structural Failure Probability," *Finite Elements in Analysis and Design*, Vol. 39, No. 5, 2003, pp. 473–485.

[16] Rosenblatt, M., "Remarks on a Multivariate Transformation," *The Annals of Mathematical Statistics*, Vol. 23, No. 3, 1952, pp. 470-472.

[17] Todd, J., *Survey of Numerical Analysis*, McGraw Hill, NY, 1962.

[18] Tvedt, L., "Two Second-order Approximations to the Failure Probability," *Section on Structural Reliability*, A/S Vertas Research, Hovik, Norway, 1984.

[19] Tvedt, L., "Distribution of Quadratic Forms in Normal Space Applications to Structural Reliability," *Journal of the Engineering Mechanics Division*, ASCE, Vol. 116, 1990, pp. 1183-1197.

[20] Wang, L.P., and Grandhi, R.V., "Improved Two-point Function Approximation for Design Optimization," *AIAA Journal*, Vol. 32, No. 9, 1995, pp. 1720-1727.

5

Reliability-based Structural Optimization

This chapter discusses the importance of multidisciplinary optimization and of including reliability-based constraints in design. Before addressing reliability issues, we first present a brief introduction to mathematical programming techniques, algorithms, sensitivity analysis, design variables linking, and function approximations. The main goal of this chapter is to introduce the power of design optimization tools for risk minimization. Whenever design modification for reliability improvement is involved, one way of solving the problem is as a nested optimization. As demonstrated in the previous chapter, the reliability index calculation is itself an iterative process, potentially employing an optimization technique to find the shortest distance from the origin to the limit-state boundary in a standard normal space. This optimization loop provides just the β value. At a higher level, the design engineer would like to modify the geometries, shapes, sizes, material properties, and topology to reduce the failure probability for the critical limit states.

5.1 Multidisciplinary Optimization

Figure 5.1 shows a design scheme representing various segments involved in multidisciplinary optimization. This design procedure is widely used in aerospace and automotive fields where weight, cost, and performance are crucial and constant improvements are demanded to meet the challenges posed by ambitious goals. Multidisciplinary optimization is most effective during the preliminary stage of structural design. At this stage, the configuration has been defined and the materials have been selected. The design task is to determine the structural sizes that will provide an optimal structure while satisfying the numerous requirements that multiple disciplines impose on the structure. Mathematical optimization tools shorten the design cycle time and provide locally optimal designs. A computer-aided design system brings the requirements of diversified disciplines into a design framework and simultaneously considers all the goals before reaching an acceptable and improved solution.

154   Reliability-based Structural Design

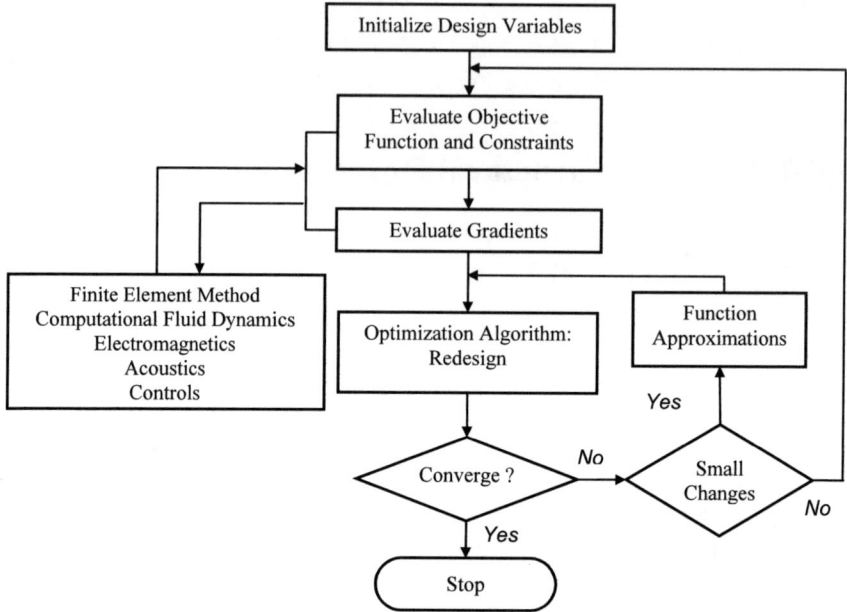

Figure 5.1. Multidisciplinary Optimization Segments

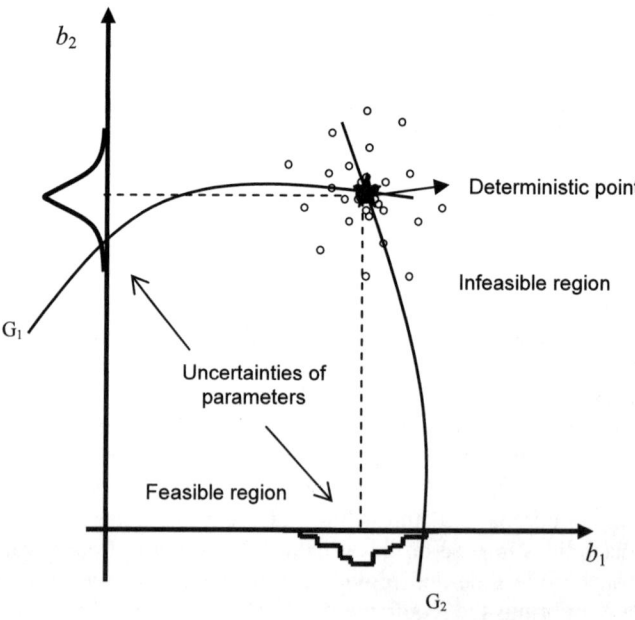

Figure 5.2. Need for Reliability-based Design Optimization

Figure 5.2 shows an optimum for a constrained minimization problem. In traditional optimization problems, all the input information, design variables, and constraints are deterministic, and a deterministic solution is found as shown. In real world applications, the design variables are uncertain, perhaps with distributions as shown in Figure 5.2, and the resulting optimum could be scattered around the single deterministic solution. Some of these scattered solutions are feasible and some are not, resulting in premature failures and unusable designs or products. In order to account for the uncertainties in variables, operating environments, physical models, and solution algorithms, a reliability-based design that considers the randomness of the variables must be carried out to minimize risk. This chapter examines the inclusion of reliability-based constraints, their sensitivities with respect to random and design variables, and incorporation of surrogate models for cost-effective solution schemes.

5.2 Mathematical Problem Statement and Algorithms

A general optimization problem is defined in a mathematical form as

Minimize: the objective function, $f(b)$ (5.1)

Subject to:

$G_j(b) \leq 0 \qquad j = 1,...,m \qquad$ inequality constraints (5.2)

$h_k(b) = 0 \qquad k = 1,...,s \qquad$ equality constraints (5.3)

$b_i^l \leq b_i \leq b_i^u \qquad i = 1,...,n \qquad$ bound constraints (5.4)

where $b = \{b_1, b_2,...,b_n\}^T$ are design variables. The optimizer searches for the best design within the design space defined by the above problem statement.

As shown in Figure 5.3, there are many algorithms that can solve the above-stated mathematical problem (e.g. gradient projection method, feasible directions algorithm, Lagrange multiplier's method, interior and exterior penalty methods, sequential quadratic programming, *etc.*). Each algorithm has certain merits when solving a specific problem; however, all these methods must face the implicit nature of objective and constraint functions. Figure 5.3 presents a short list of solution techniques for a nonlinear programming problem. These methods are classified into two broad categories: the direct approach and the indirect approach. In the direct approach, the constraints are handled in an explicit manner, whereas in the indirect approach, the constrained problem is solved as a sequence of unconstrained minimization problems. We present some of the popular methods with an example problem.

156   Reliability-based Structural Design

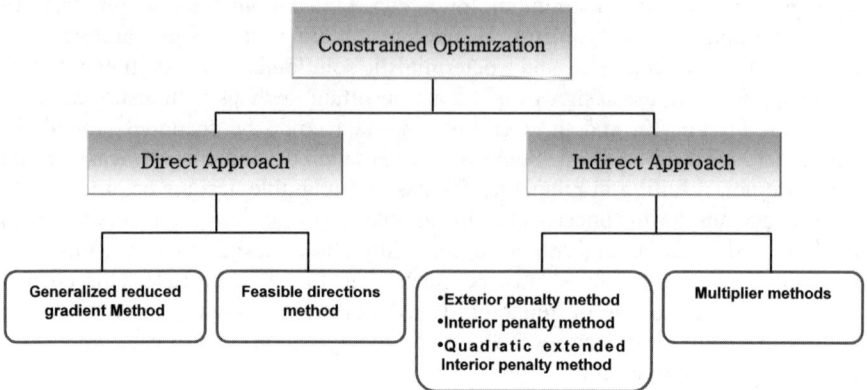

Figure 5.3. Classification of Optimization Algorithm

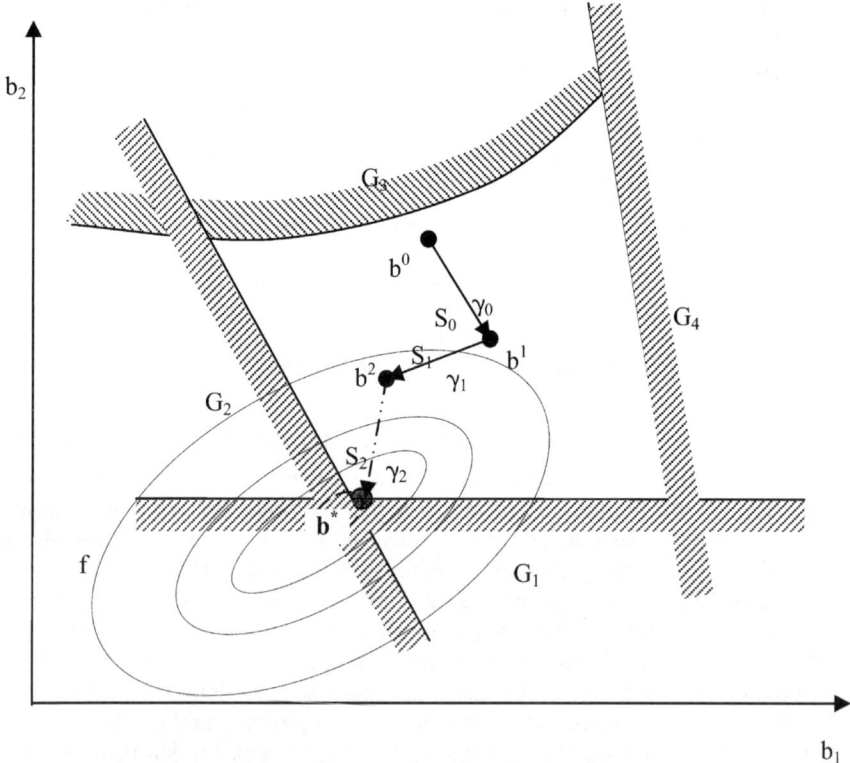

Figure 5.4. Iterative Optimization

For the most part, these are nonlinear functions and require computationally intensive finite element analysis, computational fluid mechanics, or acoustic analysis. Since the optimization scheme is essentially iterative (Figure 5.4), it involves a large number of function re-analyses. Therefore, the computational cost often becomes prohibitive when large-scale systems are optimized with multidisciplinary requirements. To make the design problem tractable, various approximation concepts are utilized throughout the design process. These include design variable linking, temporary deletion of unimportant constraints, and the generation of high quality explicit approximations for the implicit functions. The following section offers a brief description of these effective tools.

5.3 Mathematical Optimization Process

As shown in Figure 5.4, the problem stated in Equations 5.1 ~ 5.4 are solved iteratively by starting from an initial point b^0 in n-dimensional space. The design variables vector is updated as

$$b^q = b^{q-1} + \gamma^q S^q \tag{5.5}$$

where q is the iteration number, b is the vector of design variables, S is the search direction vector, and γ is the step size along the search direction.

The two main steps involved in constrained optimization are (i) determine search direction by considering the cost function and constraints, and (ii) determine travel distance along the selected direction. There are many algorithms for addressing these two steps, and there are many excellent books covering a wide spectrum of algorithms ([1], [7], [13], and [17]) with elaborate details and examples. Since the emphasis of this textbook is on reliability-based design, representative methods from the design optimization field are selected to introduce the integrated design approach. This chapter addresses the procedure for incorporating reliability-based constraints and shows how various segments are integrated to solve large-scale, complex, multidisciplinary and computationally intensive problems.

5.3.1 Feasible Directions Algorithm

In the feasible directions algorithm, we basically choose a feasible starting point and move to a better point. This is classified as the direct approach for the constrained problem. The typical iteration starts at the boundary of the feasible domain. If no constraint is active or violated, an unconstrained minimization technique is used to find the required search direction.

A vector S is defined as a feasible direction if at least a small step taken along it does not immediately leave the feasible region (Figure 5.5):

$$S^T \nabla G_j(b) \le 0, \quad j \in J \tag{5.6}$$

where J represents the set of active constraints, $G_j(b) = 0$. For numerical considerations, an active constraint is defined as one with a value between a small negative number and a small positive number. Any S vector which reduces the objective function is called a usable direction. Mathematically, it is written as

$$S^T \nabla f(b) \le 0 \tag{5.7}$$

There are many ways to choose usable feasible directions. We have two criteria for selecting a direction, (i) reducing the objective function as much as possible, and (ii) staying away from the constraint boundary as much as possible. These two goals are accomplished by solving the following optimization problem:

Maximize: τ

Subject to:

$$S^T \nabla G_j + \theta_j \tau \le 0 \qquad \theta_j \ge 0, \; j \in J \qquad \text{for active constraints}$$

$$S^T \nabla f + \tau \le 0 \tag{5.8}$$

$$-1 \le s_i \le 1, \quad i = 1, 2, \ldots n$$

where s_i is the ith component of search direction S, θ_j are positive numbers called "push-off" factors that determine how far b will move from the constraint boundaries, and τ is an additional design variable.

In this problem, θ_j are arbitrary positive scalar constants, and for simplicity, $\theta_j = 1$. Any solution of this problem with $\tau > 0$ is a usable feasible direction. The maximum value of τ keeps the direction S away from the active nonlinear constraint boundaries.

The objective function τ and the constraint equations are linear in terms of the optimization variables, $s_1, s_2, \ldots, s_n$, and τ. The linear optimization problem can be solved using the simplex algorithm. Once a search direction is found, the travel distance (step size) is typically determined based on a prescribed reduction in the objective function using a first-order Taylor series expansion:

$$f(b^{q+1}) \cong f(b^q) + \gamma \, S^T \nabla f \tag{5.9}$$

or

$$f(b^{q+1}) - f(b^q) \cong \gamma \, S^T \nabla f$$

Considering a target percentage reduction p in the cost function,

$$f(b^q) - f(b^{q+1}) = pf(b^q) \tag{5.10}$$

Substituting Equation 5.9 into Equation 5.10, we obtain the step size as

$$\gamma^* = \frac{-p\, f(b^q)}{S^T \nabla f}$$

Here $S^T \nabla f$ is negative and p is, say, 0.1, for a 10% reduction.

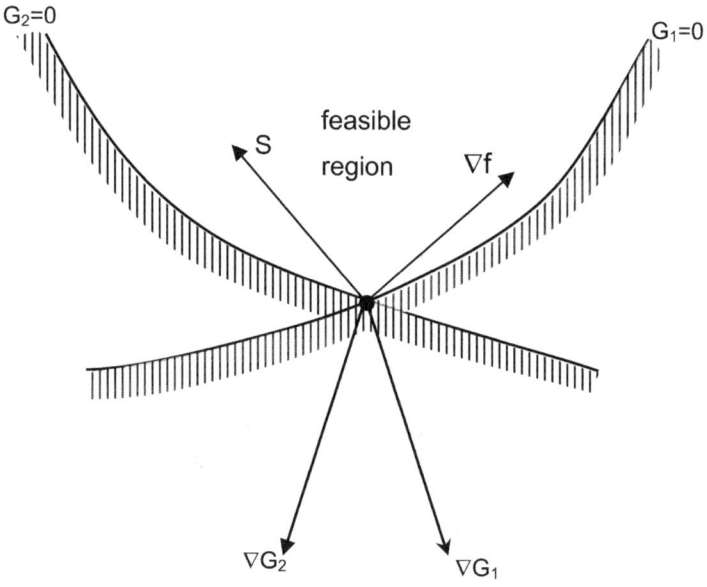

Figure 5.5. Usable-feasible Search Direction

5.3.2 Penalty Function Methods

The penalty function methods transform the constraint problem into a sequence of unconstrained minimization problems; these are classified as indirect approaches. The optimization problem is defined using only inequality constraints:

Minimize: $f(b)$

Subject to:

$$G_j(b) \leq 0 \quad j=1,...,m \tag{5.11}$$

and side bounds

$$b_i^l \leq b_i \leq b_i^u \quad i=1,...,n \tag{5.12}$$

This problem is converted into an unconstrained problem by augmenting the penalty terms:

$$F(b,r_q) = f(b) + r_q \sum_{j=1}^{m} p(G_j(b)) \tag{5.13}$$

where $p(G_j(b))$ is a penalty function based on constraints, r_q is a positive constant, known as the penalty parameter, and $F(b,r_q)$ is a pseudo-objective function, which is minimized in each iteration. Since the unconstrained minimization of $F(b,r_q)$ is repeated for a sequence of penalty parameter values, these methods are known as Sequential Unconstrained Minimization Techniques (SUMT).

The penalty function formulations for inequality-constrained problems can be divided into two categories: interior and exterior methods. We present brief descriptions of each with merits and drawbacks and an improved method, the extended interior penalty function method.

Interior Penalty Function Method

The pseudo-objective function is defined as

$$F(b,r_q) = f(b) - r_q \sum_{j=1}^{m} \frac{1}{G_j(b)} \tag{5.14}$$

$$r = r_1, r_2, ..., \quad r_q \to 0; \quad r_q > 0$$

In this method, the unconstrained minima of $F(b,r_q)$ stays in the feasible region. Once the unconstrained minimization of $F(b,r_q)$ is started from any feasible point, the subsequent points generated will always lie within the feasible domain, since the constraint boundaries act as barriers during the minimization process. If any constraint $G_j(b)$ is satisfied critically, the value of $F(b,r_q)$ tends to infinity. Critically satisfied means that the current design point is on the constraint boundary $(G_j(b)=0)$. The penalty term in Equation 5.14 is not defined in the infeasible region, and the search is confined to the feasible region.

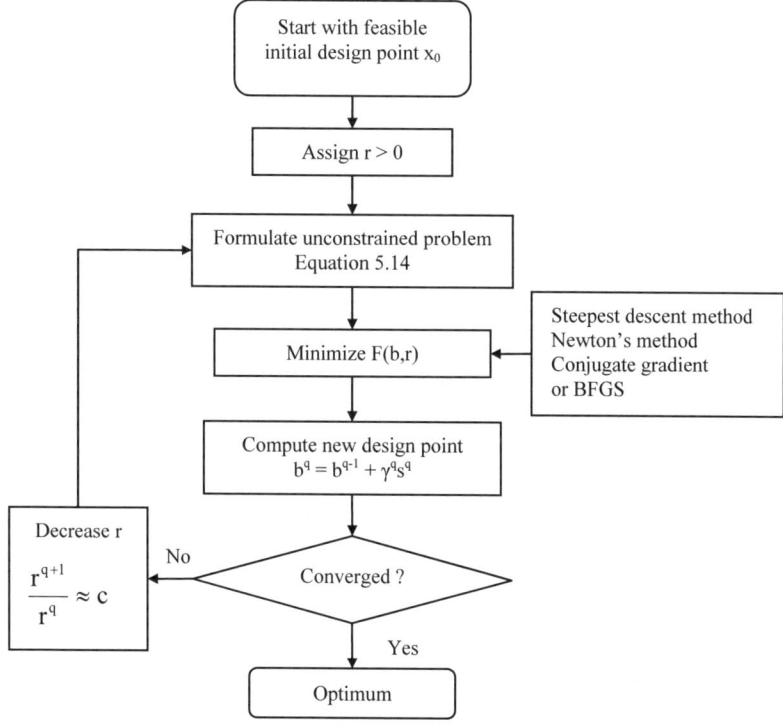

Figure 5.6. Interior Penalty Method Algorithm

The unconstrained function given in Equation 5.14 is solved by using methods such as steepest descent, conjugate gradient, Newton's method, *etc.* ([1], [7], [13], and [17]). An initial value of r is selected, typically based on the $F(b,r_q)$ value, and it is approximately equal to 1.1 to 2.0 times the $f(b)$ value at the initial design. Once the initial value r is selected, the subsequent values are chosen according to the relation

$$r_{q+1} = cr_q$$

where c is a constant. Example values of c are 0.1, 0.2, or 0.5. The unconstrained minimization is in terms of n variables and also uses one-dimensional search such as the Golden Section algorithm. Figure 5.6 shows the algorithms flow-chart.

Exterior and Quadratic Extended Interior Penalty Functions

The exterior penalty function associates a penalty with a violation of a constraint. The term 'exterior' refers to the fact that penalties are applied only in the exterior of the feasible domain.

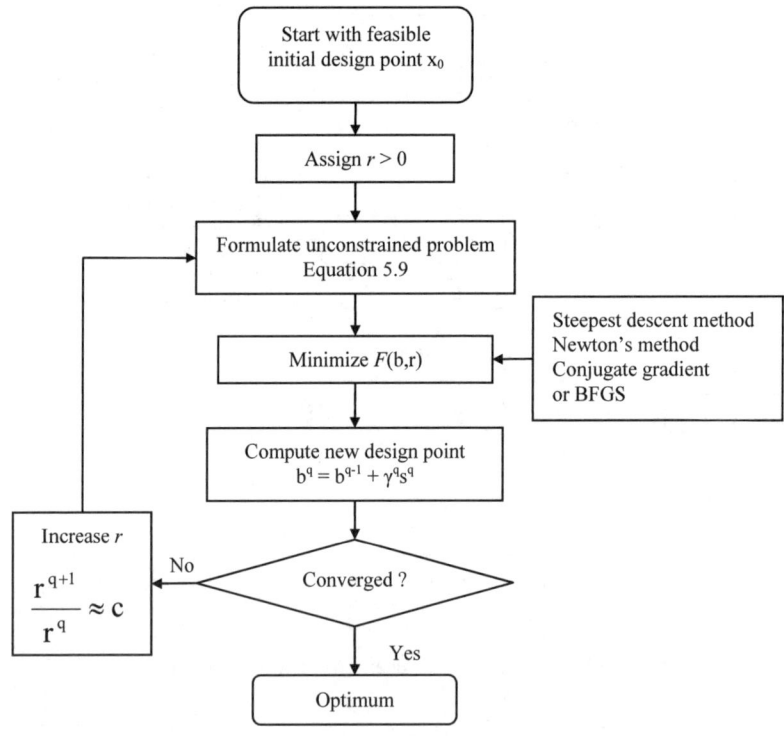

Figure 5.7. Exterior Penalty Method Algorithm

The pseudo-objective function is defined as

$$F(b,r) = f(b) + r_q \sum_{j=1}^{m} \left\{ \max\left[0, G_j(b)\right] \right\}^2 \qquad (5.15)$$

$r = r_1, r_2, ..., \quad r_q \to \infty;$

We begin with a small r_q and minimize $F(b, r_q)$. Then, r_q is gradually increased to $r_{q+1} = cr_q$, where the constant c is typically 5. Figure 5.7 shows the algorithm flow-chart. Here $G_j^+(b)$ is used to compute $\max[0, G_j(b)]$.

Quadratic Extended Interior Penalty Functions Method

In the exterior penalty function, the penalty terms are added only when constraints are violated. As a result, the design typically moves into the infeasible domain. If the iterative process is terminated due to design cycle times and costs, the resulting design is not useful. The interior penalty function method produces a sequence of improving feasible designs. This method requires a feasible starting design, and the search is confined to a feasible domain. As shown in Figure 5.1, the iterative scheme uses many levels of approximations and may lead to infeasible designs. The drawbacks of these two methods demand an approach that is continuous everywhere and provides a sequence of improving feasible designs. Haftka and Gurdal [7] present the genesis of extended interior penalty function methods. Here the quadratic extended interior penalty function is presented in which a combination of interior and exterior penalty features are incorporated.

$$F(b,r) = f(b) + r_q \sum_{j=1}^{m} p(G_j) \quad (5.16)$$

$r = r_1, r_2, ... \quad r_q \to 0$

where

$$p(G_j) = \begin{cases} -\dfrac{1}{G_j} & \text{for } G_j \leq G_0 \\ -\dfrac{1}{G_j}\left[\left(\dfrac{G_j}{G_0}\right)^2 - 3\left(\dfrac{G_j}{G_0}\right) + 3\right] & \text{for } G_j > G_0 \end{cases} \quad (5.17)$$

G_0 is the transition parameter that defines the boundary between the interior and exterior parts of the penalty terms, and it is selected as

$$G_0 = -k \, r^{\frac{1}{2}} \quad (5.18)$$

where k is a constant. A reasonable value of G_0 is -0.1. For a defined r value, k is computed. At the beginning of each unconstrained minimization, G_0 is defined by the above equation, and it is kept same throughout the unconstrained minimization.

In the following section, example problems are presented to demonstrate the complete details of solving optimization algorithms. Detailed steps are presented for finding search directions and for one-dimensional step-size algorithms. Multiple iterations are presented with detailed steps. Finally, figures are included to show convergence to an optimum and a pictorial representation of design point movements.

Example 5.1

Using the feasible directions method, find the minimum weight design of the 4-bar truss structure, subject to the following stress and displacement constraints:

a) Allowable stresses in tension = $8.74 \times 10^{-4} E$
b) Allowable stresses in compression = $4.83 \times 10^{-4} E$
c) Tip vertical displacement should be $\leq 3 \times 10^{-3} \ell$

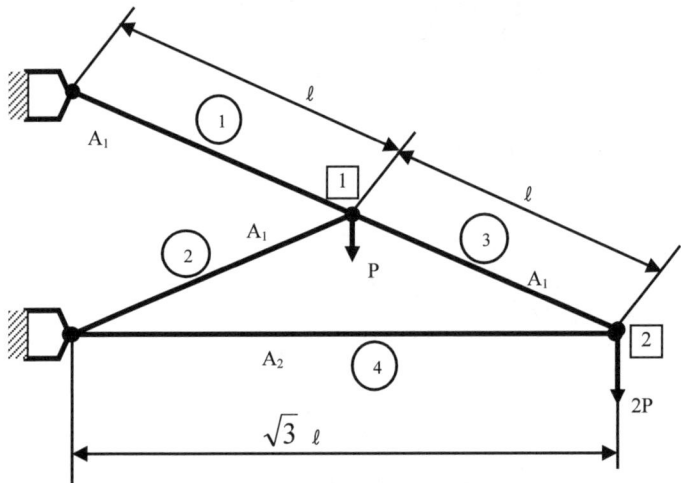

Figure 5.8. Four-bar Statically Determined Truss

Solution:

In order to demonstrate the problem using closed-form expression, we assume members 1 through 3 have the same cross-sectional area, A_1, and the member 4 has area A_2. For this statically determinate structure, the member forces and the vertical displacement at tip are

$$F_1 = 5P, \; F_2 = -P, \; F_3 = 4P, \; F_4 = -2\sqrt{3}P,$$

and $\delta_{tip} = \dfrac{6P\ell}{E}\left(\dfrac{3}{A_1} + \dfrac{\sqrt{3}}{A_2}\right)$

where a negative sign for the forces denotes compression. The maximum tensile stress occurs in member 1 as $5P/A_1$ and the maximum compressive stress occurs in member 4 as $-2\sqrt{3}P/A_2$. The displacement and two stress constraints using normalized (non-dimensional) design variables, $b_1 = A_1 E/1000P$ and $b_2 = A_2 E/1000P$, are written as the mathematical optimization statement:

minimize $\quad f(b) = 3b_1 + \sqrt{3}b_2$
subject to
$$G_1(b) = \dfrac{18}{b_1} + \dfrac{6\sqrt{3}}{b_2} - 3 \leq 0$$
$$G_2(b) = 5.73 - b_1 \leq 0$$
$$G_3(b) = 7.17 - b_2 \leq 0$$

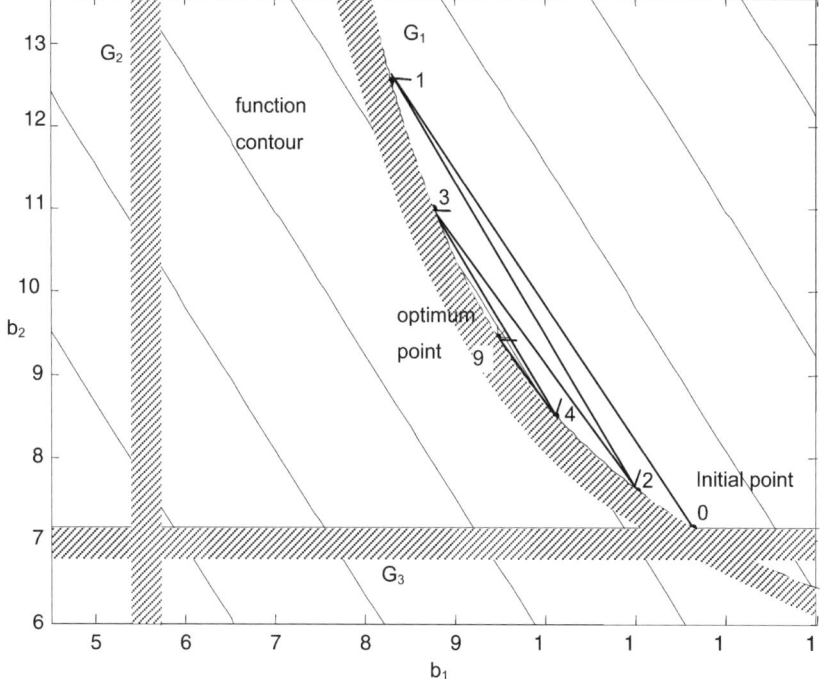

Figure 5.9. Feasible Directions Algorithm Iteration History

Since this problem is 2-dimensional and easy to visualize, we start the search at the intersection of $G_1(b) = 0$ and $G_3(b) = 0$. As shown in Figure 5.9, the initial design point b^0 is:

$$b^0 = \begin{Bmatrix} 11.61 \\ 7.17 \end{Bmatrix}$$

At this point, $G_2(b^0) = 5.73 - b_1 = -5.88$ and $f(b^0) = 47.25$. The objective function gradient is $\nabla f(b^0) = \begin{bmatrix} 3 \\ \sqrt{3} \end{bmatrix}$. The active constraint gradients are

$$\nabla G_1(b^0) = \begin{bmatrix} 0.1335 \\ 0.2021 \end{bmatrix} \text{ and } \nabla G_3(b^0) = \begin{bmatrix} 0 \\ 1 \end{bmatrix}$$

Iteration 1

Find the best search direction by solving the following maximization problem:

maximize τ
subject to $S^T \nabla G_j + \theta_j \tau \leq 0$ for $G_1(b)$ and $G_3(b)$
$S^T \nabla f + \tau \leq 0$
$|s_i| \leq 1$

By substituting the information, we get

maximize τ
subject to $-0.1335 s_1 - 0.2021 s_2 + \tau \leq 0$
$-s_2 + \tau \leq 0$
$3 s_1 + \sqrt{3} s_2 + \tau \leq 1$
$-1 \leq s_1 \leq 1;$ $-1 \leq s_2 \leq 1$

For simplicity, the push-off factors $\theta_1 = \theta_2 = 1$.

The solution of this linear problem using a simplex algorithm is

$$S^0 = \begin{Bmatrix} -0.6712 \\ 1.0000 \end{Bmatrix}$$

$b^1 = b^0 + \gamma S^0$ → $b^1 = \begin{Bmatrix} 11.61 \\ 7.17 \end{Bmatrix} + \gamma \begin{Bmatrix} -0.6172 \\ 1.0000 \end{Bmatrix}$

Here, γ can be calculated using a one-dimensional search along S^1. But to obtain an active constraint at the end of the step, an active constraint is selected, and the corresponding step size is used to demonstrate the procedure, similar to the strategy presented by Haftka and Gurdal [7].

A condition of not violating $G_2(b^1)$ leads to $\gamma = 9.527$ and a new point:

$$b^1 = \begin{Bmatrix} 11.61 \\ 7.17 \end{Bmatrix} + 9.527 \begin{Bmatrix} -0.6172 \\ 1.0 \end{Bmatrix} = \begin{Bmatrix} 5.73 \\ 16.7 \end{Bmatrix}$$

At this new point, $G_1(b^1)$ is violated. If $\gamma > 5.385$, the design point violates $G_1(b^1)$; hence the step size is limited to this value.

$$\text{New point}: b^1 = \begin{Bmatrix} 11.61 \\ 7.17 \end{Bmatrix} + 5.385 \begin{Bmatrix} -0.6172 \\ 1.0 \end{Bmatrix} = \begin{Bmatrix} 8.29 \\ 12.56 \end{Bmatrix}$$

Objective function: $f(b^1) = 46.62$

Iteration 2:

We start the iteration with the point $b^1 = \begin{Bmatrix} 8.29 \\ 12.56 \end{Bmatrix}$

Objective function gradient:

$$\nabla f(b^1) = \begin{bmatrix} 3 \\ \sqrt{3} \end{bmatrix}$$

$G_1(b^1)$ is the only active constraint, and its gradient

$$\nabla G_1(b^1) = \begin{bmatrix} 0.2619 \\ 0.0659 \end{bmatrix}$$

Find the best search direction by solving the following maximization problem:

maximize τ
subject to $S^T \nabla G_j + \theta_j \tau \leq 0 \quad \text{for } G_j(b)$
$\quad\quad\quad\quad\; S^T \nabla f + \tau \leq 0; \quad\quad |s_i| \leq 1$

By substituting the information, we get

maximize τ
subject to $-0.2619 s_1 - 0.0659 s_2 + \tau \leq 0$
$\quad\quad\quad\quad\; 3 s_1 + \sqrt{3} s_2 + \tau \leq 1$
$\quad\quad\quad\quad\; -1 \leq s_1 \leq 1; \quad -1 \leq s_2 \leq 1$

The solution of this linear problem is

$$S^1 = \begin{Bmatrix} 0.5512 \\ -1.0000 \end{Bmatrix}$$

New point:

$$b^2 = b^1 + \gamma S^1 \rightarrow b^2 = \begin{Bmatrix} 8.29 \\ 12.56 \end{Bmatrix} + \gamma \begin{Bmatrix} 0.5512 \\ -1.0000 \end{Bmatrix}$$

the lower limit of b^2 dictates that $\gamma \leq 5.35$, and $G_1(b)$ is violated if $\gamma > 4.957$

Hence $\gamma = 4.957$,

new point:

$$b^2 = \begin{Bmatrix} 8.29 \\ 12.56 \end{Bmatrix} + 4.957 \begin{Bmatrix} 0.5512 \\ -1.0 \end{Bmatrix} = \begin{Bmatrix} 11.02 \\ 7.603 \end{Bmatrix},$$

$$f(b^2) = 46.23$$

Repeat this procedure. After 9 iterations, the optimum point is obtained using MATLAB®:

$$\text{optimum point} = b^9 = \begin{Bmatrix} 9.464 \\ 9.464 \end{Bmatrix}$$

objective function: $f(b^9) = 44.786$

Figure 5.9 shows the iteration history and the final optimum. All these iterations occurred with in a narrow band of cost function values.

Example 5.2

Solve the following constrained optimization problem using the interior penalty function method:

Minimize $f(b_1, b_2) = \frac{1}{3}(b_1 + 1)^3 + b_2$

Subject to

$$G_1 = -b_1 + 1 \leq 0$$
$$G_2 = -b_2 \leq 0$$

Solution:

Combine the constraints with the objective function and develop a pseudo-objective function as follows:

$$F(b_1,b_2,r) = \frac{1}{3}(b_1+1)^3 + b_2 - r\left(\frac{1}{1-b_1} - \frac{1}{b_2}\right)$$

A penalty term is added even when constraints are satisfied.

Start with a feasible design point: $b^0 = \begin{Bmatrix} b_1 \\ b_2 \end{Bmatrix} = \begin{Bmatrix} 2 \\ 1 \end{Bmatrix}$, (here both constraints are satisfied)

Iteration 1:

Let $r^0 = 10$

$$F = \frac{1}{3}(2+1)^3 + 1 - 10\left(\frac{1}{1-2} - \frac{1}{1}\right) = 30$$

Solve the unconstrained problem using the steepest descent method, where the search direction S is simply the opposite direction of the function gradients at the current design point.

Pseudo-function gradients:

$$\frac{\partial F}{\partial b_1} = (b_1+1)^2 - \frac{r^0}{(1-b_1)^2} = (2+1)^2 - \frac{10}{(1-2)^2} = -1$$

$$\frac{\partial F}{\partial b_2} = 1 - \frac{r^0}{b_2^2} = 1 - \frac{10}{1} = -9$$

Direction: $S^0 = -\nabla F = \begin{bmatrix} 1 \\ 9 \end{bmatrix}$

New point b^1:

$$b^1 = b^0 + \gamma S^0 = \begin{bmatrix} 2 \\ 1 \end{bmatrix} + \gamma \begin{bmatrix} 1 \\ 9 \end{bmatrix} = \begin{bmatrix} 2+\gamma \\ 1+9\gamma \end{bmatrix}$$

Substitute the result b^1 into $F = \frac{1}{3}(b_1+1)^3 + b_2 - r^0\left(\frac{1}{1-b_1} - \frac{1}{b_2}\right)$ yields

$$F = \frac{1}{3}(2+\gamma+1)^3 + 1 + 9\gamma - \frac{10}{1-2-\gamma} + \frac{10}{1+9\gamma}$$

An optimum step γ is calculated by using analytical differentiation of F with respect to γ (necessary condition for an optimum).

Differentiate with respect to γ

$$\frac{\partial F}{\partial \gamma} = (3+\gamma)^2 + 9 - \frac{10}{(-1-\gamma)^2} + \frac{90}{(1+9\gamma)^2} = 0$$

One of the roots with a lowest value is $\gamma = 0.1914$

$$b^1 = b^0 + \gamma S^0 = \begin{bmatrix} 2 + 1 \times 0.1914 \\ 1 + 9 \times 0.1914 \end{bmatrix} = \begin{bmatrix} 2.1914 \\ 2.7227 \end{bmatrix}$$

b^1 becomes the starting point for the next iteration. This procedure is repeated until convergence on the unconstrained function F for the given r^0 value. After 16 steepest descent iterations, the design is converged to $b^1 = \begin{bmatrix} 2.0402 \\ 3.1623 \end{bmatrix}$. The iteration history for this unconstraint problem is presented in Table 5.1, along with step size and function values.

Table 5.1. Unconstrained Problem Convergence History for $r^0 = 10$

b_1	b_2	$\partial F/\partial x$	$\partial F/\partial y$	γ	F	f
2.0000	1.0000	-1.0000	-9.0000	0.1914	30.0000	10.0000
2.1914	2.7227	3.1403	-0.3489	0.0487	25.6239	13.5578
2.0386	2.7397	-0.0369	-0.3323	0.9050	25.3700	12.0918
2.0720	3.0404	0.7360	-0.0818	0.0438	25.3215	12.7044
2.0398	3.0440	-0.0088	-0.0792	1.0402	25.3094	12.4069
2.0490	3.1264	0.2077	-0.0231	0.0428	25.3061	12.5742
2.0401	3.1274	-0.0025	-0.0224	1.0769	25.3052	12.4928
2.0427	3.1515	0.0614	-0.0068	0.0425	25.3049	12.5418
2.0401	3.1518	-0.0007	-0.0066	1.0876	25.3048	12.5179
2.0409	3.1591	0.0184	-0.0020	0.0425	25.3048	12.5325
2.0402	3.1591	-0.0002	-0.0020	1.0908	25.3048	12.5254
2.0404	3.1613	0.0055	-0.0006	0.0424	25.3048	12.5298
2.0402	3.1613	-0.0001	-0.0006	1.0917	25.3048	12.5277
2.0402	3.1620	0.0017	-0.0002	0.0424	25.3048	12.5290
2.0402	3.1620	0.0000	-0.0002	1.0920	25.3048	12.5283
2.0402	3.1622	0.0005	-0.0001	0.0424	25.3048	12.5287
2.0402	3.1623	0.0000	-0.0001	1.0921	25.3048	12.5286

Iteration 2

Let $r^1 = 0.1$, $r^0 = 1$

$$b^1 = \begin{bmatrix} 2.0402 \\ 3.1623 \end{bmatrix}$$

$$F = \frac{1}{3}(b_1+1)^3 + b_2 - r^1\left(\frac{1}{1-b_1} - \frac{1}{b_2}\right)$$

$$F = \frac{1}{3}(2.0402+1)^3 + 3.1623 - \frac{1}{1-2.0402} + \frac{1}{3.1623} = 13.8065$$

Solve the unconstrained problem using the steepest descent method.

Pseudo-objective function gradients:

$$\frac{\partial F}{\partial b_1} = (b_1+1)^2 - \frac{r^1}{(1-b_1)^2} = (2.0402+1)^2 - \frac{1}{(1-2.0402)^2} = 8.3186$$

$$\frac{\partial F}{\partial b_2} = 1 - \frac{r^1}{b_2^2} = 1 - \frac{1}{3.1623^2} = 0.9$$

Direction: $S = -\nabla F = -\begin{bmatrix} 8.3186 \\ 0.9000 \end{bmatrix}$

New point b^2:

$$b^2 = b^1 + \gamma\, S^1 = \begin{bmatrix} 2.0402 \\ 3.1623 \end{bmatrix} + \gamma\left(-\begin{bmatrix} 8.3186 \\ 0.9000 \end{bmatrix}\right) = \begin{bmatrix} 2.0402 - 8.3186\gamma \\ 3.1623 - 0.9000\gamma \end{bmatrix}$$

Substitute in: $F = \frac{1}{3}(b_1+1)^3 + b_2 - r^1\left(\frac{1}{1-b_1} - \frac{1}{b_2}\right)$

$$F = \frac{1}{3}(2.0402 - 8.3186\gamma + 1)^3 + 3.1623 - 0.9\gamma - \frac{1}{1 - 2.0402 + 8.3186\gamma} + \frac{1}{3.1623 - 0.9\gamma}$$

Differentiate with respect to γ:

$\frac{\partial F}{\partial \gamma} = 0$, results in $\gamma = 0.0756$

172 Reliability-based Structural Design

$$b^2 = b^1 + \gamma S^1 = \begin{bmatrix} 2.0402 + 0.0756 \times (-8.3186) \\ 3.1623 + 0.0756 \times (-0.9) \end{bmatrix} = \begin{bmatrix} 1.4113 \\ 3.0943 \end{bmatrix}$$

After steepest descent iterations, the design is converged to $b^2 = \begin{bmatrix} 1.4142 \\ 1.0000 \end{bmatrix}$; here, the pseudo-objective function $F = 9.1046$.

Iteration 3

Let $r^2 = 0.1$, $r^1 = 0.1$

$$b^2 = \begin{bmatrix} 1.4142 \\ 1.0000 \end{bmatrix}$$

$$F = \frac{1}{3}(b_1 + 1)^3 + b_2 - r^2 \left(\frac{1}{1-b_1} - \frac{1}{b_2} \right) = 6.0317$$

Solve the unconstrained problem using the steepest descent method.

Pseudo-objective function gradients:

$$\frac{\partial F}{\partial b_1} = (b_1 + 1)^2 - \frac{r^2}{(1-b_1)^2} = 0.9, \quad \frac{\partial F}{\partial b_2} = 1 - \frac{r^2}{b_2^2} = 0.0513$$

Direction: $S = -\nabla F = -\begin{bmatrix} 0.9000 \\ 0.0513 \end{bmatrix}$

New point b^3:

$$b^3 = b^2 + \gamma S^2 = \begin{bmatrix} 1.4142 \\ 1.0000 \end{bmatrix} + \gamma \left(-\begin{bmatrix} 0.9000 \\ 0.0513 \end{bmatrix} \right) = \begin{bmatrix} 1.412 - 0.9\gamma \\ 1.0 - 0.0513\gamma \end{bmatrix}$$

Substitute in: $F = \frac{1}{3}(b_1 + 1)^3 + b_2 - r^2 \left(\frac{1}{1-b_1} - \frac{1}{b_2} \right)$

Differentiate with respect to γ:

$\dfrac{\partial F}{\partial \gamma} = 0$, results in $\gamma = 0.0513$, $b^3 = b^2 + \gamma S^2 = \begin{bmatrix} 1.1450 \\ 0.9538 \end{bmatrix}$

After steepest descent iterations, the design is converged to: $b^3 = \begin{bmatrix} 1.1473 \\ 0.3162 \end{bmatrix}$

Table 5.2. Interior Penalty Convergence History

R	b_1	b_2	γ	F	f
Initial point	2.0000	1.0000	0.1914	30.0000	10.0000
10	2.0402	3.1623	1.0921	25.3048	12.5286
1	1.4142	1.0000	0.0307	9.1046	5.6904
0.1	1.1473	0.3162	0.0153	4.6117	3.6164
0.01	1.0488	0.1000	0.0059	3.2716	2.9667
0.001	1.0157	0.0316	0.002	2.8569	2.7615
0.0001	1.0050	0.0100	0.0007	2.7267	2.6967
0.00001	1.0016	0.0032	0.0002	2.6856	2.6762
0.000001	1.0005	0.0010	0.0001	2.6727	2.6697
0.0000001	1.0002	0.0003	0.0000	2.6686	2.6676
0.0000000001	1.0000	0.0000	0.0000	2.6667	2.6667

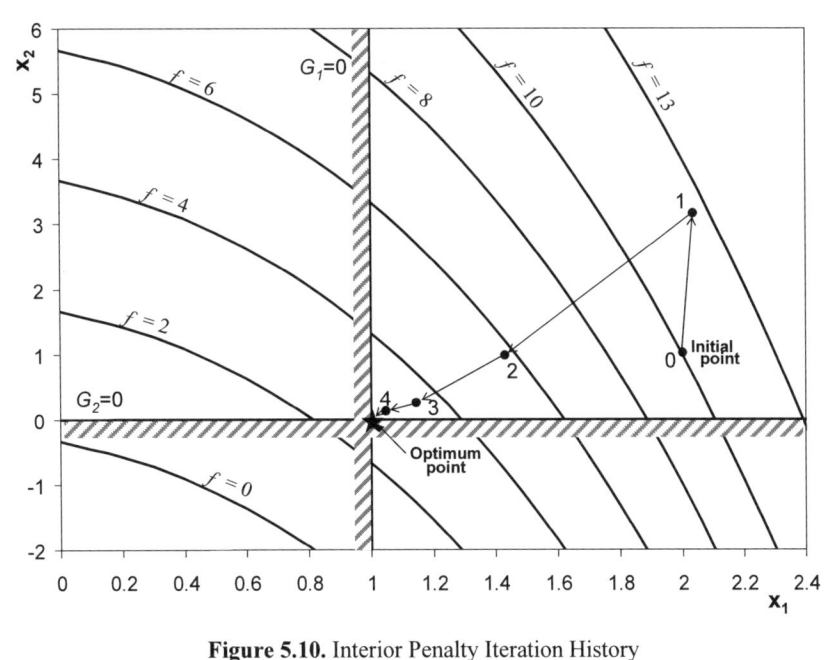

Figure 5.10. Interior Penalty Iteration History

Similarly, the r value is multiplied by a factor 0.1 in subsequent iterations until the solution is converged. As r approaches 0, the design point converges to an optimum point. The design iteration history is shown in Figure 5.10, indicating the design point movement in design space. The corresponding design variables and objective function values are given in Table 5.2. As r approaches 0, the design converges to (1,0).

Example 5.3

Solve example 5.1 using the extended interior penalty function method.

Mathematical statement for optimization:

$$\text{minimize} \quad f(b) = 3b_1 + \sqrt{3}b_2$$
subject to
$$G_1(b) = \frac{18}{b_1} + \frac{6\sqrt{3}}{b_2} - 3 \leq 0$$
$$G_2(b) = 5.73 - b_1 \leq 0$$
$$G_3(b) = 7.17 - b_2 \leq 0$$

The solution pseudo-objective function $F(b,r) = f(b) + r\sum_{j=1}^{m} p(G_j(b))$.

As described earlier, the starting point in the extended interior penalty method can be anywhere in the feasible or infeasible region. For this example we start in the infeasible region with $b^0 = \begin{bmatrix} 5 \\ 5 \end{bmatrix}$.

Iteration 1:

Taking the penalty parameters $r = 10000$ and $k = 0.001$ gives the extended interior penalty constant $G_0 = -0.001 \times (10000)^{1/2} = -0.1$. At the initial design, $G_1(b^0) = 2.6785$, $G_2(b^0) = 0.730$, and $G_3(b^0) = 2.170$. The pseudo-objective function is

$$F(b,r) = 3b_1 + \sqrt{3}b_2 + 10000[p(G_1(b)) + p(G_2(b)) + p(G_2(b))]$$

All constraint values are greater than the constant G_0; the associated penalties are

$$p(G_1(b)) = \frac{1}{G_0}\left[\left(\frac{G_1}{G_0}\right)^2 - 3\left(\frac{G_1}{G_0}\right) + 3\right] = 8007.91$$

$$p(G_2(b)) = \frac{1}{G_0}\left[\left(\frac{G_2}{G_0}\right)^2 - 3\left(\frac{G_2}{G_0}\right) + 3\right] = 781.90$$

$$p(G_3(b)) = \frac{1}{G_0}\left[\left(\frac{G_3}{G_0}\right)^2 - 3\left(\frac{G_3}{G_0}\right) + 3\right] = 5389.90$$

Substituting the information, we get

$$f(b^0) = 23.6603$$
$$F(b^0, 10000) = 23.6603 + 10000[8007.91 + 781.90 + 5389.90]$$
$$= 1.4179 \times 10^8$$

Using Newton's method for search direction vector, where the search direction is given by $S = -H^{-1}\nabla F$,

$$b^1 = b^0 - \gamma H^{-1} \nabla F$$

where ∇F is the gradient of F, H is the second derivative matrix of F, and γ is the step size obtained by one-dimensional search. The details of the calculations are:

$$\nabla F = -10^7 \begin{Bmatrix} 5.8330 \\ 6.9915 \end{Bmatrix}$$

$$H_{ij} = \frac{\partial^2 F}{\partial x_i \partial x_j}$$

$$H = 10^7 \begin{bmatrix} 4.6600 & 0.5986 \\ 0.5986 & 3.2862 \end{bmatrix}$$

$$S = -H^{-1}\nabla F = \begin{Bmatrix} 1.0005 \\ 1.9453 \end{Bmatrix}$$

$$b^1 = \begin{Bmatrix} 5 \\ 5 \end{Bmatrix} + \gamma \begin{Bmatrix} 1.0005 \\ 1.9453 \end{Bmatrix}$$

Performing one-dimensional optimization for the minimum value of $F(b,r)$ by keeping γ as a variable, we obtain optimum step size, $\gamma = 1.3$.

The new design vector $b^1 = \begin{Bmatrix} 6.3007 \\ 7.5289 \end{Bmatrix}$. This completes the first iteration of the extended interior penalty method.

Iteration 2:

$$r_2 = \frac{r_1}{10} = 1000$$

$$b^1 = \begin{Bmatrix} 6.3007 \\ 7.5289 \end{Bmatrix}$$

The function value $f(b^1) = 31.9425$. At this design point, $G_1(b^1) = 1.2371$, $G_2(b^1) = -0.5707$, and $G_3(b^1) = -0.3589$.

The pseudo-objective function is

$$F(b,r) = 3b_1 + \sqrt{3}b_2 + 1000[p(G_1) + p(G_2) + p(G_2)]$$

Since only the first constraint is violated, $G_1(b) < G_0$. The penalty associated with $G_1(b)$ is

$$p(G_1(b)) = \frac{1}{G_0}\left[\left(\frac{G_1}{G_0}\right)^2 - 3\left(\frac{G_1}{G_0}\right) + 3\right] = 1931.7$$

$G_2(b)$ and $G_3(b)$ are satisfied; hence

$$p(G_2(b)) = \frac{1}{G_2(b^1)} = 1.8$$

$$p(G_3(b)) = \frac{1}{G_3(b^1)} = 2.8$$

$$F(b^1, 1000) = 1.9363 \times 10^6$$

Using Newton's method for search direction vector, the next design point

$$b^2 = b^1 - \gamma \, H^{-1} \, \nabla F$$

The pseudo-objective function gradient $\nabla F = -10^6 \begin{Bmatrix} 1.2610 \\ 0.5164 \end{Bmatrix}$ and the Hessian matrix

$$H = 10^5 \begin{bmatrix} 8.2121 & 1.6626 \\ 1.6626 & 2.4562 \end{bmatrix},$$ gives the new search direction as

$$S = -\begin{Bmatrix} 1.2861 \\ 1.2319 \end{Bmatrix}.$$

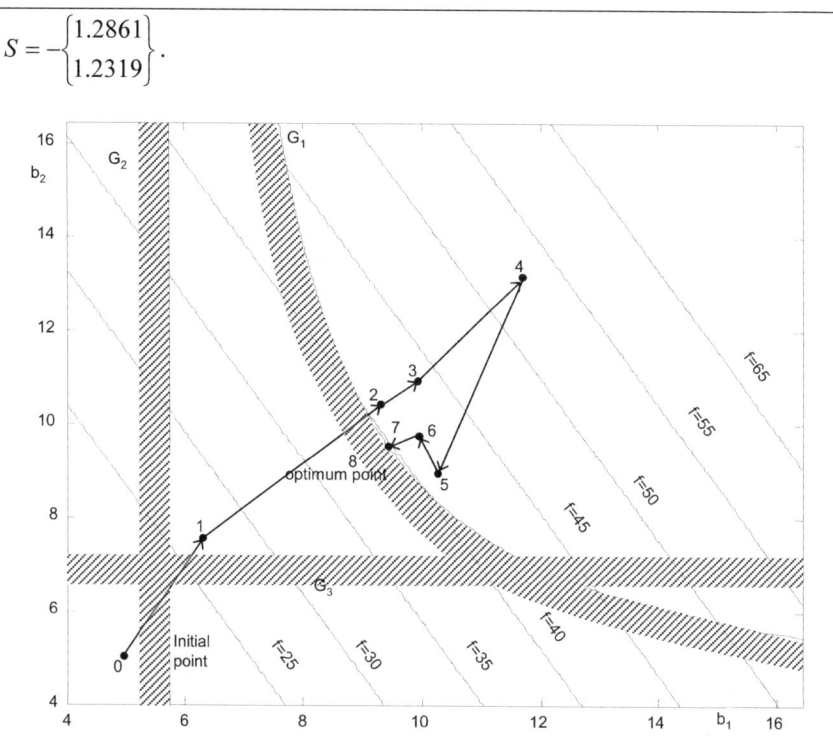

Figure 5.11. Extended Interior Penalty Iteration History

Table 5.3. Extended Interior Penalty Results Convergence

Iteration #	r	b_1	b_2	γ	$F(b,r)$	$f(b)$
0	10000	5.00	5.00	1.3	1.42E8	23.66
1	1000	6.30	7.53	2.4	2.17E5	31.94
2	100	9.39	10.49	1.1	11554.63	46.32
3	10	9.58	10.72	9	851.92	47.32
4	1	11.69	13.15	0.2	76.12	57.84
5	0.1	9.85	9.13	1.13	76.44	45.34
6	0.01	9.77	9.58	0.1	47.32	45.90
7	0.001	9.47	9.48	0.1	45.14	44.84
8	0.0001	9.46	9.46	0.0	44.90	44.78

> The one-dimensional search gives an optimum step size, $\gamma = 2.4$. This gives the new design point $b^2 = \begin{Bmatrix} 9.3873 \\ 10.4855 \end{Bmatrix}$. This completes the second iteration. At the end of the second iteration, $f(b^2) = 46.3233$, $G_1(b^2) = -0.0914$, $G_2(b^2) = -3.6573$, and $G_3(b^2) = -3.3155$.
>
> Now we have moved into the feasible region and the pseudo-objective function is modified accordingly. This gives $F(b^2, 100) = 2.6454 \times 10^3$. The remaining iterations are shown in Table 5.3. As r reaches 0, the solution is converged to the optimum. The pseudo-objective function and the original objective function approach the same value. The optimum realized in this method and that of the feasible directions algorithm give similar solution. Figure 5.11 presents the iteration region and finally, convergence to the optimum.

5.4 Sensitivity Analysis

An important step in optimal design is to obtain sensitivity derivatives, which are used to study the effect of parametric modifications, to calculate search directions to find an optimum design, and to construct function approximations. References [3] and [8] present sensitivity analysis formulations for probabilistic problems. An analytical approach for sensitivity analysis of the safety index based on the advanced first-order second-moment (AFOSM) method for evaluating probability is presented in [8]. The resulting sensitivity formula contains only the first-order derivatives. Detailed steps to derive a sensitivity formula for the safety index and failure probability are given below.

When the AFOSM method is used to evaluate the failure probability, a single mode of failure probability constraint can be written as

$$P_f = 1 - \Phi(\beta) \tag{5.19}$$

where P_f is the failure probability and $\Phi(\beta)$ is the cumulative standard normal distribution, defined in Equation 3.9.

Hence, the sensitivity of failure probability (P_f) is obtained as follows:

$$\frac{\partial P_f}{\partial b} = -\phi(\beta) \frac{\partial \beta}{\partial b} \tag{5.20}$$

Here $\phi(\beta)$ is the standard normal probability distribution and $\partial \beta / \partial b$ is the safety index sensitivity with respect to the design variables b.

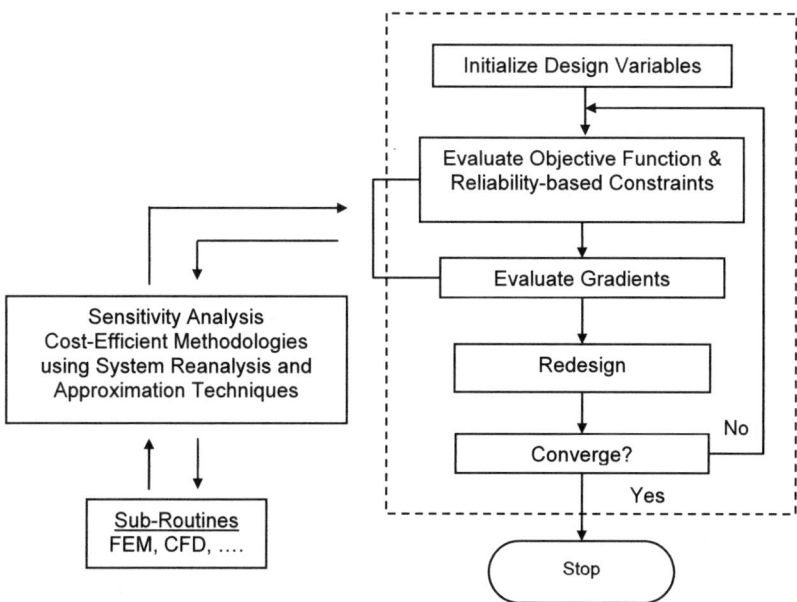

Figure 5.12. Reliability-based Optimization

The safety index is written as

$$\beta = \min_{U \in A}(U^T U)^{1/2} \qquad (5.21)$$

where $A = \{U \mid G(b,U) = 0\}$ and U denotes a vector of standard, uncorrelated and normal variables defined by

$$U = \frac{x - \mu_x}{\sigma_x} = \{U_1 \ U_2 \ \ldots \ U_n\}^T \qquad (5.22)$$

The minimization problem in Equation 5.21 is solved for the random variables U^* to obtain the safety index sensitivity $\partial \beta / \partial b$. Then, the perturbations for the most probable failure state U^* are stated as follows:

$$\delta \beta = \nabla_U L(U) \delta U \qquad (5.23)$$

where $\delta \beta$ is the change in safety index due to perturbations in design variables, δU is the change in random variables due to the change in design variables, $L(U) = (U^T U)^{1/2}$, and $\nabla_U L(U)$ represents the derivatives vector of $L(U)$ with respect to the normalized random variables U. Assuming that the minimum point U^* for the Equation 5.21 is unique, the variation can be taken in the context of the minimization problem (Equation 5.23). Now, the perturbations for U^* must satisfy the following condition:

$$\delta G = \nabla_b G \, \delta b + \nabla_U G \, \delta U = 0 \qquad (5.24)$$

where

$$\nabla_b G = \left\{ \frac{\partial G}{\partial b_1} \quad \frac{\partial G}{\partial b_2} \quad .. \quad .. \quad \frac{\partial G}{\partial b_n} \right\}^T \text{ and } \nabla_U G = \left\{ \frac{\partial G}{\partial U_1} \quad \frac{\partial G}{\partial U_2} \quad .. \quad .. \quad \frac{\partial G}{\partial U_n} \right\}^T$$

are the derivatives that are obtained at U^* and b^0. Kuhn-Tucker necessary conditions must be satisfied at b^0, since U^* and b^0 are optimal solutions for the minimization problem. This can be given as follows by introducing the Lagrangian multiplier λ:

$$\nabla_U L + \lambda \nabla_U G = 0 \qquad (5.25)$$
$$G(b^0, U^*) = 0 \qquad (5.26)$$

One Lagrangian multiplier is needed for each constraint and is added in a similar manner. From Equation 5.24, the gradient of G with respect to U is written as

$$\nabla_U G \, \delta U = -\nabla_b G \, \delta b \qquad (5.27)$$

By substituting Equation 5.27 into Equation 5.25, we get the gradient of L as follows:

$$\nabla_U L \, \delta U = \lambda \nabla_b G \, \delta b \qquad (5.28)$$

Therefore, the sensitivity of the safety index can be obtained from Equations 5.23 and 5.28 as follows:

$$\frac{\partial \beta}{\partial b} = \lambda \nabla_b G(U^*, b^0) = \lambda \frac{\partial G}{\partial U} \frac{\partial U}{\partial b} \qquad (5.29)$$

The Lagrange multiplier λ of the above equation can be obtained from Kuhn-Tucker conditions: $\lambda \nabla_U G = -\nabla_U L$, noting that

$$\nabla_U L = \frac{1}{2(U^T U)^{1/2}} 2U = \frac{1}{2(U^T U)^{1/2}} 2(U^T U)^{1/2} = 1$$

And the expression for λ is obtained as follows:

$$\lambda = -\frac{1}{\left|\nabla_U G(U^*, b^0)\right|} = -\frac{1}{\left|\frac{\partial G}{\partial U}\right|} \tag{5.30}$$

From Equations 5.29 and 5.30, when U^* is a real optimal solution of the minimization problem stated in Equation 5.21–that is, if the β value is exact–the exact derivatives of safety index β with respect to design variable b can be obtained

In many practical problems, the design variables b are modeled with uncertain means and uncertain (or fixed) standard deviations. For this case, $\partial U / \partial b$ in Equation 5.29 can be derived easily, since U is a function of mean and standard deviation, and λ and $\partial G / \partial U$ in Equation 5.29 are by-products from the reliability analysis. Thus, no additional analysis is needed for the sensitivity calculation.

It is important to point out that Equation 5.29 can only be used when the design variables are either the means or standard deviations of the random variables. If the design variables are not modeled as the uncertain means and standard deviations, other approaches are suggested to compute the sensitivity. For example, a forward-difference approach can be used. Given the deterministic variable b and perturbation δb, the approximate derivative $\partial \beta / \partial b$ is computed as

$$\frac{\partial \beta}{\partial b} = \frac{\beta(b + \delta b) - \beta(b)}{\delta b} \tag{5.31}$$

β varies as b changes, although b is not an uncertain variable, so the sensitivity of β with respect to b will not be zero.

5.4.1 Sensitivity with Respect to Means

In Equation 5.29, if the design variables b are the means of the random variables (μ_x), the sensitivity of the standard variable U with respect to the mean values can be obtained from the standard normal variable definition, $U_i = (x_i - \mu_x) / \sigma_x$, which is written as

$$\frac{\partial U}{\partial \mu_x} = -\frac{1}{\sigma_x} \tag{5.32}$$

In Equation 5.29, the constraint function gradient $\partial G / \partial U$, evaluated at (U^*, b^0), is expanded in terms of random variables as

$$\frac{\partial G}{\partial U} = \frac{\partial G}{\partial x}\frac{\partial x}{\partial U} = \frac{\partial G}{\partial x}\sigma_x \tag{5.33}$$

From Equations 5.29, 5.30, 5.32, and 5.33, the safety index sensitivity with respect to the mean is obtained as

$$\frac{\partial \beta}{\partial \mu_x} = -\frac{\dfrac{\partial G}{\partial x}\sigma_x}{\left|\dfrac{\partial G}{\partial x}\sigma_x\right|} \times \left(-\frac{1}{\sigma_x}\right) \tag{5.34}$$

By simplifying,

$$\frac{\partial \beta}{\partial \mu_x} = \frac{\dfrac{\partial G}{\partial x}}{\left|\dfrac{\partial G}{\partial x}\sigma_x\right|} \tag{5.35}$$

where μ_x and σ_x are the mean and standard deviation of the random variables.

5.4.2 Sensitivity with Respect to Standard Deviations

From Equation 5.29, the gradient of the standard normal variable with respect to σ_x is written as

$$\frac{\partial U}{\partial \sigma_x} = \frac{-(x-\mu_x)}{\sigma_x^2} \tag{5.36}$$

From Equations 5.29, 5.30, 5.33, and 5.36, the safety index sensitivity is obtained as

$$\frac{\partial \beta}{\partial \sigma_x} = -\frac{\dfrac{\partial G}{\partial x}\sigma_x}{\left|\dfrac{\partial G}{\partial x}\sigma_x\right|} \times \frac{-(x-\mu_x)}{\sigma_x^2} \tag{5.37}$$

By simplifying,

$$\frac{\partial \beta}{\partial \sigma_x} = \frac{\dfrac{\partial G}{\partial x}}{\left|\dfrac{\partial G}{\partial x}\sigma_x\right|} \times \frac{(x-\mu_x)}{\sigma_x} \tag{5.38}$$

As stated earlier, the safety index is the shortest distance from the origin to the limit surface. The coordinates of this most probable point on the limit surface are computed as:

$$U^* = \beta \cos\theta_u \tag{5.39}$$

Here, $\cos\theta_u$ (or $\cos\theta_x$) is the direction cosine of the unit outward normal vector, also known as the sensitivity factor (α_x) and given as

$$\cos\theta_x = \alpha_x = -\frac{\dfrac{\partial G}{\partial x}\sigma_x}{\left|\dfrac{\partial G}{\partial x}\sigma_x\right|} \tag{5.40}$$

From Equations 5.39 and 5.40, U^* can be written as

$$U^* = -\beta\frac{\dfrac{\partial G}{\partial x}\sigma_x}{\left|\dfrac{\partial G}{\partial x}\sigma_x\right|} \tag{5.41}$$

By substituting U^* (Equation 5.41) in (Equation 5.38), the safety index sensitivity with respect to σ_x is obtained as

$$\frac{\partial \beta}{\partial \sigma_x} = \frac{\dfrac{\partial G}{\partial x}}{\left|\dfrac{\partial G}{\partial x}\sigma_x\right|} \times -\beta\frac{\dfrac{\partial G}{\partial x}\sigma_x}{\left|\dfrac{\partial G}{\partial x}\sigma_x\right|} \tag{5.42}$$

By simplifying,

$$\frac{\partial \beta}{\partial \sigma_x} = -\frac{\left(\dfrac{\partial G}{\partial x}\right)^2 \sigma_x}{\left|\dfrac{\partial G}{\partial x}\sigma_x\right|^2}\beta \tag{5.43}$$

So far, we have presented the sensitivity of the safety index with respect to the mean and standard deviation. These calculations are needed to obtain the failure possibility sensitivities.

5.4.3 Failure Probability Sensitivity in Terms of β

From Equation 5.20, the failure probability sensitivities with respect to the mean and standard deviation are obtained as

$$\frac{\partial P_f}{\partial \mu_x} = -\phi(\beta)\frac{\partial \beta}{\partial \mu_x} \qquad (5.44)$$

$$\frac{\partial P_f}{\partial \sigma_x} = -\phi(\beta)\frac{\partial \beta}{\partial \sigma_x} \qquad (5.45)$$

where $\phi(\beta)$ is the standard normal probability distribution.

Two example problems are presented to demonstrate the sensitivity analysis calculations. The analytical methods and finite difference approaches are presented with cost and accuracy issues. Finally, a general problem where some variables are random, some are both random and design, and some are just design variables is thoroughly investigated.

Example 5.4

In this example, we consider a cylinder problem to explain how to calculate the sensitivity information. The height and diameter of the cylinder are denoted as H and D. H and D are both normally distributed with a mean of 7 and a standard deviation of 2. The cylinder volume is required to be less than 311 units. The limit-state function is given as

$$G(D,H) = 311 - V = 311 - \frac{\pi}{4}D^2 H$$

where V is the volume of the cylinder. Compute the safety index and failure probability gradients with respect to the mean and standard deviations of D and H.

Solution:

The gradients of $G(D, H)$ with respect to D and H at the mean value point (7, 7) are

$$\frac{\partial G}{\partial D} = -\frac{\pi}{2}DH = -76.969$$

$$\frac{\partial G}{\partial H} = -\frac{\pi}{4}D^2 = -38.485$$

The mean value of the limit-state function μ_G is calculated as

$$\mu_G = 311 - \frac{\pi}{4}\mu_D^2 \mu_H = 311 - \frac{\pi}{4}\times 7^2 \times 7 = 41.608$$

The standard deviation of the limit-state function, σ_G, is computed as

$$\sigma_G = |\nabla G_x(x^*) \cdot \sigma_x|$$
$$= \sqrt{(\frac{\partial G}{\partial D}\sigma_D)^2 + (\frac{\partial G}{\partial H}\sigma_H)^2}$$
$$= \sqrt{(-76.969 \times 2)^2 + (-38.485 \times 2)^2}$$
$$= 172.1080$$

From the definition, the safety index (β) can be obtained as

$$\beta = \frac{\mu_G}{\sigma_G} = \frac{41.608}{172.108} = 0.2418$$

Therefore, the probability of failure (P_f) is

$$P_f = 1 - \Phi(\beta) = 0.4045$$

The diameter and height mean values (μ_D and μ_H), and standard deviations (σ_D and σ_H) are design variables. Safety index and failure probability sensitivities with respect to these design variables are computed by using the analytical sensitivity formulations given in Equations 5.35, 5.43, 5.44 and 5.45.

Sensitivities with respect to Means (Equations 5.35 and 5.44):

The safety index sensitivity formula with respect to the mean is given as

$$\frac{\partial \beta}{\partial \mu_x} = \frac{\frac{\partial G}{\partial x}}{\left|\frac{\partial G}{\partial x}\sigma_x\right|}$$

Hence, the safety index sensitivity with respect to the diameter mean μ_D can be written as

$$\frac{\partial \beta}{\partial \mu_D} = \frac{\frac{\partial G}{\partial D}}{\sqrt{\left(\frac{\partial G}{\partial D}\sigma_D\right)^2 + \left(\frac{\partial G}{\partial H}\sigma_H\right)^2}}$$

which simplifies to

$$\frac{\partial \beta}{\partial \mu_D} = \frac{-76.969}{172.108} = -0.4461$$

The failure probability sensitivity is obtained from the safety index sensitivity

$\dfrac{\partial \beta}{\partial \mu_x}$ and standard normal probability distribution function $\phi(\beta)$ as follows:

$$\frac{\partial P_f}{\partial \mu_x} = -\phi(\beta)\frac{\partial \beta}{\partial \mu_x}$$

Therefore, the failure probability sensitivity with respect to diameter mean is

$$\frac{\partial P_f}{\partial \mu_D} = -\phi(\beta)\frac{\partial \beta}{\partial \mu_D}$$

Substitution gives

$$\frac{\partial P_f}{\partial \mu_D} = -0.3874 \times -0.4461 = 0.1728$$

Similarly, the safety index sensitivity with respect to the mean height μ_H is calculated as

$$\frac{\partial \beta}{\partial \mu_H} = \frac{\dfrac{\partial G}{\partial H}}{\sqrt{\left(\dfrac{\partial G}{\partial D}\sigma_D\right)^2 + \left(\dfrac{\partial G}{\partial H}\sigma_H\right)^2}}$$

which simplifies to

$$\frac{\partial \beta}{\partial \mu_H} = \frac{-38.4845}{172.108} = -0.2236$$

The failure probability sensitivity with respect to mean height is

$$\frac{\partial P_f}{\partial \mu_H} = -\phi(\beta)\frac{\partial \beta}{\partial \mu_H}$$

which simplifies to

$$\frac{\partial P_f}{\partial \mu_H} = -0.3874 \times -0.2236 = 0.0866$$

Sensitivities with respect to Standard Deviations (Equations 5.43 and 5.45):

From Equation 5.43, the safety index sensitivity with respect to the standard deviation of a random variable is written as

$$\frac{\partial \beta}{\partial \sigma_x} = -\frac{\left(\frac{\partial G}{\partial x}\right)^2 \sigma_x}{\left|\frac{\partial G}{\partial x}\sigma_x\right|^2}\beta$$

The safety index sensitivity with respect to the diameter standard deviation is

$$\frac{\partial \beta}{\partial \sigma_D} = -\frac{\left(\frac{\partial G}{\partial D}\right)^2 \sigma_D}{(\sigma_G)^2}\beta$$

which simplifies to

$$\frac{\partial \beta}{\partial \sigma_D} = -\frac{(-76.969)^2 \times 2}{172.108^2} \times 0.2418 = -0.0967$$

The failure probability sensitivity with respect to the diameter standard deviation is obtained from Equation 5.45 as follows:

$$\frac{\partial P_f}{\partial \sigma_D} = -\phi(\beta)\frac{\partial \beta}{\partial \sigma_D}$$

Substitution of values gives

$$\frac{\partial P_f}{\partial \sigma_D} = -0.3874 \times -0.0967 = 0.0375$$

Similarly, the safety index sensitivity with respect to the height standard deviation is calculated as

$$\frac{\partial \beta}{\partial \sigma_H} = -\frac{\left(\frac{\partial G}{\partial H}\right)^2 \sigma_H}{(\sigma_G)^2}\beta$$

which simplifies to

$$\frac{\partial \beta}{\partial \sigma_H} = -\frac{(-38.4845)^2 \times 2}{(172.108)^2} \times 0.2418 = -0.0242$$

And failure probability sensitivity with respect to height standard deviation is

$$\frac{\partial P_f}{\partial \sigma_H} = -\phi(\beta) \frac{\partial \beta}{\partial \sigma_H}$$

Substitution of values gives

$$\frac{\partial P_f}{\partial \sigma_H} = -0.3874 \times -0.0242 = 0.0094$$

These analytical sensitivities are often compared with finite difference sensitivities to validate the development of analytical methods. For demonstration, the forward finite difference method is used to compute sensitivities. In the forward finite difference method, the safety index and failure probability sensitivities are given as follows:

$$\frac{\partial \beta}{\partial b} = \frac{\beta(b + \delta b) - \beta(b)}{\delta b}$$

$$\frac{\partial P_f}{\partial b} = \frac{P_f(b + \delta b) - P_f(b)}{\delta b}$$

In the finite difference method, the safety index and failure probability are computed by making a small perturbation in the corresponding design variable while keeping other design variables constant. In this example, a 1% perturbation is provided in each of the design variables, and the results are tabulated in Table 5.4.

Table 5.4. Safety Index and Probability of Failure Values

Perturbation (1%)	M	σ	β	$\Phi(\beta)$	P_f
$\delta\mu_D$=0.0	μ_D=7.0; μ_H=7.0	σ_D=σ_H = 2.0	0.2418	0.5955	0.4045
$\delta\mu_D$=0.07	μ_D=7.07; μ_H=7.0	σ_D=σ_H = 2.0	0.2078	0.5823	0.4177
$\delta\mu_H$=0.07	μ_D=7.0; μ_H=7.07	σ_D=σ_H = 2.0	0.2243	0.5887	0.4113
$\delta\sigma_D$=0.02	μ_D=7.0; μ_H=7.0	σ_D=2.02; σ_H = 2.0	0.2398	0.5948	0.4052
$\delta\sigma_H$=0.02	μ_D=7.0; μ_H=7.0	σ_D=2.0; σ_H = 2.02	0.2413	0.5953	0.4047

From the definition, the safety index and failure probability sensitivities with respect to diameter mean (μ_D) are calculated as follows:

$$\frac{\partial \beta}{\partial \mu_D} = \frac{\beta(\mu_D + \delta\mu_D) - \beta(\mu_D)}{\delta\mu_D}$$

which simplifies to

$$\frac{\partial \beta}{\partial \mu_D} = \frac{0.2078 - 0.2418}{0.07} = -0.4857$$

$$\frac{\partial P_f}{\partial \mu_D} = \frac{P_f(\mu_D + \delta\mu_D) - P_f(\mu_D)}{\delta\mu_D}$$

which simplifies to

$$\frac{\partial P_f}{\partial \mu_D} = \frac{0.4177 - 0.4045}{0.07} = 0.1886$$

Similarly, the safety index and failure probability sensitivities with respect to the height mean (μ_H) are computed as

$$\frac{\partial \beta}{\partial \mu_H} = \frac{\beta(\mu_H + \delta\mu_H) - \beta(\mu_H)}{\delta\mu_H}$$

which simplifies to

$$\frac{\partial \beta}{\partial \mu_H} = \frac{0.2243 - 0.2418}{0.07} = -0.25$$

$$\frac{\partial P_f}{\partial \mu_H} = \frac{P_f(\mu_H + \delta\mu_H) - P_f(\mu_H)}{\delta\mu_H}$$

Substitution of information gives

$$\frac{\partial P_f}{\partial \mu_H} = \frac{0.4113 - 0.4045}{0.07} = 0.0971$$

The safety index and failure probability sensitivities with respect to the diameter standard deviation (σ_D) are

$$\frac{\partial \beta}{\partial \sigma_D} = \frac{\beta(\sigma_D + \delta\sigma_D) - \beta(\sigma_D)}{\delta\sigma_D}$$

which simplifies to

$$\frac{\partial \beta}{\partial \sigma_D} = \frac{0.2398 - 0.2418}{0.02} = -0.1$$

$$\frac{\partial P_f}{\partial \sigma_D} = \frac{P_f(\sigma_D + \delta\sigma_D) - P_f(\sigma_D)}{\delta\sigma_D}$$

Substitution of values gives

$$\frac{\partial P_f}{\partial \sigma_D} = \frac{0.4052 - 0.4045}{0.02} = 0.035$$

The safety index and failure probability sensitivities with respect to the height standard deviation (σ_H) are

$$\frac{\partial \beta}{\partial \sigma_H} = \frac{\beta(\sigma_H + \delta\sigma_H) - \beta(\sigma_H)}{\delta\sigma_H}$$

Substitution of values gives

$$\frac{\partial \beta}{\partial \sigma_H} = \frac{0.2413 - 0.2418}{0.02} = -0.025$$

$$\frac{\partial P_f}{\partial \sigma_H} = \frac{P_f(\sigma_H + \delta\sigma_H) - P_f(\sigma_H)}{\delta\sigma_H}$$

which simplifies to

$$\frac{\partial P_f}{\partial \sigma_H} = \frac{0.4047 - 0.4045}{0.02} = 0.01$$

Table 5.5. Analytical and Finite Difference Sensitivities Comparison

B sensitivity	FFD	Analytical	P_f sensitivity	FFD	Analytical
$\frac{\partial \beta}{\partial \mu_D}$	-0.486	-0.446	$\frac{\partial P_f}{\partial \mu_D}$	0.1886	0.1728
$\frac{\partial \beta}{\partial \mu_H}$	-0.250	-0.2236	$\frac{\partial P_f}{\partial \mu_H}$	0.0971	0.0866
$\frac{\partial \beta}{\partial \sigma_D}$	-0.100	-0.0967	$\frac{\partial P_f}{\partial \sigma_D}$	0.0350	0.0375
$\frac{\partial \beta}{\partial \sigma_H}$	-0.025	-0.0242	$\frac{\partial P_f}{\partial \sigma_H}$	0.0100	0.0094

The analytical and finite difference sensitivity values are tabulated in Table 5.5. The table shows that finite difference sensitivities match closely with analytical sensitivities. Moreover, finite difference sensitivities require $(n+1)$ function evaluations for n design variables. This number of function evaluations increases rapidly with increases in design variables, whereas no additional function evaluations are required for analytical sensitivity computations. Finally, the finite difference sensitivities are highly dependent on the nonlinearity of the function and perturbation size.

Example 5.4

In a multidisciplinary design optimization environment, some parameters could be both random and optimization variables, some are only random and pre-assigned design variables, and some are deterministic design variables. To demonstrate these possibilities, the following four-bar truss example (shown in Figure 5.8) is considered.

Assume members 1 through 3 have same cross-sectional area A_1, and member 4 has an area of A_2. Under the specified loads, the vertical displacement at joint 2 is defined as

$$d = \frac{6P\ell}{E}\left(\frac{3}{A_1} + \frac{\sqrt{3}}{A_2}\right)$$

Here P is denoted as a load that is normally distributed with a mean of 20 MN and 10% COV, and E is Young's modulus, which varies normally with a mean of 200 GPa and 10% COV. A_1 and A_2 are cross-sectional areas that vary with mean 1 m^2 and 1% COV, and ℓ is a deterministic variable with a length of 1 m. The limit on the vertical displacement is 0.003 m at joint 2. Calculate the safety index and probability of failure and their sensitivities.

Solution:

The limit-state function is defined as the vertical displacement. It should be less than 0.003 m and is written as

$g(A_i, \ell) : d \leq 0.003$
$g : 0.003 - d \geq 0$

Therefore, the limit-state function becomes

$$g = 0.003 - \frac{6P\ell}{E}\left(\frac{3}{A_1} + \frac{\sqrt{3}}{A_2}\right)$$

Table 5.6. Variables Characteristics

Parameter	Random variable	Design variable
E	×	—
P	×	—
A_1 and A_2	×	×
ℓ	—	×

Observe that the limit-state is a function of load (P), Young's modulus (E), length (ℓ), and areas (A_1 and A_2). The characteristics of each variable are tabulated in Table 5.6. The statistical properties of these variables are defined in Table 5.7.

Table 5.7. Variables Properties

Variable	Mean(μ)	Standard deviation (σ)
E	200×10^9 PA	20×10^9
P	20×10^6 N	2×10^6
A_1 and A_2	1 m²	0.01
ℓ	1 m	0

The gradients of $g(A_i, \ell)$ with respect to E, P, A_i, and ℓ at the mean values are calculated as

$$\frac{\partial g}{\partial E} = \frac{6P\ell}{E^2}\left(\frac{3}{A_1} + \frac{\sqrt{3}}{A_2}\right) = \frac{6 \times 20 \times 10^6 \times 1}{(200 \times 10^9)^2}\left(\frac{3}{1} + \frac{\sqrt{3}}{1}\right) = 1.42 \times 10^{-14}$$

$$\frac{\partial g}{\partial P} = -\frac{6\ell}{E}\left(\frac{3}{A_1} + \frac{\sqrt{3}}{A_2}\right) = -\frac{6 \times 1}{200 \times 10^9}\left(\frac{3}{1} + \frac{\sqrt{3}}{1}\right) = -1.42 \times 10^{-10}$$

$$\frac{\partial g}{\partial \ell} = -\frac{6P}{E}\left(\frac{3}{A_1} + \frac{\sqrt{3}}{A_2}\right) = -\frac{6 \times 20 \times 10^6}{200 \times 10^9}\left(\frac{3}{1} + \frac{\sqrt{3}}{1}\right) = -0.00283$$

$$\frac{\partial g}{\partial A_1} = -\frac{6P\ell}{E}\left(-\frac{3}{A_1^2}\right) = -\frac{6 \times 20 \times 10^6 \times 1}{200 \times 10^9}\left(-\frac{3}{1^2}\right) = 0.0018$$

$$\frac{\partial g}{\partial A_2} = -\frac{6P\ell}{E}\left(-\frac{\sqrt{3}}{A_2^2}\right) = -\frac{6 \times 1000 \times 1}{200 \times 10^9}\left(-\frac{\sqrt{3}}{1^2}\right) = 0.00104$$

The mean value of the limit-state function is

$$\mu_g = 0.003 - \frac{6\mu_P\mu_\ell}{\mu_E}\left(\frac{3}{\mu_{A_1}} + \frac{\sqrt{3}}{\mu_{A_2}}\right) = 0.003 - \frac{6 \times 20 \times 10^6 \times 1}{200 \times 10^9}\left(\frac{3}{1} + \frac{\sqrt{3}}{1}\right) = 1.61 \times 10^{-4}$$

The standard deviation of the limit-state function σ_g, is

$$\sigma_g = |\nabla_x g \times \sigma_x|$$

$$\sigma_g = \sqrt{\left(\frac{\partial g}{\partial E}\sigma_E\right)^2 + \left(\frac{\partial g}{\partial P}\sigma_P\right)^2 + \left(\frac{\partial g}{\partial l}\sigma_l\right)^2 + \left(\frac{\partial g}{\partial A_1}\sigma_{A_1}\right)^2 + \left(\frac{\partial g}{\partial A_2}\sigma_{A_2}\right)^2}$$

$$= 4.021 \times 10^{-4}$$

The safety index is obtained as

$$\beta = \frac{\mu_g}{\sigma_g} = \frac{1.61 \times 10^{-4}}{4.021 \times 10^{-4}} = 0.39986$$

The probability of failure is

$$P_f = 1 - \Phi(\beta) = 0.34463$$

Sensitivities with respect to Means:

The safety index and failure probability sensitivities with respect to the mean values are computed using Equation 5.35 and Equation 5.43, as follows:
Using Equation 5.35, the safety index sensitivity with respect to the mean is written as

$$\frac{\partial \beta}{\partial \mu_x} = \frac{\frac{\partial G}{\partial x}}{\left|\frac{\partial G}{\partial x}\sigma_x\right|}$$

The safety index sensitivity with respect to the length mean value (μ_ℓ) is

$$\frac{\partial \beta}{\partial \mu_\ell} = \frac{\frac{\partial G}{\partial \ell}}{\sigma_g} = \frac{-0.002839}{0.000421} = -7.062$$

From Equation 5.43, the failure probability sensitivity is computed as

$$\frac{\partial P_f}{\partial \mu_\ell} = -\phi(\beta)\frac{\partial \beta}{\partial \mu_\ell} = -0.3683 \times -7.062 = 2.6009$$

The safety index sensitivity with respect to the mean of A_1 (μ_{A1}) is

$$\frac{\partial \beta}{\partial \mu_{A_1}} = \frac{\frac{\partial G}{\partial A_1}}{\sigma_g} = \frac{0.0018}{0.000421} = 4.4769$$

The failure probability sensitivity with respect to the mean of A_1 (μ_{A1}) is

$$\frac{\partial P_f}{\partial \mu_{A_1}} = -\phi(\beta)\frac{\partial \beta}{\partial \mu_{A_1}} = -0.3683 \times 4.4769 = -1.6488$$

The safety index sensitivity with respect to the mean of A_2 (μ_{A2}) is

$$\frac{\partial \beta}{\partial \mu_{A_2}} = \frac{\frac{\partial G}{\partial A_2}}{\sigma_g} = \frac{0.00104}{0.000421} = 2.5847$$

The failure probability sensitivity with respect to the mean of A_2 (μ_{A2}) is

$$\frac{\partial P_f}{\partial \mu_{A_2}} = -\phi(\beta)\frac{\partial \beta}{\partial \mu_{A_2}} = -0.3683 \times 2.5847 = -0.9519$$

Sensitivities with respect to Standard Deviations:

The safety index and failure probability sensitivities with respect to standard deviations are computed using Equation 5.44 and Equation 5.45. Using Equation 5.44, the safety index sensitivity is written as

$$\frac{\partial \beta}{\partial \sigma_x} = -\frac{\left(\frac{\partial G}{\partial x}\right)^2 \sigma_x}{\left(\sigma_g\right)^2}\beta$$

Since the length is a deterministic variable, the standard deviation is zero. Hence, the sensitivity does not exist. The safety index sensitivity with respect to the standard deviation of A_1 is calculated as

$$\frac{\partial \beta}{\partial \sigma_{A_1}} = -\frac{\left(\frac{\partial G}{\partial A_1}\right)^2 \sigma_{A_1}}{\left(\sigma_g\right)^2}\beta$$

$$\rightarrow \frac{\partial \beta}{\partial \sigma_{A_1}} = -\frac{(0.0018)^2 \times 0.01}{(4.2 \times 10^{-4})^2} \times 0.39986 = -0.08014$$

The failure probability sensitivity with respect to the standard deviation A_1 is

$$\frac{\partial P_f}{\partial \sigma_{A_1}} = -\phi(\beta)\frac{\partial \beta}{\partial \sigma_{A_1}} = -0.3683 \times -0.08014 = 0.0295$$

Similarly, the safety index sensitivity with respect to the standard deviation A_2 is obtained as

$$\frac{\partial \beta}{\partial \sigma_{A_2}} = -\frac{\left(\frac{\partial G}{\partial A_2}\right)^2 \sigma_{A_2}}{(\sigma_g)^2}\beta = -\frac{(0.00104)^2 \times 0.01}{(4.2 \times 10^{-4})^2} \times 0.39986 = -0.026714$$

The failure probability sensitivity with respect to the standard deviation of area 2 is

$$\frac{\partial P_f}{\partial \sigma_{A_2}} = -\phi(\beta)\frac{\partial \beta}{\partial \sigma_{A_2}} = -0.3683 \times -0.026714 = 0.0093$$

Table 5.8. Safety Index and Failure Probability for Design Variables Perturbations

Perturbation	μ	σ	β	$\Phi(\beta)$	P_f
$\delta\mu = \delta\sigma_{A_1} = 0.0$	$\mu_l = \mu_{A_1} = \mu_{A_2} = 1.0$	$\sigma_{A_1} = \sigma_{A_2} = 0.01$	0.39986	0.65537	0.34463
$\delta\mu_l = 0.005$	$\mu_l = 1.005, \mu_{A_1} = \mu_{A_2} = 1.0$	$\sigma_{A_1} = \sigma_{A_2} = 0.01$	0.36274	0.64160	0.35840
$\delta\sigma_{A_1} = 0.005$	$\mu_l = 1.0, \mu_{A_1} = 1.005, \mu_{A_2} = 1.0$	$\sigma_{A_1} = \sigma_{A_2} = 0.01$	0.42347	0.66402	0.33598
$\delta\sigma_{A_2} = 0.005$	$\mu_l = \mu_{A_1} = 1.0, \mu_{A_2} = 1.005$	$\sigma_{A_1} = \sigma_{A_2} = 0.01$	0.41347	0.66037	0.33963
$\delta\sigma_{A_2} = 0.001$	$\mu_l = \mu_{A_1} = \mu_{A_2} = 1.0$	$\sigma_{A_1} = 0.011, \sigma_{A_2} = 0.01$	0.39978	0.65534	0.34466
$\delta\sigma_{A_2} = 0.001$	$\mu_l = \mu_{A_1} = \mu_{A_2} = 1.0$	$\sigma_{A_1} = 0.01, \sigma_{A_2} = 0.011$	0.39983	0.65536	0.34464

These analytical sensitivities are compared with forward finite difference sensitivities, which are computed using the following expressions:

$$\frac{\partial \beta}{\partial b} = \frac{\beta(b + \delta b) - \beta(b)}{\delta b}$$

$$\frac{\partial P_f}{\partial b} = \frac{P_f(b+\delta b) - P_f(b)}{\delta b}$$

In this example, the mean values and standard deviations are perturbed by 0.005 and 0.001, respectively. The safety index and failure probability are tabulated in Table 5.8.

The safety index sensitivities with respect to the design variable mean values are

$$\frac{\partial \beta}{\partial \mu_\ell} = \frac{\beta(\mu_\ell + \delta\mu_\ell) - \beta(\mu_\ell)}{\delta\mu_\ell} = \frac{0.36274 - 0.39986}{0.005} = -7.424$$

$$\frac{\partial \beta}{\partial \mu_{A_1}} = \frac{\beta(\mu_{A_1} + \delta\mu_{A_1}) - \beta(\mu_{A_1})}{\delta\mu_{A_1}} = \frac{0.42347 - 0.39986}{0.005} = 4.722$$

$$\frac{\partial \beta}{\partial \mu_{A_2}} = \frac{\beta(\mu_{A_2} + \delta\mu_{A_2}) - \beta(\mu_{A_2})}{\delta\mu_{A_2}} = \frac{0.413468 - 0.39986}{0.005} = 2.722$$

The failure probability sensitivities with respect to the design variable mean values are

$$\frac{\partial P_f}{\partial \mu_\ell} = \frac{P_f(\mu_\ell + \delta\mu_\ell) - P_f(\mu_\ell)}{\delta\mu_\ell} = \frac{0.35840 - 0.34463}{0.005} = 2.754$$

$$\frac{\partial P_f}{\partial \mu_{A1}} = \frac{P_f(\mu_{A_1} + \delta\mu_{A_1}) - P_f(\mu_{A_1})}{\delta\mu_{A_1}} = \frac{0.33598 - 0.34463}{0.005} = -1.73$$

$$\frac{\partial P_f}{\partial \mu_{A2}} = \frac{P_f(\mu_{A_2} + \delta\mu_{A_2}) - P_f(\mu_{A_2})}{\delta\mu_{A_2}} = \frac{0.339632 - 0.34463}{0.005} = -1.02$$

The sensitivities with respect to the design variable standard deviations are obtained as

Safety Index Sensitivities:

$$\frac{\partial \beta}{\partial \sigma_{A_1}} = \frac{\beta(\sigma_{A_1} + \delta\sigma_{A_1}) - \beta(\sigma_{A_1})}{\delta\sigma_{A_1}} = \frac{0.399775 - 0.39986}{0.001} = -0.085$$

$$\frac{\partial \beta}{\partial \sigma_{A_2}} = \frac{\beta(\sigma_{A_2} + \delta\sigma_{A_2}) - \beta(\sigma_{A_2})}{\delta\sigma_{A_2}} = \frac{0.399831 - 0.39986}{0.001} = -0.029$$

Failure Probability Sensitivities:

$$\frac{\partial P_f}{\partial \sigma_{A_1}} = \frac{P_f(\sigma_{A_1} + \delta\sigma_{A_1}) - P_f(\sigma_{A_1})}{\delta\sigma_{A_1}} = \frac{0.3446613 - 0.34463}{0.001} = 0.0313$$

$$\frac{\partial P_f}{\partial \sigma_{A_2}} = \frac{P_f(\sigma_{A_2} + \delta\sigma_{A_2}) - P_f(\sigma_{A_2})}{\delta\sigma_{A_2}} = \frac{0.3446405 - 0.34463}{0.001} = 0.0105$$

The analytical and finite difference sensitivities are tabulated in Table 5.9.

Table 5.9. Comparison of FFD and analytical sensitivities

β sensitivity	FFD	Analytical	P_f sensitivity	FFD	Analytical
$\dfrac{\partial \beta}{\partial \mu_\ell}$	-7.424	-7.062	$\dfrac{\partial P_f}{\partial \mu_\ell}$	2.754	2.601
$\dfrac{\partial \beta}{\partial \mu_{A_1}}$	4.722	4.477	$\dfrac{\partial P_f}{\partial \mu_{A_1}}$	-1.73	-1.649
$\dfrac{\partial \beta}{\partial \mu_{A_2}}$	2.722	2.585	$\dfrac{\partial P_f}{\partial \mu_{A_2}}$	-1.02	-0.952
$\dfrac{\partial \beta}{\partial \sigma_{A_1}}$	-0.085	-0.08014	$\dfrac{\partial P_f}{\partial \sigma_{A_1}}$	0.0313	0.0295
$\dfrac{\partial \beta}{\partial \sigma_{A_2}}$	-0.029	-0.026714	$\dfrac{\partial P_f}{\partial \sigma_{A_2}}$	0.0105	0.0098

In summary, the sensitivities analysis that is a crucial part of reliability-based design optimization is presented for deterministic variables, random variables and for both categories. Since optimization is an iterative process, there are many challenges for large-scale problems where implicit analysis is needed. In the following, some of the practical aspects of design optimization are presented.

5.5 Practical Aspects of Structural Optimization

5.5.1 Design Variable Linking

Having an independent design variable for each free parameter or finite element gives additional degrees of freedom for solving the mathematical optimization problem. But sometimes this results in impractical or difficult to manufacture components. In addition, a problem solved with hundreds or thousands of design variables may not be tractable. Hence there are several practical advantages to reducing the number of design variables. One way to do this is to link the local design variables with global variables. The global variables b are the ones that are directly involved in the design process. The local variables are linked to the global values through a matrix relationship of the form

$$t = T b \tag{5.46}$$

where t is a vector of local optimization variables, b is a vector of global design variables, and T is the linking matrix. There are various forms of linking options possible, based on the physics of the problem. The idea is to significantly reduce the number of optimization variables using the T matrix. Linking of design variables imposes additional constraints on the problem and may not lead to the lowest possible objective function.

5.5.2 Reduction of Number of Constraints

A multidisciplinary design problem often involves a large number of inequality constraints—both behavioral and side constraints. The large number of constraints arises because it is usually necessary to guard against a wide variety of failure modes in each of several distinct maneuver and load conditions. During each stage of an iterative process, only critical and potentially critical constraints play significant roles in deriving the solution. Non-critical and redundant constraints that are not currently influencing the iterative design process are temporarily ignored. Two commonly used techniques to accomplish this are regionalization and "throw-away." In regionalization, for example under multiple static loading conditions, if the region contains various types of finite elements (e.g. bars, shear panels, quadrilaterals, beams), it may be desirable to retain the one most critical stress constraint for each load condition and element type. Reduction of constraints using the regionalization concept hinges upon the assumption that the design changes made during the redesign step are not so drastic as to result in a shift of the constraint location within a region. In the "throw away" approach, unimportant (redundant or very inactive) constraints are temporarily ignored in a particular iteration.

5.5.3 Approximation Concepts

The basic objective in the approximate structural analysis approach is to obtain high quality algebraically explicit expressions for the objective function and behavior constraints. These explicit approximations are used in place of the detailed analysis during different parts of the iterative process. The function approximations play a very significant role in reliability-based optimization, and they were extensively discussed in Chapters 3 and 4, where they were successfully applied in computing the safety index, β, and the failure probability, P_f. As shown in Chapters 3 and 4, these concepts have been used extensively in mathematical optimization and are also popular in reliability analysis and design.

5.5.4 Move Limits

The approximations constructed at a specific point are accurate within certain bounds of n-dimensional space. Once the exact problem is replaced with a surrogate problem, in order to maintain the validity of the approximations, bounds are placed on how much a design variable can change during a single design cycle. Move limits artificially restrict the design space. Due to the nonlinearity of functions, a certain percentage of change in design variables may not proportionately translate to the same percentage change in the response function. Proper selection of move limits is important for convergence to the optimum. These artificial bounds always fall within the bounds of the original design problem.

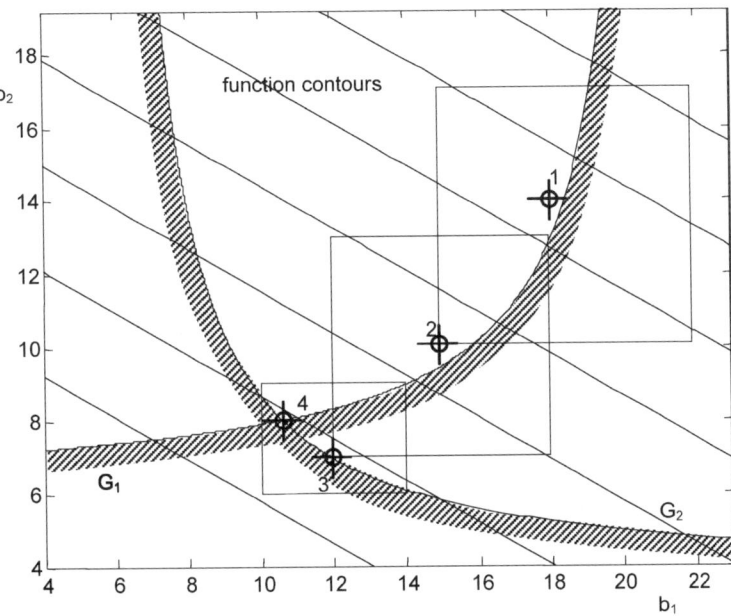

Figure 5.13. Surrogate Models and Sequential Optimization

Figure 5.13 shows a sketch of how an approximate problem converges to the true optimum. For a 2-D problem, the move limits are shown as a box around the current design point. Design freedom decreases as we move closer to the optimum where the constraint violations are not acceptable. For the first iteration, the solution to the approximate problem is found to be at a corner of the box, since only the objective is minimized in this iteration where the constraints are not active. Typically, an approximate problem is reconstructed at this solution using an exact analysis (simulation), and the optimization-seeking process is continued. By the end of the second iteration, one of the constraints is violated due to the typical nature of approximations. At the end of the third iteration, a near optimal solution is reached. Multiple types of move-limit strategies are implemented in the design optimization community.

5.6 Convergence to Local Optimum

Since numerical optimization is an iterative process, one of the steps is to define termination criteria. The convergence parameters vary from problem to problem; all of the following conditions may not be needed for every problem.

The first criterion requires that the relative change in the objective between iterations is less than a specific tolerance ε_1. Thus, the criterion is satisfied if

$$\frac{\left|F(b^q) - F(b^{q-1})\right|}{\left|F(b^{q-1})\right|} \leq \varepsilon_1 \tag{5.47}$$

The second criterion is that the absolute change in the objective between the iterations is less than a specified tolerance ε_2. This criterion is satisfied if

$$\left|F(b^q) - F(b^{q-1})\right| \leq \varepsilon_2 \tag{5.48}$$

The third criterion is that the absolute change in the design variables is less than a specified tolerance, ε_3. The criterion s satisfied if

$$\left|b^q - b^{q-1}\right| \leq \varepsilon_3 \tag{5.49}$$

5.7 Reliability-based Design Optimization

Up to this point, various segments involved in optimization have been presented, as weel as a broad classification of optimization algorithms with details about two methods, including examples. The concept of reliability analysis, presented in Chapter 4, can be incorporated into the described deterministic optimization method; this incorporated scheme is referred to as Reliability-based Design Optimization (RBDO). In RBDO, the statistical nature of constraints and design problems are defined in the objective function and probabilistic constraint. The probabilistic constraint can specifiy the required reliability level of the system.

The formation of RBDO is similar to that of deterministic optimization:

Minimize: the objective function, $f(b, \mu_X)$ \hfill (5.50)

Subject to:

$$P_j[g_j(b,X) < 0] \le P_{Rj} \qquad j = 1,\ldots,m \qquad (5.51)$$

or

$$P_j[g_j(b,X) \ge 0] \ge R_j \qquad (5.52)$$

where $g_j(\cdot)$ represents the limit-state function, b is the vector of deterministic design variables, and X is the random vector, which can be random design variables or random parameters of the system. P_{Rj} and R_j are the specified probability of failure level and the specified reliability level of the system, respectively.

Equation 5.51 or Equation 5.52 can be expressed in terms of the safety index:

$$\beta_{g_j} \ge \beta_{R_j} \qquad (5.53)$$

where β_{R_j} is the required safety index of the system, and the safety index of the probabilistic constraint, β_{g_j}, can be calculated from Equation 4.19. The relationship between the safety index and the normally distributed limit state is specified in Equation 3.9. An alternative approach for RBDO problems is the Performance Measure Approach (PMA), which can efficiently measure violations of the constraint. In PMA, the performance measure is determined after solving inverse reliability analysis problems. Details of PMA are available in [15], [16], and [19].

This chapter presented the role of multidisciplinary optimization in engineering design and how reliability of a system can be improved by incorporating probability-of-failure constraints. One important part of optimization is the determination of search direction based on function gradients, also known as sensitivity analysis. This chapter presented details about sensitivities with respect to random variables, and design variables with detailed calculations. Real-world applications have implicit functions that are computationally expensive when seeking the optimum through numerous iterations. This chapter presented tools for making problems tractable by reducing the number of design variables and constructing explicit function approximations as a function of design variables. In the last section, we discussed one effective probablistic optimization method, RBDO; details of more sophiscated RBDO methods are available in [1], [3]-[6], [11], and [14].

5.8 References

[1] Agarwal, H., and Renaud, J.E., "Reliability-based Design Optimization using Response Surfaces in Application to Multidisciplinary Systems," Engineering Optimization, Vol. 36, No.3, June 2004, pp. 291-311.

[2] Arora, J.S., *Introduction to Optimum Design*, Second Edition, Elsevier Academic Press, San Diego, CA, 2004.

[3] Belegundu, A.D., "Probabilistic Optimal Design using Second Moment Criteria", *Journal of Mechanisms*, Transmissions and Automation in Design, Vol. 110, September, 1988, pp. 324-329.

[4] Du, X., Sudjianto, A., and Chen, W., "An Integrated Framework for Optimization under Uncertainty Using Inverse Reliability Strategy," *ASME Journal of Mechanical Design*, Vol.126, No.4, pp. 561-764, 2004

[5] Enevoldsen, J., and Sorensen, J.D., "Reliability-based Optimization in Structural Engineering," *Structural Safety*, Vol. 15, 1994, pp. 169-196.

[6] Frangopol, D.M., and Maute, K., "Reliability-based Optimization of Civil and Aerospace Structural Systems," *Chapter 24 in Engineering Design Reliability Handbook* (Edited by Nikolaidis, E., Ghiocel, D.M., and Singhal, S.), CRC Press, Boca Raton, FL, 2005, 24-1 to 24-32.

[7] Haftka, R.T., and Gurdal, Z., *Elements of Structural Optimization*, Kluwer Academic Publishers, Dordrecht, Netherlands, 1993.

[8] Kwak, B.M., and Lee, T.W., "Sensitivity Analysis for Reliability-based Optimization using an AFOSM Method", *Computers and Structures*, Vol. 27, No. 3, 1987, pp. 399-406.

[9] Luo, X., and Grandhi, R.V., "ASTROS for Reliability-based Multidisciplinary Structural Analysis and Optimization", *Journal of Computers and Structures*, Vol. 62, No. 4, 1997, pp. 737.

[10] Mourelatos, Z.P., and Zhou, J., "Reliability Estimation and Design with Insufficient Data Based on Possibility Theory," *AIAA Journal*, Vol. 43, No. 8, 2005, pp. 1696-1705.

[11] Pettit, C.L.,, and Grandhi, R.,V., "Preliminary Weight Optimization of a Wing Structure for Gust Response and Aileron Effectiveness Reliability," AIAA Journal of Aircraft, Vol 40, No. 6, 2003, pp. 1185-1191.

[12] Qu, X., Haftka, R.T., Venkataraman, S., and Johnson, T.F., "Deterministic and Reliability-Based Optimization of Composite Laminates for Propellant Tanks," AIAA Journal, Vol. 41, No. 10, 2003, pp. 2029-2036.

[13] Rao, S.S., Engineering Optimization: Theory and Practice, Third Edition, *Wiley Interscience*, New York, USA, 1996.

[14] Sues, R.H., Shin, Y., and Wu, Y.-T., "Applications of Reliability-based Design Optimization," *Engineering Design Reliability Handbook*, (Edited by Nicholaidis, E., Ghiocel, D.M., and Singhal, S.) CRC Press, Boca Raton, FL, 2005, pp. 34-1 to 34-24.

[15] Tu, J., and Choi, K.K., *A Performance Measure Approach in Reliability-based Structural Optimization*, Technical Report R97-02, University of Iowa, 1997.

[16] Tu, J., and Choi, K.K., *Design Potential Concept for Reliability-based Design Optimization*, Technical Report R99-07, University of Iowa, 1999.

[17] Vanderplaats, G.N., *Numerical Optimization Techniques for Engineering with Applications*, Third Edition, VR&D Inc., Colorado Springs, CO, 1999.

[18] Yang, Y.S., and Nikolaidis, E., "Design of Aircraft Wings Subjected to Gust Loads: A Safety Index Based Approach", *AIAA Journal*, Vol. 29, No. 5, 1990, pp. 804-812.

[19] Youn, B.D. and Choi, K.K., "Selecting Probabilistic Approaches for Reliability-based Design Optimization," *AIAA Journal*, Vol. 42, No. 1, 2004, pp. 124-131.

6

Stochastic Expansion for Probabilistic Analysis

The direct use of stochastic expansions is an efficient choice for representing uncertain parameters because they provide analytically appealing convergence properties. This chapter discusses two stochastic expansions that are used to represent random processes: the PCE and the KL expansion. Basic properties of PCE and the KL transform are illustrated with simple problems. In addition, a new stochastic approximation procedure is presented that uses PCE and the KL transform. In the latter part of the chapter, the roles of both expansions in the spectral stochastic finite element method are also discussed.

6.1 Polynomial Chaos Expansion (PCE)

As discussed in Chapter 3, one of the effective choices for uncertainty analysis is the direct use of stochastic expansions of output responses and input random variables for representing uncertainty. Stochastic expansions provide analytically appealing convergence properties based on the concept of a random process. The polynomial chaos expansion can reduce computational effort of uncertainty quantification in engineering design applications where the system response is computed implicitly. The basic concepts and mathematical aspects of the PCE are outlined in the following section.

6.1.1 Fundamentals of PCE

In the latter half of the eighteenth century, Robert Brown, a botanist, observed through a microscope random motions of pollen grains in water. The particles kept changing directions, and the grains' behavior was quite random. Termed *Brownian motion*, this random behavior was explained by Einstein in 1905 to be the collision between tiny water molecules and the pollen particles. This attempt of explanations was induced to the estimation of the water molecules' weight, and Einstein won the Nobel prize. Later, Wiener, a mathematician, provided meaningful mathematical descriptions of irregularities in Brownian motion. His study of Brownian motion (also termed the *Wiener process*) led him to establish a novel stochastic analysis, introducing the multiple stochastic integral with homogeneous chaos in his 1938 paper [31]. Subsequently, Wiener's work was modified and completed by Ito, who showed that any stochastic processes can be described in terms of the Wiener process, which is the simplest stochastic process.

The PCE stemmed from both Wiener and Ito's work on mathematical descriptions of irregularities. Since Wiener introduced the concept of homogeneous chaos, the PCE has been successfully used for the uncertainty analysis in various applications, for example, in [12], [14], and [15]. A simple definition of the PCE for a Gaussian random response $u(\theta)$ as a convergent series is as follows:

$$u(\theta) = a_0 \Gamma_0 + \sum_{i=1}^{\infty} a_{i_1} \Gamma_1(\xi_{i_1}(\theta)) + \sum_{i_1=1}^{\infty} \sum_{i_2=1}^{i_1} a_{i_1 i_2} \Gamma_2(\xi_{i_1}(\theta), \xi_{i_2}(\theta))$$
$$+ \sum_{i_1=1}^{\infty} \sum_{i_2=1}^{i_1} \sum_{i_3=1}^{i_2} a_{i_1 i_2 i_3} \Gamma_3(\xi_{i_1}(\theta), \xi_{i_2}(\theta), \xi_{i_3}(\theta)) + \ldots \quad (6.1)$$

where $\{\xi_i(\theta)\}_{i=1}^{\infty}$ is a set of Gaussian random variables; $\Gamma_p(\xi_{i_1}, \ldots, \xi_{i_p})$ is the generic element of a set of multidimensional Hermite polynomials, usually called homogeneous chaos of order p; $a_{i_1}, \ldots, a_{i_p}$ are deterministic constants; and θ represents an outcome in the space of possible outcomes of a random event.

PCE is convergent in the mean-square sense [1] and the pth order PCE consists of all orthogonal polynomials of order p, including any combination of $\{\xi_i(\theta)\}_{i=1}^{\infty}$; furthermore, $\Gamma_p \perp \Gamma_q$ for $p \neq q$. This orthogonality greatly simplifies the procedure of statistical calculations, such as moments. Therefore, PCE can be used to approximate non-Gaussian distributions using a least-squares scheme: for example, in order to compare the skewness and kurtosis of distributions.

The general expression to obtain the multidimensional Hermite polynomials is given by [12]

$$\Gamma_p(\xi_{i_1}, \ldots, \xi_{i_p}) = (-1)^n \frac{\partial^n e^{-\frac{1}{2}\vec{\xi}^T \vec{\xi}}}{\partial \xi_{i_1}, \ldots, \partial \xi_{i_p}} e^{\frac{1}{2}\vec{\xi}^T \vec{\xi}} \quad (6.2)$$

where the vector $\vec{\xi}$ consists of n Gaussian random variables ($\xi_{i_1}, \ldots, \xi_{i_n}$).

Equation 6.1 can be written more simply as

$$u(\theta) = \sum_{i=0}^{P} b_i \Psi_i(\vec{\xi}(\theta)) \quad (6.3)$$

where b_i and $\Psi_i(\vec{\xi}(\theta))$ are one-to-one correspondences between the coefficients $a_{i_1},......,a_{i_p}$ and the functions $\Gamma_p(\xi_{i_1},......,\xi_{i_p})$, respectively. For example, the two-dimensional case of Equation 6.1 can be expanded as:

$$u(\theta) = a_0\Gamma_0 + a_1\Gamma_1(\xi_1) + a_2\Gamma_1(\xi_2) \tag{6.4}$$

$$+ a_{11}\Gamma_2(\xi_1,\xi_1) + a_{12}\Gamma_2(\xi_2,\xi_1) + a_{22}\Gamma_2(\xi_2,\xi_2)$$

$$+ a_{111}\Gamma_3(\xi_1,\xi_1,\xi_1) + a_{211}\Gamma_3(\xi_2,\xi_1,\xi_1) + a_{221}\Gamma_3(\xi_2,\xi_2,\xi_1)$$

$$+ a_{222}\Gamma_3(\xi_2,\xi_2,\xi_2) \ldots$$

Equation 6.3 can be recast in terms of $\Psi_i[.]$ and b_i as follows:

$$u(\theta) = b_0\Psi_0 + b_1\Psi_1 + b_2\Psi_2 + b_3\Psi_3 + b_4\Psi_4 + b_5\Psi_5 + \ldots \tag{6.5}$$

Thus, the term $a_{11}\Gamma_2(\xi_1,\xi_1)$ becomes $b_3\Psi_3$ for this two-dimensional case.

In the one-dimensional case, we can expand the random response u using orthogonal polynomials in ξ, which has a known probability distribution such as unit normal, $N[0,1]$. If u is a function of a normally distributed random variable x, which has the known mean μ_x and variance σ_x^2, ξ is a normalized variable:

$$\xi = \frac{x - \mu_x}{\sigma_x} \tag{6.6}$$

Generally, the one-dimensional Hermite polynomials are defined by

$$\Psi_n(\xi) = (-1)^n \frac{\varphi^{(n)}(\xi)}{\varphi(\xi)} \tag{6.7}$$

where $\varphi^{(n)}(\xi)$ is the nth derivative of the normal density function, $\varphi(\xi) = 1/\sqrt{2\pi}\ e^{-\xi^2/2}$. This is simply the single-variable version of Equation 6.2. From Equation 6.7, we can readily find

$$\{\Psi_i\} = \{1, \xi, \xi^2 - 1, \xi^3 - 3\xi, \xi^4 - 6\xi^2 + 3, \xi^5 - 10\xi^3 + 15\xi,...\} \tag{6.8}$$

Thus, a second-order, 2-D PCE is given by

$$u(\theta) = b_0 + b_1\xi_1(\theta) + b_2\xi_2(\theta) + b_3(\xi_1^2(\theta) - 1) \tag{6.9}$$

$$+b_4\xi_1(\theta)\xi_2(\theta)+b_5(\xi_2^2(\theta)-1)$$

where $\xi_1(\theta)$ and $\xi_2(\theta)$ are two independent random variables.

The standard normal random variable ξ and orthogonal polynomials Ψ_i satisfy

$$\Psi_0=1,\ E[\Psi_i]=0\ \text{and}\ E[\Psi_i\Psi_j]=E[\Psi_i^2]\delta_{ij}\quad \forall\ i,j \qquad (6.10)$$
$$E[\xi^0]=1,\ E[\xi^k]=0\quad \forall\ k\ odd$$

where $E[.]$ indicates the expected value operation, and δ_{ij} is the Kronecker delta, which equals 1 when $i=j$ and 0 for $i\neq j$.

If the solution is known, the generalized Fourier coefficients b_i can be evaluated from

$$b_i=\frac{E[u(\theta)\Psi_i(\vec{\xi}(\theta))]}{E[\Psi_i(\vec{\xi}(\theta))\Psi_i(\vec{\xi}(\theta))]} \qquad (6.11)$$

This approach was applied in [25] in the non-intrusive formulation procedure by using MCS to evaluate the expected values. This equation is equivalent to the coefficient calculation procedure of general regression analysis described in Section 2.3; thus, the method presented in Section 6.1.2 provides an alternative to general MCS for estimating the expansion coefficients.

As mentioned earlier, PCE can be used to approximate non-Gaussian distributions using a least-squares scheme. The following two examples demonstrate this procedure by utilizing the knowledge of statistic moments. First, we derive the expression of the moments in terms of the coefficients of PCE (Example 6.1); then, the least-squares criterion is applied to compute the coefficients (Example 6.2).

Example 6.1

Suppose we have a random variable x that is normally/non-normally distributed. This random variable x can be approximated by the first four terms of the PCE as follows:

$$x\approx z(\xi)=b_0+b_1\xi+b_2(\xi^2-1)+b_3(\xi^3-3\xi)$$

Calculate the first four central moments of z in terms of the coefficients b_i.

Solution:

Since the polynomial chaoses of order greater than one have zero mean (Equation 6.10), the mean value of z is b_0, $\mu_z = b_0$.

From the definition of the central moment (Equation 2.19), the second moment of z is

$$m_z^2 = E[(z-b_0)^2] = E[\{b_1\xi + b_2(\xi^2-1) + b_3(\xi^3-3\xi)\}^2]$$
$$= E[\xi^2 b_1^2 - 2\xi b_1 b_2 + 2\xi^3 b_1 b_2 + b_2^2 - 2\xi^2 b_2^2 + \xi^4 b_2^2 - 6\xi^2 b_1 b_3 +$$
$$2\xi^4 b_1 b_3 + 6\xi b_2 b_3 - 8\xi^3 b_2 b_3 + 2\xi^5 b_2 b_3 + 9\xi^2 b_3^2 - 6\xi^4 b_3^2 + \xi^6 b_3^2]\quad\ldots\ldots\ldots(A)$$

From Equation 6.10,

$$E[\xi^k] = 0 \quad \forall\ k\ odd,$$
$$E[\xi^0] = 1,\ E[\xi^2] = 1,\ E[\xi^4] = 3,\ E[\xi^6] = 15\ \ldots\ldots\ldots\ldots\ldots\ldots(B)$$

Thus, after substituting the values of Equation B into Equation A, the second moment of z can be computed as

$$m_z^2 = E[(z-b_0)^2] = b_1^2 + 2b_2^2 + 6b_3^2$$

Similarly, the third and fourth moments are obtained as

$$m_z^3 = E[(z-b_0)^3] = E[\{b_1\xi + b_2(\xi^2-1) + b_3(\xi^3-3\xi)\}^3]$$
$$= 6b_1^2 b_2 + 8b_2^3 + 36b_1 b_2 b_3 + 108b_2 b_3^2$$

$$m_z^4 = E[(z-b_0)^4] = E[\{b_1\xi + b_2(\xi^2-1) + b_3(\xi^3-3\xi)\}^4]$$
$$= 3b_1^4 + 60b_2^4 + 3348b_3^4 + 24b_1^3 b_3 + 60b_1^2 b_2^2$$
$$+ 252b_1^2 b_3^2 + 576b_1 b_2^2 b_3 + 1296b_1 b_3^3 + 2232b_2^2 b_3^2$$

The coefficients b_i of PCE can be determined by matching the known moments of given random variables; that is, b_i are numerically calculated through the least-square criterion. The following example continues with the remaining procedures of this moment-matching method.

Example 6.2

Suppose the first four moments of a random variable, x, are given by $m_x^1 = 1.0$, $m_x^2 = 1.5$, $m_x^3 = 3.5$, and $m_x^4 = 17.0$. Estimate the coefficients b_i of z (Example 6.1) by using the least-square criterion:

Minimize: $\sum_{j=1}^{4} f_j^2(b_i)$, $(i = 1, 2, 3)$

where $f_1(b_i) = m_z^1 - m_x^1 = 0$

$f_2(b_i) = m_z^2(b_i) - m_x^2$

$f_3(b_i) = m_z^3(b_i) - m_x^3$

$f_4(b_i) = m_z^4(b_i) - m_x^4$

Solution:

The coefficient results in Example 6.1 permit matching of the known moments of a random variable x. Although a true minimum of the least-square criterion is difficult to obtain, good approximations of the moment match can be expected. The following solutions are obtained by using an optimization solver in MATLAB® for the above problem:

$$b_0 = 1, \ b_1 = -0.93597, \ b_2 = 0.55361, \ b_3 = 0.04934$$

Thus, the PCE representation of the random variable x can be

$$x \approx z(\xi) = 1 - 0.93597\xi + 0.55361(\xi^2 - 1) + 0.04934(\xi^3 - 3\xi)$$

To check the accuracy of the obtained result, 20,000 Gaussian sampling points are generated in the Monte Carlo Sampling. The corresponding sketch of the PDF (Figure 6.1) and moments of the random variable z are

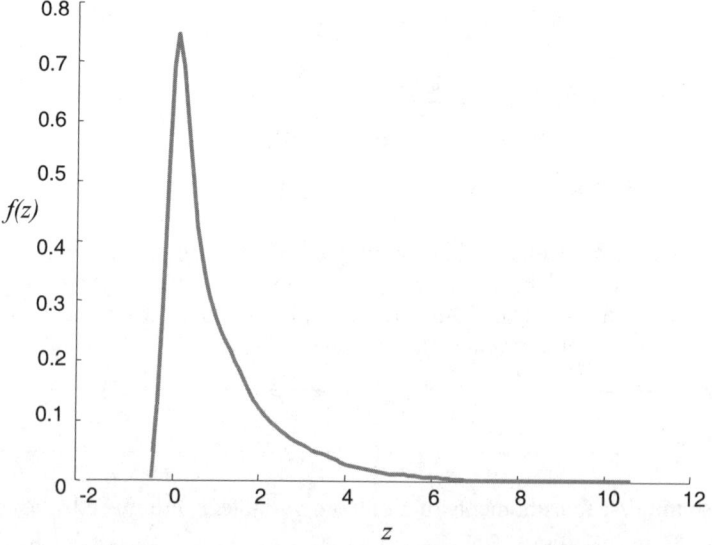

Figure 6.1. Random Variable z PDF Plot (Example 6.2)

> $m^1 = 0.99283, m^2 = 1.47433, m^3 = 3.49398, m^4 = 17.32207$
>
> As expected, adequate results for the moments are also obtained, and the PDF plot shows the positive skewness and sharp peak shape. We will discuss another possible scheme of the moment-match method in Section 6.1.4.
>
> Note: This procedure was suggested by Tatang [27]. Further details and applications are available in the corresponding reference.

6.1.2 Stochastic Approximation

As discussed in Section 3.1, one of the uses for stochastic expansion is the non-intrusive formulation to create a surrogate model of stochastic responses using PCE. The methodology is presented using a simple example. If we fit curvilinear data, the following regression model can be considered:

$$Y(x) = \beta_0 F_0(x) + \beta_1 F_1(x) + \beta_2 F_2(x) + \beta_3 F_3(x) \tag{6.12}$$

where, β_0, β_1, β_2, and β_3 represent the mean, linear, quadratic and cubic effect, respectively, of the response; Y, and $F_i(x)$ are basis polynomials.

Generally, the common polynomial models, for example $F_0(x) = 1$, $F_1(x) = x$, $F_2(x) = x^2$, and $F_3(x) = x^3$, are useful approximations of unknown or complex nonlinear relationships. However, these often are not the best choice, as discussed in Section 2.3. For instance, large positive values of x give large positive values of the response. If x is negative, all of the odd power terms yield negative values and the even power terms give positive values. Since they are not orthogonal, small changes in $F(x)$ would produce relatively large changes in the coefficients $\beta_0,...,\beta_3$. This collinearity will make the associated least-squares problem ill-conditioned. A more satisfactory solution is to use orthogonal polynomials, e.g., Chebyshev polynomials, $F_0(x) = 1$, $F_1(x) = x$, $F_2(x) = 2x^2 - 1$, and $F_3(x) = 4x^3 - 3x$.

Obviously, the use of orthogonal polynomials can eliminate collinearity and ill-conditioned problems. There are several orthogonal polynomials described in the literature whose orthogonality weighting functions match the standard probability density functions. Table 6.1 shows several PDFs with their related orthogonal polynomials and ranges [32],[33]. Hermite polynomials are uncorrelated when x is Gaussian on $(-\infty, +\infty)$, and Laguerre polynomials are uncorrelated when x is Gamma-distributed on $(0, +\infty)$. The basic idea of the stochastic approximation utilizing stochastic expansions is to select an appropriate basis function to represent the response of uncertain systems. The PCE, which employs orthogonal basis functions and is mean-square convergent, is a good choice for estimating the response variability of uncertain systems.

Table 6.1. Orthogonal Polynomials [33]

		PDF	Orthogonal Polynomial	Support Range
Continuous		Gaussian	Hermite	$(-\infty, +\infty)$
		Gamma	Laguerre	$[0, +\infty)$
		Beta	Jacobi	[a,b]
		Uniform	Legendre	[a,b]
Discrete		Poisson	Charlier	$\{0,1,2,...\}$
		Binomial	Krawtchouk	$\{0,1,2,...,N\}$
		Negative Binomial	Meixner	$\{0,1,2,...\}$
		Hypergeometric	Hahn	$\{0,1,2,...,N\}$

PCE can be used to represent the response of an uncertain system in the non-intrusive formulation. The basic idea of this approach is to project the response and stochastic system operator onto the stochastic space spanned by PCE, with the projection coefficients, b_i, being evaluated through an efficient sampling scheme. We first define vector x at a particular point $(\xi_1, \xi_2, ..., \xi_m)$ of random variables

$$x^T = [1 \; \Psi_1(\xi_1) \; \Psi_2(\xi_1)...\Psi_p(\xi_1) \; \Psi_1(\xi_2) \; \Psi_2(\xi_2) \\ ...\Psi_p(\xi_2) \; \Psi_1(\xi_m) \; \Psi_2(\xi_m)...\Psi_p(\xi_m)] \quad (6.13)$$

where p is the order of polynomial and $\Psi_j(\xi_i)$ are PCE, which were described in the previous section. The estimated response at this point is

$$y(x) = x^T \hat{\beta} \quad (6.14)$$

where $\hat{\beta}$ is a set of undetermined coefficients of PCE.

Generally, the method of least squares is used to obtain the regression coefficients for n sample values of x and y as

$$\hat{\beta} = (X^T X)^{-1} X^T Y \quad (6.15)$$

where X is a $n \times p$ matrix of the levels of the regressor variables and Y is a $n \times 1$ vector of the responses.

The fitted model $\hat{Y}$ and the residuals e are

$$\hat{Y} = X\hat{\beta} \quad (6.16)$$

and

$$e = Y - \hat{Y} \qquad (6.17)$$

Once the analyst determines, through ANOVA and residual plots (Section 2.3), that the model is acceptable, various statistics can be obtained including the mean, variance, and confidence interval of the stochastic responses. A confidence interval indicates a range of values that likely contains the analysis results. Generally, the confidence interval of any parameter includes two parts: the confidence level and margin of error. The confidence level denotes the probability with which the interval contains the true parameter value. The margin of error represents how accurate our guess of the true parameter value is. Next, we estimate the confidence interval of the PCE on the mean response at a particular point. We first define vector x_0 at a particular point ($\xi_1, \xi_2, ..., \xi_m$) of random variables from Equation 6.13. Then, the estimated mean response at this point is

$$\hat{y}(x_0) = x_0^T \hat{\beta} \qquad (6.18)$$

where $\hat{\beta}$ is a set of undetermined coefficients of PCE. A $100(1-\alpha)$ percent confidence interval at the particular point x_0 is

$$\hat{y}(x_0) - t_{\frac{\alpha}{2}, v} \sqrt{\sigma^2 x_0^T (X^T X)^{-1} x_0} \leq \mu \leq \hat{y}(x_0) + t_{\frac{\alpha}{2}, v} \sqrt{\sigma^2 x_0^T (X^T X)^{-1} x_0} \qquad (6.19)$$

where σ^2 is variance, v is degree of freedom and α indicates the $100(1-\alpha)$th percentile of the t distribution [21], and X is a $n \times p$ matrix of the random variables. Note that the point x_0 is not limited to one of the sampling points used, since the interval includes the results of random samples from the given population with mean μ. Related discussions about the confidence interval in multiple regression analysis can be found in [20],[21].

6.1.3 Non-Gaussian Random Variate Generation

The PDF of any Gaussian distribution exhibits a symmetric bell-shaped graph. To define the variety of skewed distribution shapes, we need to introduce different families of PDFs, such as gamma distributions or exponential distributions, which are widely used in engineering and science disciplines. Xiu and Karniadakis [33] introduced the Askey scheme to represent various probability distributions for use with PCE. The Askey scheme classifies the hypergeometric orthogonal polynomials and indicates the limit relations between them. In the *generalized* PCE, instead of the Hermite polynomials, any orthogonal polynomials in the Askey scheme can be used to represent either Gaussian or non-Gaussian random processes. Another possible choice for PCE of non-Gaussian cases is the transformation technique [17]. In this procedure, the random input variables, which include non-normal distributions, are transformed into standard normal random variables. This section briefly discusses two techniques, a generalized PCE algorithm and a transformation method, for non-Gaussian random variable cases in the non-intrusive formulation.

Generalized Polynomial Chaos Expansion

In order to represent various probability distributions with PCE, Xiu and Karniadakis introduced the Askey scheme, which classifies the hypergeometric orthogonal polynomials and indicates the limit relations between them. They expected that the use of the original PCE (Equation 6.1) may converge slowly for non-normal distribution cases. In the generalized PCE, the polynomials Ψ_n (Equation 6.3) are not limited to the Hermite polynomials. Any orthogonal polynomials in the Askey scheme can be used to represent both Gaussian and non-Gaussian random processes instead of the Hermite polynomials. For instance, for Gamma and Poisson distributions, the Laguerre and Charlier polynomials are more appropriate choices, respectively. Table 6.1 shows the polynomial types and their associated random variables. After selecting the type of random variables and corresponding polynomials, the non-intrusive formulation procedure can be applied to the uncertainty analysis.

Table 6.2. Representation of Various Distributions as Functionals of Normal Random Variables [17]

Distribution Type	Transformation
Uniform (a, b)	$a + (b-a)\left(\dfrac{1}{2} + \dfrac{1}{2} erf(\dfrac{\xi}{\sqrt{2}})\right)$
Normal (μ, σ)	$\mu + \sigma\xi$
Lognormal (μ, σ)	$\exp(\mu + \sigma\xi)$
Gamma (a,b)	$ab\left(\xi\sqrt{\dfrac{1}{9a}} + 1 - \dfrac{1}{9a}\right)^3$
Exponential (λ)	$-\dfrac{1}{\lambda}\log\left(\dfrac{1}{2} + \dfrac{1}{2} erf(\dfrac{\xi}{\sqrt{2}})\right)$

Transformation Technique

Devroye [9] presented useful transformations in terms of the standard normal random variables. Isukapalli introduced the transformation technique in the stochastic response surface method. According to Table 6.2, the input random variables can be represented by a function of standard normal random variables. In this table, ξ is a standard normal random variable, μ and σ are the mean and standard deviation of the random variable, respectively, while a, b, and λ are scale parameters of each distribution. By using the transformation technique, the random input variables, which include non-normal distributions, are transformed into standard normal random variables. After that, the original PCE can be employed as the response surface model for uncertain systems. Detailed descriptions of various transformation schemes can be found in [9] and [17].

Chapter 7 presents the applications of both methods with the stochastic approximation procedure from Section 6.1.2 on a three-bar truss structure and a joined-wing SensorCraft structure.

6.1.4 Hermite Polynomials and Gram-Charlier Series

Before beginning the topic of the KL transform (Section 6.2), it is useful to see several properties of the Hermite polynomial, which is the basis of the PCE. The construction of the Hermite Polynomial (also known as the Tchebycheff-Hermite or Chebyshev-Hermite polynomial) was described by Pafnuty Chebyshev (1859) and Charles Hermite (1864). The second-order differential equation (also called Hermite's differential equation) is given by

$$\frac{d^2 y}{dx^2} - x\frac{dy}{dx} + ny = 0 \tag{6.20a}$$

or

$$\frac{d^2 y}{dx^2} - 2x\frac{dy}{dx} + 2ny = 0 \tag{6.20b}$$

where n is a positive integer.

The corresponding possible solutions are

$$H_n(x) = (-1)^n e^{\frac{x^2}{2}} \frac{d^n}{dx^n} e^{-\frac{x^2}{2}} \tag{6.21a}$$

for Equation 6.20a and

$$H_n(x) = (-1)^n e^{x^2} \frac{d^n}{dx^n} e^{-x^2} \tag{6.21b}$$

for Equation 6.20b.

These polynomials are called the Hermite polynomials. Although these two equations are not equivalent, Equation 6.21b is a linear rescaling of the domain of Equation 6.21a. Since the nth derivative of the normal density function,

$\varphi(x) = 1/\sqrt{2\pi}\ e^{-x^2/2}$, is included in Equation 6.21a, the definition of Equation 6.21a is often used in probabilistic analysis. The second definition, Equation 6.21b, is often encountered in other subjects, such as physics.

In practice, the following recurrence relations are usually more efficient than the direct computation via Equations 6.21a and 6.21b:

$$H_0(x) = 1, \quad H_1(x) = x \qquad (6.22a)$$
$$H_n(x) = xH_{n-1}(x) - (n-1)H_{n-2}(x), \quad n \geq 2$$

or

$$H_0(x) = 1, \quad H_1(x) = 2x \qquad (6.22b)$$
$$H_n(x) = 2xH_{n-1}(x) - 2(n-1)H_{n-2}(x), \quad n \geq 2$$

Thus, the first several Hermite polynomials are given by

$$H_0(x) = 1, \quad H_1(x) = x, \quad H_2(x) = x^2 - 1, \qquad (6.23a)$$

$$H_3(x) = x^3 - 3x, \quad H_4(x) = x^4 - 6x^2 + 3,$$

$$H_5(x) = x^5 - 10x^3 + 15x, \quad H_6(x) = x^6 - 15x^4 + 45x^2 - 15$$

or

$$H_0(x) = 1, \quad H_1(x) = 2x, \quad H_2(x) = 4x^2 - 2, \qquad (6.23b)$$

$$H_3(x) = 8x^3 - 12x, \quad H_4(x) = 16x^4 - 48x^2 + 12,$$

$$H_5(x) = 32x^5 - 160x^3 + 120x,$$

$$H_6(x) = 64x^6 - 480x^4 + 720x^2 - 120$$

The orthogonal properties of the Hermite polynomials are given in the interval $[-\infty, \infty]$ with respect to the weight function of $e^{-x^2/2}$ or e^{-x^2}:

$$\int_{-\infty}^{\infty} e^{-\frac{x^2}{2}} H_n(x) H_m(x) dx = n!\sqrt{2\pi}\,\delta_{nm} \qquad (6.24a)$$

and

$$\int_{-\infty}^{\infty} e^{-x^2} H_n(x) H_m(x) dx = 2^n n! \sqrt{\pi} \delta_{nm} \tag{6.24b}$$

Equation 6.24a implies that the Hermite polynomials (Equation 6.21a) are orthogonal with respect to the Gaussian distribution. Also, notice that the weight functions, $e^{-x^2/2}$ or e^{-x^2}, help keep the integral from reaching infinity over the interval from $-\infty$ to ∞, since the exponential functions converge to zero much faster than the polynomials blow up when x is large.

When this orthogonal property of the Hermite polynomials is used to estimate the probability density function, the procedure is known as the *Gram-Charlier* method. The basic idea of the Gram-Charlier method is that the density function of the Gaussian distribution and its derivatives provide a series expansion to represent an arbitrary density function. The Gram-Charlier series [7] is given by

$$f(x) = b_0 \varphi(x) + b_1 \varphi'(x) + b_2 \varphi''(x) + \ldots \tag{6.25}$$

where $f(x)$ denotes the unknown probability density function, and $\varphi^{(n)}(x)$ is the nth derivative of the normal density function, $\varphi(x) = 1/\sqrt{2\pi}\, e^{-x^2/2}$. From Equation 6.21a,

$$H_n(x) = (-1)^n \frac{\varphi^{(n)}(x)}{\varphi(x)} \tag{6.26}$$

$$\varphi^{(n)}(x) = (-1)^n \varphi(x) H_n(x)$$

Substituting Equation 6.26 into Equation 6.25 yields

$$f(x) = \varphi(x)[b_0 H_0(x) - b_1 H_1(x) + b_2 H_2(x) + \ldots] \tag{6.27}$$

$$= \varphi(x) \sum_{m=0}^{\infty} (-1)^m b_m H_m(x)$$

To find the b_i coefficients in Equation 6.27, multiply both sides by $H_n(x)$ and integrate from $-\infty$ to ∞. The result is

$$\int_{-\infty}^{\infty} f(x) H_n(x) dx = \sum_{m=0}^{\infty} (-1)^m b_m \int_{-\infty}^{\infty} \varphi(x) H_n(x) H_m(x) dx \tag{6.28}$$

Because of the orthogonal property of the Hermite polynomials,

$$\int_{-\infty}^{\infty} \varphi(x) H_n(x) H_m(x) dx = \begin{cases} n! & \text{for } n = m \\ 0 & \text{for } n \neq m \end{cases} \tag{6.29}$$

This property can be used to compute the coefficients of Equation 6.28. Finally, the coefficients can be computed as

$$b_n = \frac{(-1)^n}{n!} \int_{-\infty}^{\infty} f(x) H_n(x) dx \qquad (6.30)$$

Example 6.3

Using the recurrence formula of Equation 6.22a,

(a) Generate the values of the first 20 Hermite polynomials at the arguments $x = -2.0, -1.0, 0, 1.0,$ and 2.0.
(b) Plot the first four Hermite polynomials on the same plot for the interval -2 to 2.

Solution:

(a) In practice, the recurrence relationship (Equation 6.22a) is more efficient than the direct calculation of Equation 6.8 when many degrees of the polynomials must be computed. The results are summarized in Table 6.3.

Table 6.3. Hermite Polynomials $H_n(x)$ (Example 6.3)

n \ x	-2.0	-1.0	0	1.0	2.0
0	1	1	1	1	1
1	-2	-1	0	1	2
2	3	0	-1	0	3
3	-2	2	0	-2	2
4	-5	-2	3	-2	-5
5	18	-6	0	6	-18
6	-11	16	-15	16	-11
7	-86	20	0	-20	86
8	249	-132	105	-132	249
9	190	-28	0	28	-190
10	-2621	1216	-945	1216	-2621
11	3342	-936	0	936	-3342
12	22147	-12440	10395	-12440	22147
13	-84398	23672	0	-23672	84398
14	-119115	138048	-135135	138048	-119115
15	1419802	-469456	0	469456	-1419802
16	-1052879	-1601264	2027025	-1601264	-1052879
17	-20611074	9112560	0	-9112560	20611074
18	59121091	18108928	-34459425	18108928	59121091
19	252757150	-182135008	0	182135008	-252757150
20	-1628815029	-161934624	654729075	-161934624	-1628815029

As shown in Table 6.3, the even-degree Hermite polynomials are even functions, whereas the odd-degree Hermite polynomials are odd functions. Geometrically, the even functions are symmetric with respect to the y-axis, and the odd functions are anti-symmetric with respect to the y-axis. Some of these polynomials are plotted in Figure 6.2.

(b) The first four Hermite polynomials are plotted in Figure 6.2. It is clear that $H_n(x)$ is an even function when n is even; otherwise the function is odd.

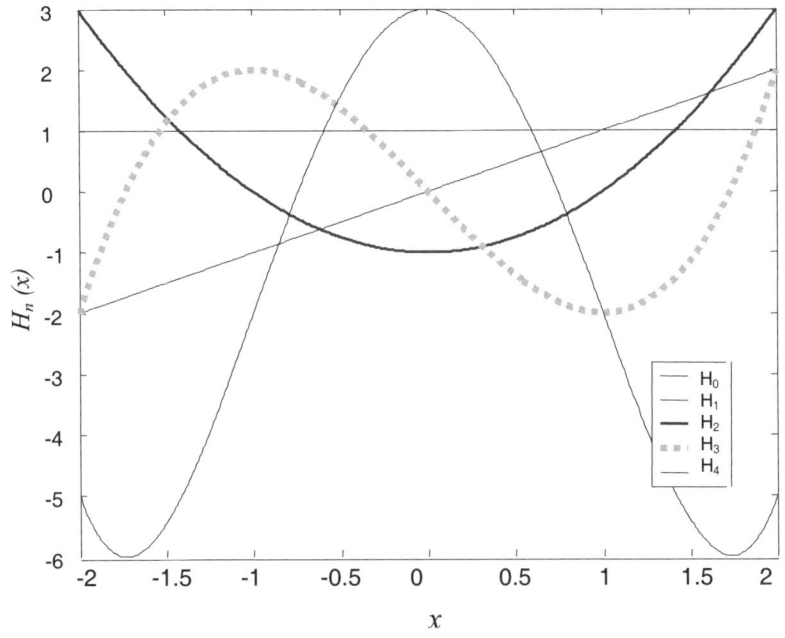

Figure 6.2. Hermite Polynomials (Example 6.3)

Example 6.4

Specify the first seven coefficients of the Gram-Charlier series by using Equation 6.30 and Equation 6.23a.

Solution:

Substituting Equation 6.23a into Equation 6.30 yields

$$b_0 = \int_{-\infty}^{\infty} f(x)dx,$$

$$b_1 = -\int_{-\infty}^{\infty} f(x)x\,dx,$$

$$b_2 = \frac{1}{2}\int_{-\infty}^{\infty} f(x)(x^2 - 1)dx,$$

$$b_3 = -\frac{1}{6}\int_{-\infty}^{\infty} f(x)(x^3 - 3x)dx,$$

$$b_4 = \frac{1}{24}\int_{-\infty}^{\infty} f(x)(x^4 - 6x^2 + 3)dx,$$

$$b_5 = -\frac{1}{120}\int_{-\infty}^{\infty} f(x)(x^5 - 10x^3 + 15x)dx,$$

$$b_6 = \frac{1}{720}\int_{-\infty}^{\infty} f(x)(x^6 - 15x^4 + 45x^2 - 15)dx$$

From the definition of the central moment (Equation 2.19), the nth order moments are given by

$$m_x^n = E[(X - \mu_x)^n] = \int_{-\infty}^{\infty}(X - \mu_x)^n f_X(x)dx$$

To keep things simple, $m^1, m^2, \ldots, m^6$ are the central moments, and let the first moment be zero ($\mu_x = m^1 = 0$). Then, replace each term by the corresponding moment. We obtain

$$b_0 = 1, \quad b_1 = -m^1 = 0, \quad b_2 = \frac{1}{2}(m^2 - 1),$$

$$b_3 = -\frac{1}{6}(m^3 - 3m^1), \quad b_4 = \frac{1}{24}(m^4 - 6m^2 + 3),$$

$$b_5 = -\frac{1}{120}(m^5 - 10m^3 + 15m^1),$$

$$b_6 = \frac{1}{720}(m^6 - 15m^4 + 45m^2 - 15)$$

This result shows that the coefficients of the Gram-Charlier series can be expressed by the Hermite polynomials in terms of the central moments, m^n.

6.2 Karhunen-Loeve (KL) Transform

The primary challenge of a stochastic analysis is to discover effective ways to represent the various types of uncertainty information and to use the information to evaluate the safety of structural systems in such a way that the computational effort of the analysis is minimized. Many engineering properties in structural analysis are distributed in space and time domains. For example, material properties, like Young's modulus and distributed dynamic loads, vary over the space or time domain of the structure. The description of such space-and-time-varying quantities can be represented by the concept of the random field. This section presents an efficient way of handling spatially-correlated data and dimensionality reduction of the random variables by using the KL transform.

6.2.1 Historical Developments of KL Transform

Uncertainty analysis often encounters situations in which a large number of variables is required to realistically describe the stochasticity of structural systems, such as the description of random fields. The accuracy and reliability of the prediction model will suffer when we consider highly-correlated variables and superfluous variables in the stochastic analysis. Therefore, finding ways to reduce dimensionality in conjunction with minimizing the loss of accuracy is essential in uncertainty analysis.

The orthogonal transform can effectively reduce the dimensionality of the correlated data set. The orthogonal transform for the variability of a random parameter is also known as the KL transform, the *Hotelling transform*, or the *principal component analysis*. Karhunen and Loeve independently represented the continuous-time random process in terms of orthonormal coordinate functions derived from the covariance function. Hotelling proposed the use of the principal components to analyze the correlation structure between many random variables [13]. A principal component analysis of an autocovariance matrix yields identical results to that of the KL transform when a continuous process is sampled periodically at a finite number of occasions [18]. Basically, the principal component analysis and the KL transform are the same for the zero mean vector. A later section (6.2.2) describes the details of the orthogonal transform and presents critical aspects of the method for generating and reducing the correlated random variables.

After showing the superiority of LHS by McKay *et al.* [19], various modifications are presented to reduce spurious correlations or to control correlation structures of sampling sets. Inman and Conover suggested the distribution-free approach to generate the dependent random variables using rank correlation [16]. Stein pointed out the inappropriateness of the distribution-free approach for large sampling sets and suggested a different modification of LHS that uses the ranks of the sample elements [26]. Another viable way to generate correlated random variables is the use of orthogonal decomposition, since a decomposition of the random process is an appropriate approach to make use of the correlations between the observations of a time series or spatially-distributed data. Various applications showed the usefulness of the orthogonal transform for generating the correlated random variables within the framework of LHS [22]-[24]. Section 6.2.2 utilizes the KL transform to generate the random fields within the framework of LHS, and Section 6.2 demonstrates the applicability of the method.

As discussed in Chapter 3, we can classify the usage of the stochastic expansions into two approaches: the non-intrusive formulation and the intrusive formulation. One type of the non-intrusive formulation procedure employs PCE to create the response surface as a surrogate model without interfering with the finite element analysis procedure. Thus, this type non-intrusive analysis is sometimes called the stochastic response surface method. By contrast, use of the KL expansion and PCE to directly modify FEM procedures or analytical formulas is referred to as the intrusive formulation, which is also known as the SSFEM or the stochastic Galerkin FEM [1]. In SSFEM, the KL expansion provides an accurate representation of input random fields and PCE represents the random process of the responses. Due to the simplicity of the response surface method, the non-intrusive formulation of PCE is prevalently applied in diverse research fields. Considering the KL expansion for the random field inputs in complex engineering problems is less prevalent in response surface methodologies, since the non-intrusive formulation is susceptible to the curse of dimensionality after the random field inputs are introduced. Choi et al. ([3]-[6]) present a robust procedure for the non-intrusive case using the KL transform, PCE, and LHS to account for large fluctuations of structural properties. In the presented method, the KL transform is introduced to represent the spectral description of uncertain quantities and to reduce the dimension of the random variables with a clear criterion for selecting the appropriate reduced dimension.

The following sections describe preliminaries of the KL transform and the KL expansion. It is also shown that the KL transform permits generation of the random fields within the framework of LHS and the dimensionality reduction of the random variables.

6.2.2 KL Transform for Random Fields

Due to the simplicity of its procedure, the most widely used method of multivariate data analysis is the orthogonal transform method. The KL transform is a viable tool with multiple uses for uncertainty analysis because it can generate correlated random variables and effectively reduce the dimensionality of the correlated data set.

The KL expansion can be viewed as part of a general orthogonal series expansion. Consider a general series expansion of $f(x)$ with a complete set of orthogonal and normalized base functions $\phi_i(x)$:

$$f(x) = \sum_{i=1}^{N} b_i \phi_i(x) \tag{6.31}$$

where the coefficients b_i represent the projection of $f(x)$ on the basis function $\phi_i(x)$ and b_i are obtained by

$$b_i = \int f(x) \phi_i(x) dx \tag{6.32}$$

The condition of uncorrelated coefficients yields

$$\langle (b_i - \mu_i)(b_j - \mu_j) \rangle = \lambda_j \delta_{ij} \tag{6.33}$$

where $\langle \cdot \rangle$ indicates the expected value operation, δ_{ij} is the Kronecker delta, and μ is the mean of the coefficients b. This restriction results in the following eigenvalue analysis of the covariance function [12],[30]:

$$\lambda_i \phi_i(x) = \int K(x, y) \, \phi_i(y) dy \tag{6.34}$$

where $\phi_i(x)$ and λ_i denote the eigenfunctions and eigenvalues of the covariance function $K(x, y)$, respectively, and x and y are the temporal or spatial coordinates:

$$K(x, y) = \langle (b(x) - \mu(x))(b(y) - \mu(y)) \rangle \tag{6.35}$$

where $\mu(x)$ is the mean of the coefficients $b(x)$.

The series of the eigenfunctions and the eigenvalues forms the KL expansion:

$$w(x) = \sum_{i=1}^{\infty} \sqrt{\lambda_i} \xi_i \phi_i(x) \tag{6.36}$$

where ξ_i is a set of uncorrelated random variables, and this expansion expresses the projection of the random process $w(x)$.

In the discrete case, Equation 6.34 can be specified by a statistical interpretation of the eigenvalue problem

$$[P][\Lambda] = [K][P] \tag{6.37}$$

where the covariance matrix $[K]$ is a symmetric and nonnegative definite matrix, and $[P]$ and $[\Lambda]$ are the orthogonal eigenvector matrix and the eigenvalue matrix, respectively.

Consequently, the orthogonal decomposition of the covariance matrix provides the product of the matrices of eigenvectors and eigenvalues:

$$[K] = [P][\Lambda][P]^T \tag{6.38}$$
$$\text{or } [K] = [A][A]^T \tag{6.39}$$

where $[A]$ is the transform matrix chosen as $[A] = [P][\Lambda]^{1/2}$.

The transform matrix $[A]$ can be employed to yield the correlated random vector T:

$$[T] = [A][X] \tag{6.40}$$

where [X] is the (n×1) matrix of uncorrelated random variables X_j, (j=1,...,n), and the transformed matrix, [T], possesses a given covariance matrix [K].

In addition to generating the dependent random variables, T, the KL transform can be used to reduce the dimension of the random variables. The main advantage of this procedure is to permit significant reduction in the number of uncorrelated random variables that represent random fields, especially for high levels of correlation [8],[18].

Suppose $z = [z_1, z_2, ..., z_n]^T$ is a stochastic observation with a zero mean and a covariance matrix [K_z], of which the orthogonal decomposition of [K_z] provides a compact description, y:

$$y = [P_z]z \qquad (6.41)$$

where [P_z] is a principal eigenvector matrix, which is arranged in descending order corresponding to its largest eigenvalues of [K_z]. Furthermore, the dimension of [P_z] can be reduced as $m \times n$, where $m < n$ (often $m << n$), so that a small number of independent variables can adequately describe strongly correlated observations. It is important to note that [P_z] should be arranged in size order before its dimension is reduced, while [P] of Equation 6.37 is not constrained to be arranged in descending order to its largest eigenvalues. The elements of the random vector y are known as the principal components of z.

Example 6.5

Suppose a target covariance matrix (Example 2.4) is given by

$$[C] = \begin{bmatrix} 1 & 0.95 & 0.83 & 0.65 \\ 0.95 & 1 & 0.95 & 0.83 \\ 0.83 & 0.95 & 1 & 0.95 \\ 0.65 & 0.83 & 0.95 & 1 \end{bmatrix}$$

Generate the correlated random variables ($\mu = 0$, $\sigma = 1$) and compare the sample covariance matrix $[\tilde{C}]$ with the target covariance matrix [C]

Solution:

From Equation 6.37, the eigenvalues and eigenvectors of [C] are calculated as

$$[\Lambda] = \begin{bmatrix} 3.5863 & 0 & 0 & 0 \\ 0 & 0.3921 & 0 & 0 \\ 0 & 0 & 0.0137 & 0 \\ 0 & 0 & 0 & 0.0079 \end{bmatrix}$$

$$[P] = \begin{bmatrix} -0.4785 & 0.6672 & 0.5206 & 0.2341 \\ -0.5206 & 0.2341 & -0.4785 & -0.6672 \\ -0.5205 & -0.2340 & -0.4785 & 0.6672 \\ -0.4785 & -0.6672 & 0.5406 & -0.2341 \end{bmatrix}$$

The transformation matrix [A] is obtained as

$$[A] = [P][\Lambda]^{1/2} = \begin{bmatrix} -0.9063 & 0.4178 & 0.0609 & 0.0208 \\ -0.9858 & 0.1466 & -0.0560 & -0.0593 \\ -0.9858 & -0.1466 & -0.0560 & 0.0593 \\ -0.9063 & -0.4178 & 0.0609 & -0.0208 \end{bmatrix}$$

With this transformation matrix, 50, 100, 500, 1,000, 2,000, 5,000, 10,000, and 20,000 correlated random vectors were generated for four standard normal random variables ($\mu = 0$, and $\sigma = 1$). For instance, 5,000 independent normal variates are generated using LHS, and these input vectors are transformed by the matrix [A]. The covariance matrix of 5,000 sample points was

$$[\tilde{C}] = \begin{bmatrix} 1.0085 & 0.9576 & 0.8365 & 0.6585 \\ 0.9576 & 1.0072 & 0.9565 & 0.8397 \\ 0.8365 & 0.9565 & 1.0065 & 0.9603 \\ 0.6585 & 0.8397 & 0.9603 & 1.0151 \end{bmatrix}$$

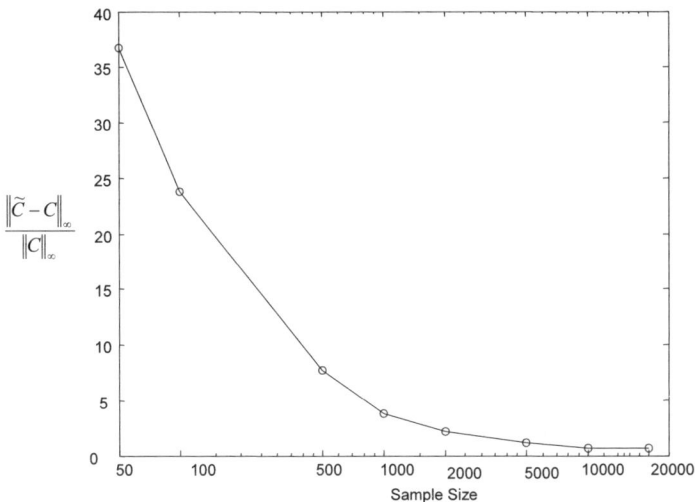

Figure 6.3. Error Estimation of KL Transform (Example 6.5)

The norm comparison between the target covariance matrix, $[C]$, and the sample covariance matrix, $[\tilde{C}]$, of the generated random vectors is shown in Figure 6.3. During the simulation, asymptotically correct joint distributions are obtained for large samples of 2,000 through 20,000. The larger the sample size, the closer it gets to the target covariance matrix $[C]$. Although the pattern of the sample covariance structure is similar to the target covariance in smaller-sized samples, deviations from the target covariance do exist, as shown in Figure 6.3. To accurately match joint distributions of the input random vectors, a large number of realizations, such as 10,000 or 20,000, should be used, but if computational budget is limited, at least several hundred realizations are required to obtain an approximate pattern of joint distributions for the target covariance matrix.

Example 6.6

This example is modified from [28]. The accompanying figure shows a simply supported beam ($L = 9$ m) loaded with two concentrated loads, P_1 and P_2. Both P_1 and P_2 are assumed to be normal distributions with mean values of 10 kN and 15 kN and standard deviations of 2 kN and 3 kN, respectively. Determine the probability of failure for both cases of the correlated (the correlation coeff. $\rho_{12} = 0.25$) and uncorrelated P_1 and P_2 when the allowable maximum moment M_a in the beam is assumed to be normally distributed with a mean of 50 kNm and a standard deviation of 5 kNm.

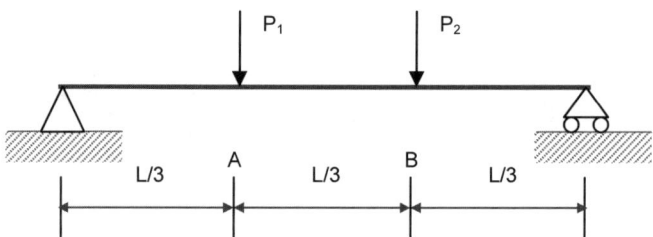

Solution:

(a) Uncorrelated Case

The maximum moment, M, caused by two loads, can be obtained at the point B:

$$M = \frac{L}{9}(P_1 + 2P_2)$$

The limit-state function can be expressed as

$$g(\{M_a, P_1, P_2\}) = M_a - \frac{L}{9}(P_1 + 2P_2) = M_a - P_1 - 2P_2$$

The basic variables, M_a, P_1, and P_2 are normalized by $\xi_1 = (M_a - \mu_{M_a})/\sigma_{M_a}$, $\xi_2 = (P_1 - \mu_{P_1})/\sigma_{P_1}$ and $\xi_3 = (P_3 - \mu_{P_3})/\sigma_{P_3}$.

Thus, the limit-state function can be normalized as

$$g(\{\xi_1, \xi_2, \xi_3\}) = \mu_{M_a} + \sigma_{M_a}\xi_1 - \mu_{P_1} - \sigma_{P_1}\xi_2 - 2\mu_{P_2} - 2\sigma_{P_2}\xi_3$$
$$= 50 + 5\xi_1 - 10 - 2\xi_2 - 30 - 6\xi_3 = 10 + 5\xi_1 - 2\xi_2 - 6\xi_3 \quad \ldots\ldots\ldots\ldots \text{(A)}$$

From Equation A, the probability of failure can be determined by FORM, SORM, or MCS. The probability of failure is obtained using MCS with 100,000 simulations:

$$P_f = 0.1073$$

(b) Correlated Case ($\rho_{12} = 0.25$)

$$\text{Cov}(P_1, P_2) = \rho_{12}\sigma_{P_1}\sigma_{P_2} = 0.25 \times 2 \times 3 = 1.5$$

Thus, the covariance matrix is given by

$$[C] = \begin{bmatrix} 4 & 1.5 \\ 1.5 & 9 \end{bmatrix}$$

From Equation 6.38, the eigenvalues and eigenvectors are calculated as

$$[\Lambda] = \begin{bmatrix} 9.4155 & 0 \\ 0 & 3.5845 \end{bmatrix}, \quad [P] = \begin{bmatrix} 0.2669 & 0.9637 \\ 0.9637 & -0.2669 \end{bmatrix}$$

From Equation 6.41, the transformed variable Y_1, and Y_2, equivalent to P_1 and P_2, are

$$[Y] = \begin{bmatrix} 0.2669 & 0.9637 \\ 0.9637 & -0.2669 \end{bmatrix} \begin{bmatrix} P_1 \\ P_2 \end{bmatrix} \quad \ldots\ldots\ldots\ldots\ldots\ldots\ldots\ldots\ldots \text{(B)}$$

Thus,

$$Y_1 = 0.2669 P_1 + 0.9637 P_2 \quad\quad Y_2 = 0.9637 P_1 - 0.2669 P_2$$

and

$$E[Y_1] = 0.2669\, E[P_1] + 0.9637\, E[P_2] = 17.1245$$
$$\text{with } \sigma_{Y_1} = \sqrt{9.4155} = 3.0685$$

> $E[Y_2] = 0.9637\ E[P_1] - 0.2669\ E[P_2] = 5.6335$
> with $\sigma_{Y_2} = \sqrt{3.5845} = 1.8933$
>
> From the inverse of Equation B, P_1 and P_2 can be written as:
>
> $P_1 = 0.2669\ Y_1 + 0.9637\ Y_2$
> $P_2 = 0.9637\ Y_1 - 0.2669\ Y_2$
>
> Thus, the limit-state function can be written in uncorrelated variables by
>
> $g(\{M_a, Y_1, Y_2\}) = M_a - (0.2669 Y_1 + 0.9637 Y_2) - 2(0.9637 Y_1 - 0.2669 Y_2)$
> $= M_a - 2.1943 Y_1 - 0.4299 Y_2$
>
> Again, the limit-state function can be normalized as
>
> $g(\{\xi_1, \xi_2, \xi_3\}) = \mu_{M_a} + \sigma_{M_a}\xi_1 - 2.1943(\mu_{Y_1} + \sigma_{P_1}\xi_2) - 0.4299(\mu_{P_2} + \sigma_{P_2}\xi_3)$
> $= 10.0019 + 5\xi_1 - 6.7332\xi_2 - 0.8139\xi_3$
>
> The probability of failure is obtained using the MCS with 100,000 simulations:
>
> $P_f = 0.1171$
>
> The correlation increases the probability of failure for this problem.

6.2.3 KL Expansion to Solve Eigenvalue Problems

The KL expansion can be derived based on the analytical properties of its covariance function. In this section, we discuss details of solving the associated integral eigenvalue problem for the KL expansion.

Let the covariance function of Equation 6.35 be specified by the exponential covariance (Equation 2.74) with a variance of C_0, correlation length of $1/h$, and two different locations of x_1 and x_2 defined in $[-a \le x_1, x_2 \le a]$:

$$K(x_1, x_2) = C_0 e^{-h|x_1 - x_2|}, \quad -a \le x_1, x_2 \le a \tag{6.42}$$

Then, Equation 6.34 can be written as

$$\lambda \phi(x_1) = \int_{-a}^{a} C_0 e^{-h|x_1 - x_2|} \phi(x_2) dx_2 \tag{6.43}$$

We need to solve Equation 6.43 by converting the integral equation to a differential equation, and then substituting the solution back into the integral equation. To eliminate the absolute magnitude sign, Equation 6.43 can be rewritten as

$$\lambda\phi(x_1) = \int_{-a}^{x_1} C_0 e^{-h(x_1-x_2)}\phi(x_2)dx_2 + \int_{x_1}^{a} C_0 e^{h(x_1-x_2)}\phi(x_2)dx_2 \tag{6.44}$$

Differentiating Equation 6.44 with respect to x_1, yields

$$\lambda\phi'(x_1) = -C_0 h e^{-hx_1} \int_{-a}^{x_1} e^{hx_2}\phi(x_2)dx_2 + C_0 h e^{hx_1} \int_{x_1}^{a} e^{-hx_2}\phi(x_2)dx_2 \tag{6.45}$$

Differentiating once more with respect to x_1, yields

$$\begin{aligned}\lambda\phi''(x_1) &= C_0 h^2 \int_{-a}^{x_1} e^{-h(x_1-x_2)}\phi(x_2)dx_2 - 2C_0 h\phi(x_1) \\ &= h^2\lambda\phi(x_1) - 2C_0 h\phi(x_1) = (h^2\lambda - 2C_0 h)\phi(x_1)\end{aligned} \tag{6.46}$$

Defining $\omega = (2C_0 h - h^2\lambda)/\lambda$, Equation 6.46 becomes

$$\phi''(x_1) + \omega^2\phi(x_1) = 0, \quad -a \leq x_1 \leq a \tag{6.47}$$

Letting $x_1 = t$, the solution of Equation 6.47 is given by

$$\phi(t) = c_1 e^{j\omega t} + c_2 e^{j\omega t}, \quad \omega^2 \geq 0 \tag{6.48}$$

where, c_1 and c_2 are constants.

Applying the boundary condition (substituting Equation 6.48 into Equation 6.44), yields

$$(h - \omega\tan(\omega a))(\omega + h\tan(\omega a)) = 0$$

$$= h - \omega\tan(\omega a) = 0 \tag{6.49a}$$

$$\text{or} \quad \omega + h\tan(\omega a) = 0 \tag{6.49b}$$

The values of ω can be determined graphically or numerically, and the corresponding eigenvalues are

$$\lambda_i = \frac{2C_0 h}{h^2 + \omega_i^2}, \quad i = 1,2,3,\ldots,n \tag{6.50}$$

where the solutions of Equation 6.49 are arranged as $\omega_1 < \omega_2 < \ldots < \omega_n$, and this results in the corresponding eigenvalues as $\lambda_1 > \lambda_2 > \ldots > \lambda_n$.

The resulting eigenfunctions are

$$\phi_i(t) = \frac{\cos\omega_i t}{\sqrt{a + \frac{\sin(2\omega_i a)}{2\omega_i}}} \qquad \text{(for, } i = \text{odd)}, \ -a \le t \le a \qquad (6.51\text{a})$$

$$\phi_i(t) = \frac{\sin\omega_i t}{\sqrt{a - \frac{\sin(2\omega_i a)}{2\omega_i}}} \qquad \text{(for, } i = \text{even)}, \ -a \le t \le a \qquad (6.51\text{b})$$

After graphical or numerical solution of transcendental equations (Equation 6.49) for ω_i, the eigenfunctions can be given as a set of periodic sines and cosines at approximately $(i-1)\pi/2$ [29]:

$$\phi_i(t) = \frac{\cos\left[\frac{(i-1)\pi}{2a} t\right]}{\sqrt{a + \frac{\sin(2\omega_i a)}{2\omega_i}}} \qquad \text{(for, } i = \text{odd)}, \ -a \le t \le a \qquad (6.52\text{a})$$

$$\phi_i(t) = \frac{\sin\left[\frac{(i-1)\pi}{2a} t\right]}{\sqrt{a - \frac{\sin(2\omega_i a)}{2\omega_i}}} \qquad \text{(for, } i = \text{even)}, \ -a \le t \le a \qquad (6.52\text{b})$$

Thus, the random process, which has exponential covariance, can be represented by Equation 6.36 in terms of Equation 6.50 and Equation 6.51 or Equation 6.52.

Example 6.7

Consider the exponential covariance function with $C_0 = 0.01$, $h = 4$, and $-0.5 \le t \le 0.5$. Numerically determine the first five solutions of Equation 6.49a and Equation 6.49b.

Solution:

To obtain the eigenvalues of Equation 6.50, we need to find the solutions of both

$$\frac{h}{\omega_n} = \tan(\omega_n a) \qquad \text{from (6.49a)}$$

and $\qquad -\dfrac{\omega_n}{h} = \tan(\omega_n a) \qquad$ from (6.49b);

the graphical solutions of both equations are given in Figure 6.4. As seen in the figure, the intersections occur in the intervals of $[(i-1)\pi/2a, i\pi/2a]$ for $i=1,2,\ldots n$, namely $[0, \pi/2a]$, $[\pi/2a, \pi/a]$, $[\pi/a, 3\pi/2a]$, etc.

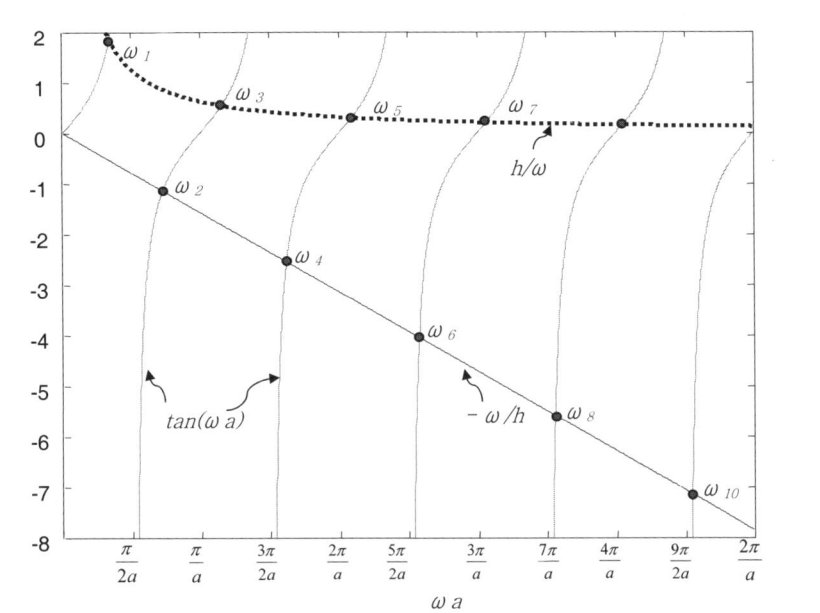

Figure 6.4. Graphical Solutions (Example 6.6)

The numerical tools of MATLAB® (*fzero*(.)) or Mathematica® (*FindRoot*[.]) require initial guesses or interval information to find the solutions. The interval information of $(i-1)\pi/2a$ is used to find the solutions, using the *fzero*(.) function in MATLAB®. The obtained solutions, ω_i, are

$\omega_1 = 2.1537$, $\omega_2 = 4.5779$, $\omega_3 = 7.2868$, $\omega_4 = 10.1732$, $\omega_5 = 13.1567$
$\omega_6 = 16.1921$, $\omega_7 = 19.2590$, $\omega_8 = 22.3454$, $\omega_9 = 25.4449$, $\omega_{10} = 28.5546$

Substitution of this information into Equation 6.50 and Equation 6.51 constructs the KL expansion (Equation 6.36).

6.3 Spectral Stochastic Finite Element Method (SSFEM)

As mentioned in Chapter 3, PCE is used to represent stochastic responses, and the KL expansion is used to represent the input of random fields in the intrusive formulation procedure. This method is also known as the SSFEM and yields appropriate results for a wide range of random fluctuations. The following sections briefly discuss the roles of the KL expansion and PCE in SSFEM.

6.3.1 Role of KL Expansion in SSFEM

Recall that in the KL expansion (Equation 6.36) a series of eigenfunctions and eigenvalues with a set of random variables ξ_i represent the random process. The eigenvalues and eigenfunctions can be obtained from Equation 6.34. Let $w(x,\theta)$ denote a random process, so that the function can be expanded in the following form, truncated to M terms:

$$w(x,\theta) = \overline{w}(x) + \sum_{i=1}^{M} \sqrt{\lambda_i}\, \xi_i(\theta) \phi_i(x) \tag{6.53}$$

where $\overline{w}(x)$ denotes the expected value of the random process, and θ represents an outcome in the space of possible outcomes of a random event.

Suppose the Young's modulus is a Gaussian random field. Then, the elasticity matrix D can be written as

$$D(x,\theta) = w(x,\theta) D_0 \tag{6.54}$$

where D_0 is a constant matrix similar to the one in deterministic finite element analysis.

The element stiffness matrix is

$$K^e(\theta) = K_0^{(e)} + \sum_{i=1}^{M} K_i^{(e)} \xi_i(\theta) \tag{6.55}$$

where $K_0^{(e)}$ is the mean element stiffness matrix and

$$K_i^{(e)} = \sqrt{\lambda_i} \int_{\Omega_e} \phi(x) B_e^T D_0 B_e\, d\Omega^e \tag{6.56}$$

where B_e is the matrix determined from the shape functions and geometric condition of the finite element.

Assembling the above element contributions in the finite element analysis procedure eventually gives

$$[K_0 + \sum_{i=1}^{M} K_i \xi_i(\theta)] u(\theta) = f \tag{6.57}$$

$$K_0 [I + \sum_{i=1}^{M} K_0^{-1} K_i \xi_i(\theta)] u(\theta) = f \tag{6.58}$$

Multiplying Equation 6.58 by K_0^{-1},

$$[I + \sum_{i=1}^{M} K_0^{-1} K_i \xi_i(\theta)] u(\theta) = u_0 \qquad (6.59)$$

where $u_0 = K_0^{-1} f$.

Equation 6.59 leads to

$$u(\theta) = [I + \sum_{i=1}^{M} K_0^{-1} K_i \xi_i(\theta)]^{-1} u_0 \qquad (6.60)$$

Now, the displacement vector can be obtained by the Neumann series [12]:

$$u(\theta) = \sum_{i=0}^{\infty} (-1)^i \left(\sum_{n=1}^{M} K_0^{-1} K_n \xi_n(\theta) \right)^i u_0 \qquad (6.61)$$

Applying the expected value operator, the mean of the response yields [12]

$$E[u] = \sum_{i=0}^{\infty} (-1)^i E[(\sum_{n=1}^{M} K_0^{-1} K_n \xi_n)^i u_0] \qquad (6.62)$$

In a general case, the covariance matrix yields

$$Cov[u,u] = \sum_{i=0}^{\infty} \sum_{j=0}^{\infty} (-1)^{i+j} E[(\sum_{n=1}^{M} K_0^{-1} K_n \xi_n)^i K_0^{-1} f \qquad (6.63)$$
$$\times f^T K_0^{-T} (\sum_{m=1}^{M} K_m^T K_0^{-T} \xi_m)^j]$$

Obviously, the KL expansion requires known covariance functions to obtain the eigenvalues and eigenfunctions. Since the covariance function of stochastic responses often is not known, PCE is used to represent stochastic responses in SSFEM. The following section will describe the role of the PCE in the intrusive formulation procedure.

6.3.2 Role of PCE in SSFEM

Recalling the definition of PCE from Equation 6.3, $u(\theta)$ can be projected on the expansion:

$$u(\theta) = \sum_{j=0}^{\infty} b_j \Psi_j(\theta) \qquad (6.64)$$

232 Reliability-based Structural Design

Substituting the expansion into Equation 6.57 yields

$$\left(\sum_{i=0}^{\infty} K_i \xi_i(\theta)\right)\left(\sum_{j=0}^{\infty} b_j \Psi_j(\theta)\right) = f \qquad (6.65)$$

Truncating the KL expansion after M terms and PCE after P terms results in

$$\sum_{i=0}^{M} \sum_{j=0}^{P} \xi_i(\theta)\Psi_j(\theta)K_i b_j - f = \varepsilon \qquad (6.66)$$

Minimization of the residual, ε, leads to an accurate approximation of the solution $u(\theta)$. This requires the residual to be orthogonal to the approximating space spanned by the PCE [10], [11]. Orthogonality requires the inner product be equal to zero, namely, $E[\varepsilon \cdot \Psi_k] = 0$. Thus, the expected value of Equation 6.66 becomes

$$\sum_{i=0}^{M} \sum_{j=0}^{P} E[\xi_i(\theta)\Psi_j(\theta)\Psi_k(\theta)]K_i b_j = E[f\ \Psi_k(\theta)],\ k=0,\ldots,P \qquad (6.67)$$

which can be rewritten as

$$\sum_{j=0}^{P} K_{jk} b_j = f_k \qquad (6.68)$$

where

$$K_{jk} = \sum_{i=0}^{M} C_{ijk} K_i\ ,\ C_{ijk} = E[\xi_i(\theta)\Psi_j(\theta)\Psi_k(\theta)]\text{ and } f_k = E[f\ \Psi_k(\theta)]$$

or

$$K^{(j,k)} b^{(j)} = f^{(k)} \qquad (6.69)$$

which we may rewrite as

$$\begin{bmatrix} K^{(0,0)} & K^{(0,1)} & \ldots & K^{(0,P)} \\ K^{(1,0)} & K^{(1,1)} & \ldots & K^{(1,P)} \\ \vdots & \vdots & \vdots & \vdots \\ K^{(P,0)} & K^{(P,1)} & \ldots & K^{(P,P)} \end{bmatrix} \begin{Bmatrix} b^{(0)} \\ b^{(1)} \\ \vdots \\ b^{(P)} \end{Bmatrix} = \begin{Bmatrix} f^{(0)} \\ f^{(1)} \\ \vdots \\ f^{(P)} \end{Bmatrix}$$

There is a $P+1$ N-dimensional matrix. Each b_j is a N-dimensional vector that is identical to the number of physical degrees of freedom. Thus, the final K matrix is of order $N(P+1)$.

Once the system is computed with the coefficient vectors b_j, the statistics of the solution can be readily obtained. The mean and covariance matrix of $u(\theta)$ can be obtained as

$$E[u(\theta)] = b_0 \tag{6.70}$$

and

$$Cov[u,u] = E[(u-u_0)(u-u_0)^T] = \sum_{j=1}^{P} E[\Psi_j^2] b_j b_j^T \tag{6.71}$$

Thus far, we have developed and demonstrated the important aspects of the PCE and the KL transform and expansion with respect to uncertainty analysis. In the following chapter, we present relatively novel procedures for the uncertainty analysis, utilizing these stochastic expansions in large-scale engineering problems.

6.4 References

[1] Babuska, I., Tempone, R., and Zouraris, G.E., "Galerkin Finite Element Approximations of Stochastic Elliptic Differential Equations," *SIAM Journal on Numerical Analysis*, Vol. 42, 2004, pp. 800-825.
[2] Cameron, R.H. and Martin, W.T., "The Orthogonal Development of Nonlinear Functionals in Series of Fourier-Hermite Functionals," *Annals of Mathematics*, Vol. 48, 1947, pp. 385-392.
[3] Choi, S., Canfield, R.A., and Grandhi, R.V., "Estimation of Structural Reliability for Gaussian Random Fields," *Structure & Infrastructure Engineering*, Nov. 2005 (in press).
[4] Choi, S., Grandhi, R.V., and Canfield, R.A., "Structural Reliability under non-Gaussian Stochastic Behavior," *Computers and Structures Journal*, Vol. 82, 2004, pp.1113-1121.
[5] Choi, S., Grandhi, R.V., and Canfield, R.A., "Robust Design of Mechanical Systems via Stochastic Expansion," *International Journal of Materials and Product Technology*, Vol. 25, 2006, pp. 127-143.
[6] Choi, S., Grandhi, R.V., Canfield, R.A., and Pettit, C.L., "Polynomial Chaos Expansion with Latin Hypercube Sampling for Estimating Response Variability," *AIAA Journal*, Vol. 42, (6), 2004, pp. 1191-1198.
[7] Cramer, H., *Mathematical Methods of Statistics*, Princeton University Press, New Jersey, 1958.
[8] Cureton, E., and D'Agostinao, R., *Factor Analysis: An Applied Approach*, Lawrence Erlbaum Associates, NJ, 1983.
[9] Devroye L., *Non-Uniform Random Variate Generation*, Springer-Verlag, New York, 1986.
[10] Ghanem R., "Ingredients for a General Purpose Stochastic Finite Elements Formulation," *Comput. Methods Appl. Mech. Eng.*, Vol. 168, 1999, pp. 19-34.

[11] Ghanem, R., and Kruger R., "Numerical Solution of Spectral Stochastic Finite Element Systems," *Comput. Methods Appl. Mech. Eng.*, Vol. 129, 1996, pp. 289–303.
[12] Ghanem, R., and Spanos, P.D., *Stochastic Finite Elements: A Spectral Approach*, Springer-Verlag, NY, 1991.
[13] Hotelling, H., "Analysis of a Complex of Statistical Variables into Principal Components," *Journal of Educational Psychology*, Vol. 24, 1933, pp. 498-520.
[14] Imamura, T., and Meecham, W., "Wiener-Hermite Expansion in Model Turbulence in the Late Decay Stage," *Journal of Mathematical Phyisics.*, Vol. 6, (5), 1965, pp.707-721.
[15] Imamura, T., Meecham, W., and Siegel, A., "Symbolic Calculus of the Wiener Process and Wiener-Hermite Expansion," *Journal of Mathematical Phyisics*, Vol.6, (5), 1965, pp. 695-706.
[16] Iman, R. L., and Conover, W. J., "A Distribution-Free Approach to Inducing Rank Correlation among Input Variables," *Communications in Statistics: Simulation and Computation*, B11, 1982, pp. 311-334.
[17] Isukapalli, S.S., *Uncertainty Analysis of Transport-Transformation Models*, Ph.D. Dissertation, Rutgers, the State University of New Jersey, New Brunswick, NJ., 1999.
[18] Jackson, J., *A User's Guide to Principal Components*, Wiley, NY, 1991.
[19] McKay, M.D., Beckman, R.J. and Conover, W.J., "A Comparison of Three Methods for Selecting Values of Input Variables in the Analysis of Output from a Computer Code," *Technometrics*, Vol. 21, (2), 1979, pp. 239-245.
[20] Montgomery, D.C, and Peck, E.A., *Introduction to Linear Regression Analysis*, Wiley, New York, 1992.
[21] Montgomery, D.C., *Design and Analysis of Experiments*, Wiley, New York, 1997.
[22] Novák, D., Lawanwisut, W., and Bucher, C., "Simulation of Random Fields Based on Orthogonal Transform of Covariance Matrix and Latin Hypercube Sampling," *Proceedings of International Conference on Monte Carlo Simulation MC 2000*, Monte Carlo, Monaco, June 2000.
[23] Olsson, A., Sandberg, G., and Dahlblom, O., "On Latin Hypercube Sampling for Structural Reliability Analysis," *Structural Safety*, Vol. 25, (1), 2003, pp. 47-68.
[24] Pebesma, E. and Heuvelink, G., "Latin Hypercube Sampling of Gaussian Random Fields," *Technometrics*, Vol. 41, (4), 1999, pp. 303-312.
[25] Pettit, C.L., Canfield, R.A., and Ghanem, R., "Stochastic Analysis of an Aeroelastic System," presented at *15th ASCE Engineering Mechanics Conference*, Columbia University, New York, June 2-5, 2002.
[26] Stein, M., "Large Sample Properties of Simulations Using Latin Hypercube Sampling," *Technometrics*, Vol. 29, (2), 1987, pp.143-151.
[27] Tatang, M.A., Direct Incorporation of Uncertainty in Chemical and Environmental Engineering Systems, Ph.D. Dissertation, Massachusetts Institute of Technology, Cambridge, MA, 1995.
[28] Thoft-Christensen, P. and Murotsu, P., *Application of Structural Systems Reliability Theory*, Springer-Verlag, NY, 1986.
[29] Van Trees, H., *Detection, Estimation, and Modulation Theory Part I*, Wiley, New York, 1971.
[30] Watanabe, S., "Karhunen-Loève Expansion and Factor Analysis," *In Transactions of the 4th Prague Conference on Information Theory*, Prague, 1965, pp. 635-660.
[31] Wiener, N., "The Homogeneous Chaos," *American Journal of Mathematics*, Vol. 60, 1938, pp.897-936.
[32] Xiu, D., and Karniadakis, G. E., "Modeling Uncertainty in Flow Simulations via Generalized Polynomial Chaos," *Journal of Computational Physics*, Vol. 187, 2003, pp. 137-167.

[33] Xiu, D., and Karniadakis, G., "The Wiener-Askey Polynomial Chaos for Stochastic Differential Equations," *SIAM Journal on Scientific Computing*, Vol. 24, (2), 2002, pp. 619-644.

7

Probabilistic Analysis Examples via Stochastic Expansion

This chapter applies the stochastic simulation procedure introduced in Chapter 6 to Gaussian and non-Gaussian problems with examples involving the input of random fields. The technique is applied to both basic problems and practical engineering problems. Geometrically nonlinear structural models of a joined-wing aircraft and a supercavitating torpedo are considered to demonstrate the efficiency and applicability of the procedure. In addition, the stochastic simulation procedure is incorporated into stochastic optimization problems. Implementation of the method is demonstrated for a three-bar truss structure and a joined-wing aircraft.

7.1 Gaussian and Non-Gaussian Distributions

In Chapter 6, we presented important characteristics of the PCE and KL transform that are important to help quantify uncertainty and find significant parameters of uncertainty models. In this section we present a stochastic analysis procedure [6] that utilizes PCE with Latin Hypercube Sampling (LHS) to represent the response of an uncertain system. Also, several analytical examples and a large finite element model of a joined-wing aircraft are used to demonstrate the effectiveness of the procedure. The next section (7.2) discusses how the KL transform can deal with the random nature of input parameters of structural models.

7.1.1 Stochastic Analysis Procedure

The underlying concept for the stochastic alaysis procedure for estimating PCE coefficients is to build approximations for the response model as a PCE of the uncertain parameters. This procedure comprises the following steps:
    1) Select experimental designs using Latin Hypercube Sampling (LHS)
    2) Simulate system response at each design point
    3) Construct approximate model using PCE
    4) Conduct ANOVA and residual analysis

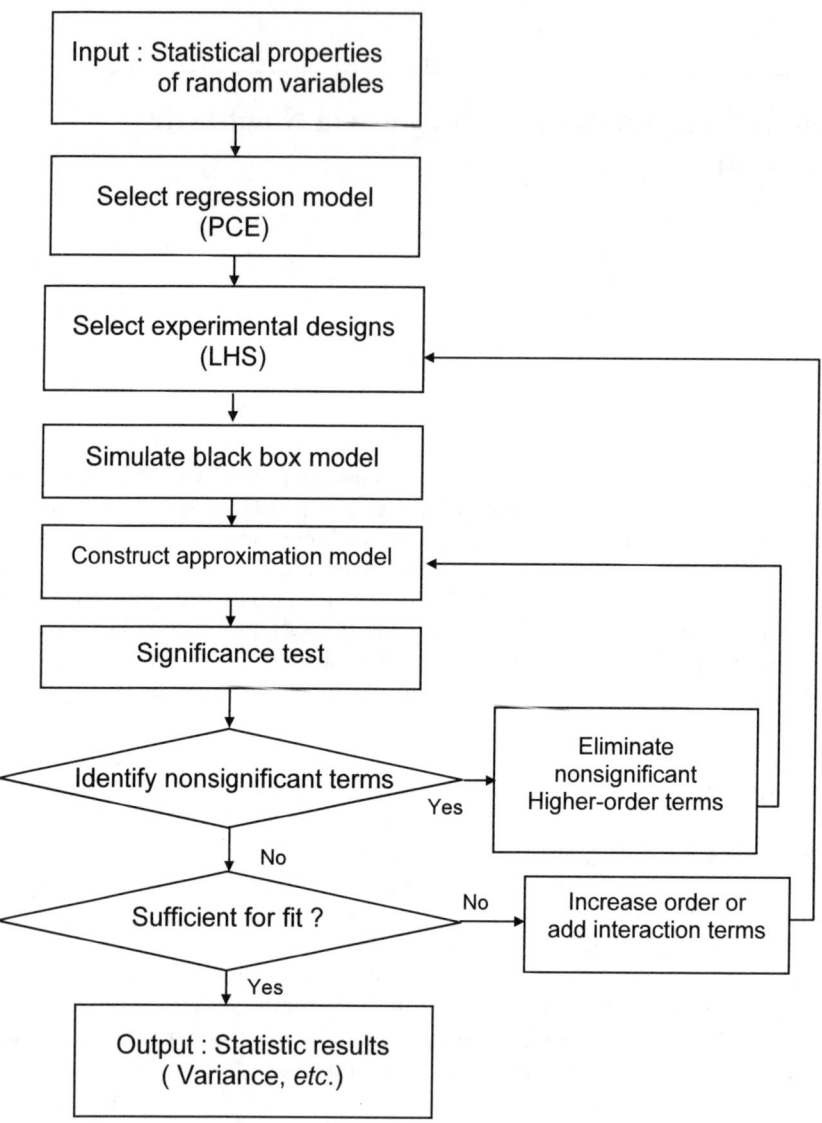

Figure 7.1. Backward Elimination Procedure

In this procedure, Equation 6.3 can be used as a surrogate model to represent the response of the uncertain system. The basic idea of this approach is to project the response and stochastic system operator onto the stochastic space spanned by PCE, evaluating the projection coefficients through an efficient sampling scheme. As mentioned earlier (Section 2.3), the forward selection and backward elimination procedures [14] – which can minimize regression errors according to *F*-statistics – are used to calculate the unknown projection coefficients b_i in Equation 6.3 using LHS. A detailed procedure for the latter approach is shown in Figure 7.1.

In this procedure, we expect that utilization of the stratified sampling technique known as LHS will decrease the number of simulations needed. The stochastic responses of the structural system are projected onto the PCE, and undetermined coefficients are estimated using LHS. Next, an approximate model of uncertainty in a system is constructed. Then, the significance test and residual analysis are performed to check the sufficiency of the fit. ANOVA is employed to determine the significant contributors of the model. ANOVA can estimate the lack of fit and confidence interval on the mean response. Residual analysis can detect model inadequacies with little additional effort. An abnormality of the residual plots indicates that the selected model is inadequate or that an error exists in the analysis. These two analyses are principal factors for measuring the model adequacy in stochastic approximations.

7.1.2 Gaussian Distribution Examples

The framework of the stochasic analysis procedure consists of four parts: i) sampling design points, ii) simulation, iii) approximating the regression model, and iv) checking the model adequacy. This section gives two simple examples to demonstrate these four steps and then applies the concepts to a more complex model of a joined-wing structure (see Figure 7.4).

Demonstration Examples

To understand the hybrid procedure of PCE employed here, consider a simple model that has one normal random variable *x*, which has a mean of 2.0 and unit standard deviation. This random variable is used to define a lognormal random variable *Y*; i.e., *ln Y = x*, or

$$Y = e^x \qquad (7.1)$$

Generally, in regression analysis, a lower-order model is preferred over a higher-order model, because an arbitrary fit of higher-order polynomials may create considerable errors. Therefore, choosing the order of an approximation is an important part of regression analysis. To accomplish this, the significance test and residual analysis are introduced. In this example, we will use the forward selection procedure to fit models of increasing order until the significance test for the highest-order term is nonsignificant. Set the initial approximate model of *Y* by introducing the second-order polynomials,

$$\hat{Y} = \beta_0 p_0(\xi) + \beta_1 p_1(\xi) + \beta_2 p_2(\xi) \qquad (7.2)$$

where ξ has a standard normal distribution, *N(0,1)*.

For an efficient approach, we choose orthogonal polynomials such as the PCE,

$$p_0(\xi) = 1, \; p_1(\xi) = \xi, \; p_2(\xi) = \xi^2 - 1 \tag{7.3}$$

The random variable x is transformed as

$$x = \mu_x + \sigma_x p_1(\xi) \tag{7.4}$$

Thus, from the given distribution parameters, x can be written as

$$x = 2 + p_1(\xi) \tag{7.5}$$

To find the unknown coefficients (β_0 through β_2) of the approximate model, LHS is used to identify the input design points. The number of input points must be higher than the number of unknown coefficients; generally, about twice the number of input points is sufficient for a simple case like this. For example, suppose that the five input points $\{x_1, x_2, ..., x_5\}$ are $\{1.3598, 4.0560, -0.3177, 2.5972, 1.9640\}$, with corresponding $\xi = \{-0.6402, 2.0560, -2.3177, 0.5972, -0.0360\}$. The corresponding responses are $Y = \{3.8954, 57.7439, 0.7278, 13.4263, 7.1274\}$. The unknown coefficients β_0 through β_2, found using regression analysis, give the following regression function:

$$\hat{Y} = 12.2538 + 13.8439 \times p_1(\xi) + 4.9133 \times p_2(\xi) \tag{7.6}$$

Table 7.1. ANOVA Result of 1-D Example (2nd order PCE)

Source of variance	Sum of squares	Degrees of freedom	Mean square	F_o
Regression	3492.90	3	1164.30	26.40
$\beta_0 p_0(\xi)$	750.78	1	750.78	17.02
$\beta_1 p_1(\xi)$	1986.84	1	1986.84	45.04
$\beta_2 p_2(\xi)$	755.29	1	755.29	17.12
Residual	88.22	2	44.11	
Total	3581.12	5		

The exact and approximate responses are shown in Figure 7.2. The second-order approximation is very poor in the interval [0, 2]. In contrast, the cubic polynomials produce more accurate results over the entire interval. The ANOVA result is summarized in Table 7.1. In this case, the residual plots were not used because a visual comparison of the response is quite easy and more intuitive for this simple case. ANOVA shows that all β terms are significant when we select a $F_{0.10,1,2}$ value of 8.53 (Appendix D), which is smaller than the F_o values in Table 7.1. Therefore, the effect of the higher-order terms of the regression model should be checked.

Table 7.2. ANOVA Result of 1-D Example (3rd order PCE)

Source of variance	Sum of squares	Degrees of freedom	Mean square	F_o
Regression	4291.16	4	1072.79	8.60
$\beta_0 p_0(\xi)$	1070.53	1	1070.53	8.58
$\beta_1 p_1(\xi)$	1773.05	1	1773.05	14.22
$\beta_2 p_2(\xi)$	1277.46	1	1277.46	10.24
$\beta_3 p_3(\xi)$	170.12	1	170.12	1.36
Residual	374.14	3	124.71	
Total	4665.30	7		

Figure 7.2 also shows the approximation results of a third-order PCE model, and Table 7.2 illustrates the ANOVA of the third-order case when seven simulations are conducted using the same procedure as in the second-order case. The regression coefficients (β_0 through β_3) of the third-order case are {12.3667, 11.8221, 5.6240, 1.5042}. If we choose the $F_{0.10,1,3}$ value of 5.54 (Appendix D), the regression coefficients β_0 through β_2 and the total regression are significant, and β_3 is nonsignificant. Based on the F-value of the coefficient β_3, we can expect that the other higher terms should have trivial effects on the current model. Thus, checking the effect of higher-order terms in the regression is not necessary. Since the contribution of the highest-order term is not significant, the third-order PCE is sufficient for fitting, as shown in Figure 7.2. In this example, the 90% ($\alpha = 0.10$) F-value is used as a rule to determine when to stop including additional the polynomial orders in the forward selection procedure. Thus, the results of the F-statistics can be used to identify a reasonable choice of polynomial orders in the regression model. If users want to use the F-statistic as a cutoff criterion, they must keep in mind that valuable regressors can be removed by relatively large values of F-statistic. Choosing the cutoff values of F-statistics in a stepwise regression procedure has been criticized in the literature, because it often depends on the personal preference of the analyst. Thus, inexperienced analysts must consider cutoff rules more carefully in the stepwise procedure. Further discussions of this procedure and alternative algorithms are available in [8], [14] and [15].

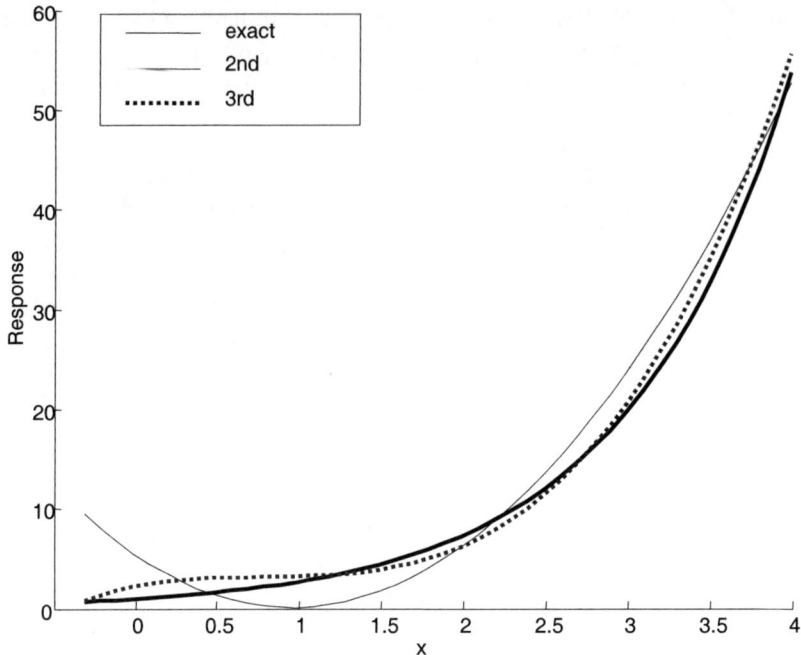

Figure 7.2. Plot of Approximate Versus Exact Response for $Y = e^x$

In this simple system, $Y = e^x$, we can check the acceptable error by plotting the response and comparing it to the Maclaurin polynomial. Since such visual comparisons are not available in practical problems, a residual analysis will be applied to the procedure in the next examples. After finding the tolerable order term of the regression model, we can proceed to check for a sufficient fit using residual analysis. The following examples illustrate the procedure. Now, the same procedure is applied to a two-dimensional case, $Y = e^{x_1 + x_2}$, where x_1 and x_2 are independent variables that are normally distributed, with a mean of 2.0 and a standard deviation of 1.0. The approximate model of Y is considered by introducing polynomials that include the interaction of x_1 and x_2:

$$\hat{Y} = \beta_0 + \beta_1 p_1(\xi_1) + \beta_2 p_1(\xi_2) + \beta_3 p_2(\xi_1) \qquad (7.7)$$
$$+ \beta_4 p_2(\xi_2) + \beta_5 p_1(\xi_1) p_1(\xi_2)$$

where $p_i(\xi)$, $i = 1, 2, \ldots, n$, are n^{th}-order PCE.

Table 7.3 summarizes the ANOVA analysis of the two-dimensional case results. When we choose the $F_{0.10,1,5}$ value of 4.06 with respect to the given degree of freedom of the regression model, the F_o column values are greater than the F value, as shown in Table 7.3. Thus, the effect of higher-order terms of the given model should be considered. Figure 7.3 illustrates the results of third and fourth-order cases. Though every coefficient is significant in the third-order case, the highest term is nonsignificant in the fourth-order case. Hence, we do not need to exploit the effects of higher order terms based on the F-statistics. The plot of residuals versus $\hat{y}$, or y versus $\hat{y}$, can provide a visual assessment of model effectiveness in regression analysis. Since both residual plots in Figure 7.3 yield points around the 45° line, the estimated regression function gives accurate predictions of the values that are actually observed. Therefore, the fourth-order PCE model is sufficient for fitting the given data, and useful statistical properties can be obtained using an approximation model.

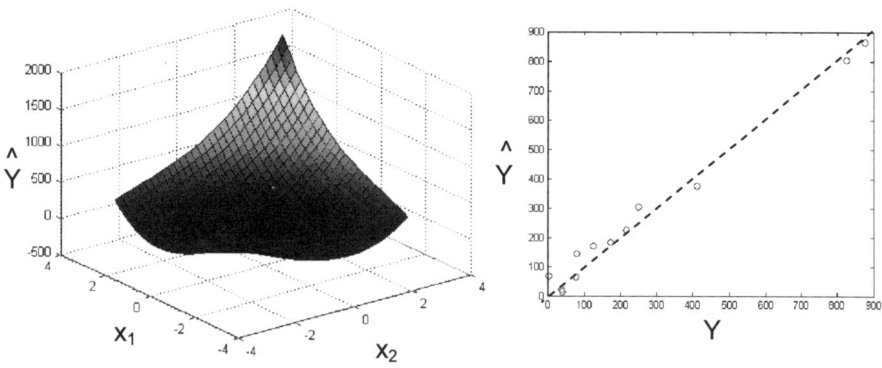

(a) 3rd Order PCE with 17 Simulations

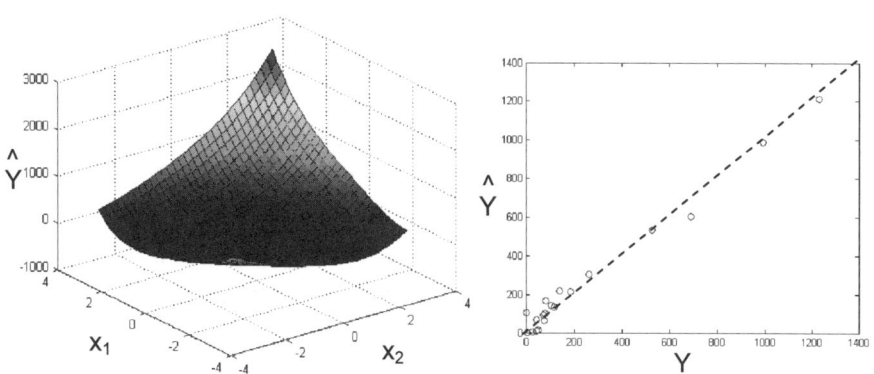

(b) 4th Order PCE with 25 Simulations

Figure 7.3. Response and Residual Plots for 2-D Example

Table 7.3. ANOVA Result of 2-D Example (2nd order PCE)

Source of variance	Sum of squares	Degrees of freedom	Mean square	F_o
Regression	987900.92	6	164650.15	43.13
β_0	180947.67	1	180947.67	47.40
$\beta_1 p_1(\xi_1)$	326465.17	1	326465.17	85.53
$\beta_2 p_1(\xi_2)$	187782.45	1	187782.45	49.20
$\beta_3 p_2(\xi_1)$	177967.54	1	177967.54	46.62
$\beta_4 p_2(\xi_2)$	93736.60	1	93736.60	24.56
$\beta_5 p_1(\xi_1) p_1(\xi_2)$	21001.49	1	21001.49	5.50
Residual	19085.47	5	3817.09	
Total	1006986.39	11		

Joined-wing Example

In this section, using the same procedure discussed in the previous section, structural reliability analysis of an aircraft wing structure is used to address the probability of failure due to buckling. In the design of the joined-wing (Figure 7.4), the backward- and forward-swept lifting surfaces combine to replace the traditionally separate wing and horizontal tail. This non-traditional design offers potential weight savings, reduces maneuver drag, and improves stability, but the nontraditional structural layout also introduces the possibility of uncommon failure modes.

An important design issue in the joined-wing is overcoming the potential buckling of the aft wing in the presence of compressive loads. This unconventional configuration demands investigation of the coupling between buckling and aeroelastic instabilities to diagnose the variability of the response. Traditional design tools do little to capture the uncertain nature of such unconventional designs and their environment. This section demonstrates that the current procedure can quantify variability of the joined-wing structural response, which is modeled by assuming stochasticity in the wing joint and the wing roots because of their importance in determining the response characteristics of the coupled structure.

The joined-wing structure model consists of 1,562 grid points and 3,013 elements in a geometrically nonlinear finite element model in NASTRAN, as shown in Figure 7.5. In the current model, 2173 QUAD4, 156 TRIA3, and 684 SHEAR elements compose the wing skins, ribs, and webs, respectively, in each wing section. The roots of the fore and aft wings have fully constrained boundary conditions, and all other grid points have translational and rotational degrees of freedom. The distributed steady aerodynamic pressures are applied as inputs to the buckling analysis in the baseline model, which has 30 degrees of leading-edge sweep, a semi-span of 26.0 m, and a uniform chord length of 2.50 m. A more detailed description of the joined-wing design can be found in [3].

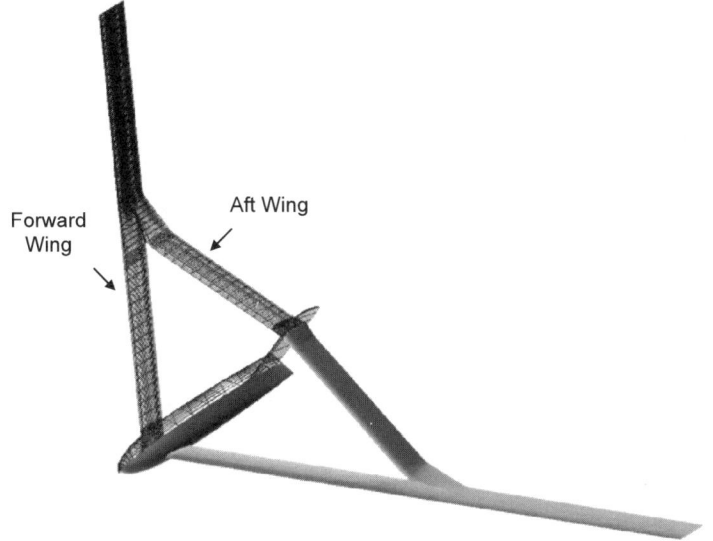

Figure 7.4. Joined-wing SensorCraft

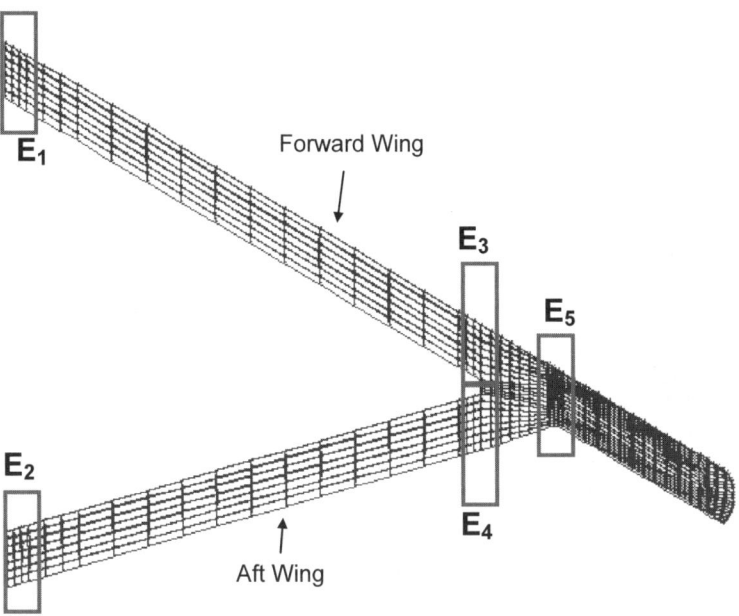

Figure 7.5. Joined-wing Model and the Locations of Five Random Variable Parts

In the buckling analysis procedure, the buckling eigenvalue is the relevant output of the joined-wing system. In the current model, five locations were chosen to include uncertain material properties, which are limited to the Young's modulus of the skin, spar, and rib elements in the vicinity of the wing joint and the two wing roots. Figure 7.5 depicts the five corresponding groups of elements: 1) forward wing root, 2) aft wing root, 3) forward wing joint, 4) aft wing joint, and 5) outboard wing joint. The associated Young's moduli are assumed to be Gaussian and uncorrelated. The correlated case of the same model will be discussed in later sections. The Young's moduli, which have a coefficient of variation (COV) of 0.1 and a mean of 6.9×10^{10} Pa, are denoted by E_1 through E_5, respectively, as shown in Figure 7.5.

The eigenvalue problem can be expressed as a function of ξ because Young's modulus, which is computed as $E = \mu_E + \sigma_E \xi$, enters through the stiffness matrices:

$$[K(\xi) + \lambda(\xi)\Delta K(\xi)]\phi(\xi) = 0 \qquad (7.8)$$

where ΔK represents the geometric stiffness matrix. The parameter λ represents a multiplier on the fixed-load distribution, so that the loads must be multiplied by the lowest eigenvalue λ to induce buckling. For a random structure, λ must also be random. The eigenvalues are projected onto the PCE, $\lambda = \sum_{i=0}^{P} \beta_i \Psi_i$; then the method discussed in Section 6.1.2 is applied to obtain the undetermined coefficients of the PCE. The associated moments of the eigenvalue can be obtained by the following expressions:

$$<\lambda> = \beta_0, \quad Var(\lambda) = \sum_{i=1}^{P} <\Psi_i^2> \beta_i^2 \qquad (7.9)$$

where β_i are the regression coefficients of PCE Ψ_i, and $<\cdot>$ indicates the expected value operation.

In the current model, we use the backward elimination procedure (Figure 7.1) with a fourth-order PCE; therefore, we initially have twenty-one unknown coefficients in the regression model without considering the interaction effects. If the selected regression model does not satisfy model adequacy, as discussed in the previous section, we should consider the interaction terms or the other-order terms. To obtain the unknown regression coefficients of this model, NASTRAN is used for 40 simulations based on LHS. According to the result of ANOVA (90% ($\alpha = 0.10$) F value), the higher order terms (3rd and 4th) have no effect on our regression model since the values of these coefficients are almost zero (-4.87×10^{-5} to 1.62×10^{-5}). Thus, these higher order terms can be eliminated. The current joined-wing model has first and second mean buckling eigenvalues of $\lambda_1 = 1.3445$ and $\lambda_2 = 1.8071$, with standard deviations of 0.00503 and 0.00519, respectively. The relatively small changes observed in the current model are natural since the buckling eigenvalue exhibits marked sensitivity only when applying large stiffness reductions.

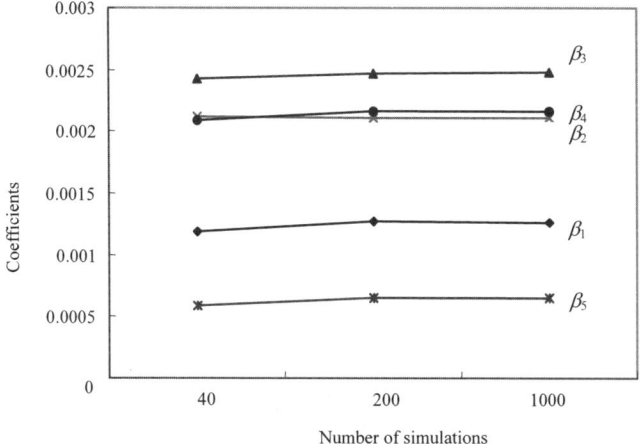

Figure 7.6. Coefficients Convergence Check with Increased Number of Simulations

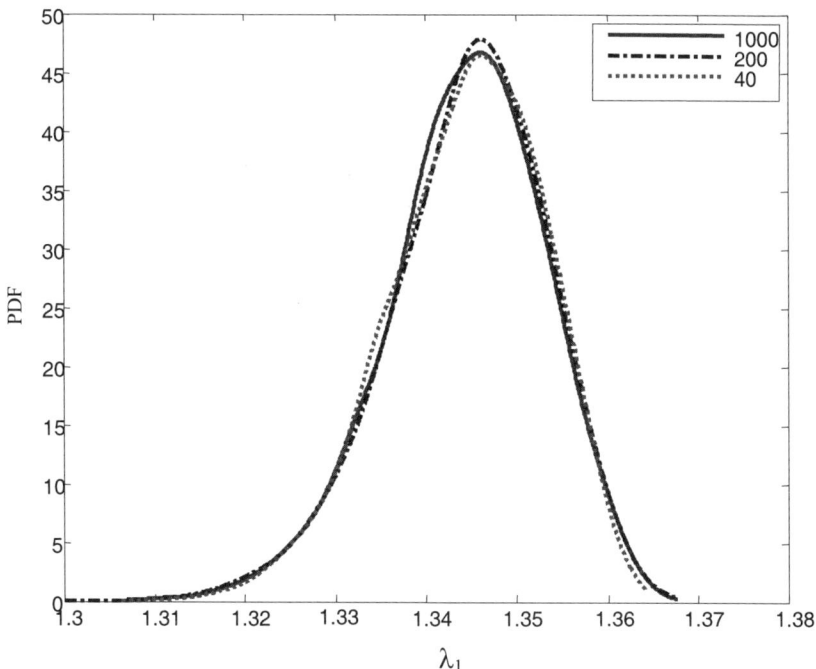

Figure 7.7. PDF of the First Buckling Eigenvalue as a Function of the Number of Simulations

The outer wing joint shows the lowest effect on the eigenvalues. The convergence of each coefficient was checked by comparing them with coefficients from 200 and 1,000 simulations, as shown in Figure 7.6. After 200 simulations, all coefficients are converged to within 0.1 to 0.5 percent of the 1,000 simulation results. In the case of 40 simulations, the mean term, β_0, has exactly the same value, and the other coefficients have 0.33 to 8.28 percent differences when compared with 1,000 simulation results. Figure 7.7 depicts the PDFs resulting from the MCS for the first buckling eigenvalue in each of the simulation results. All significant coefficients of the surrogate model are used to obtain the PDF by running 10,000 simulations in MCS. The fifth coefficient of the 40 simulation results has the maximum error, but Figure 7.7 shows that this error does not severely alter the probability density function.

7.1.3 Non-Gaussian Distribution Examples

This section presents the reliability analysis for non-Gaussian distributions to demonstrate the accuracy and efficiency of two existing techniques (the generalized PCE algorithm and the transformation method), which were described in Section 6.1.3. Assessments of the statistical characteristics, including the confidence interval, are demonstrated. Also, reliability analysis is applied to the joined-wing aircraft to demonstrate the procedure's effectiveness in real engineering problems.

Pin-connected Three-bar Truss Structure

An indeterminate, asymmetric system consisting of a three-pin-connected truss structure is illustrated in Figure 7.8. The unloaded length, L_m, and orientation, α_m, of each member are deterministic. Young's modulus, E_m, of each member is also assumed to be deterministic. The load has random magnitude (P) and direction (θ). The cross-sectional area A for all members is also random. The random quantities are initially considered normally distributed as a baseline for comparison to a non-Gaussian distribution:

$A \sim N(1 \text{ in}^2, 0.1 \text{ in}^2)$
$P \sim N(1000 \text{ lb}, 250 \text{ lb})$
$\theta \sim N(45°, 7.5°)$

where the symbol $x \sim N(\mu_x, \sigma_x)$ denotes that the random variable x is treated as a normal distribution and has the mean of μ_x and standard deviation of σ_x.

The principle of virtual work is used to calculate the displacement vector $[\bar{u}, \bar{v}]^T$ of the joint at which the load is applied, and it is found by solving the following system of equations:

$$P\cos\theta = \sum_{m=1}^{3}(\bar{u}\cos^2\alpha_m + \bar{v}\cos\alpha_m\sin\alpha_m)\frac{E_m A_m}{L_m} \qquad (7.10)$$

$$P\sin\theta = \sum_{m=1}^{3}(\bar{v}\sin^2\alpha_m + \bar{u}\cos\alpha_m\sin\alpha_m)\frac{E_m A_m}{L_m}$$

The horizontal deflection of the structure should be $\bar{u} < 0.001$ in.

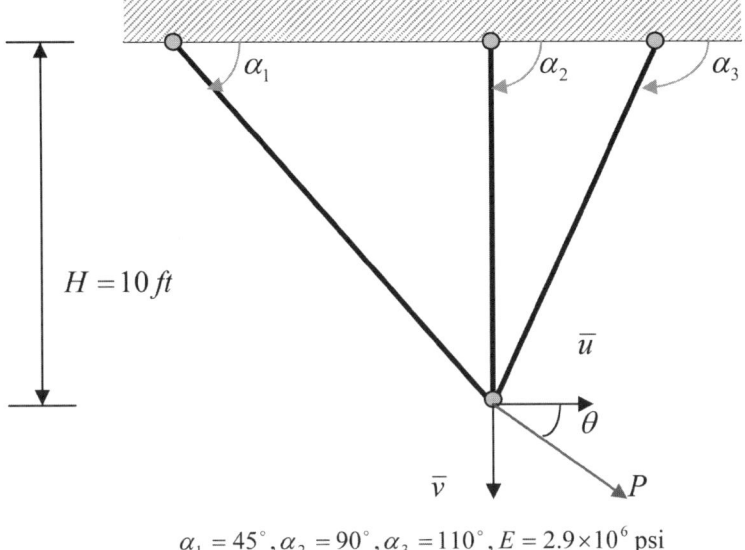

$\alpha_1 = 45°, \alpha_2 = 90°, \alpha_3 = 110°, E = 2.9\times 10^6$ psi

Figure 7.8. Pin-connected Three-bar Truss Structure

Table 7.4. Comparison of Methods for Reliability Analysis

	P_f	Difference (%)
MCS	0.0045	-
PCE	0.0044	2.2
FORM	0.0051	13.3

The results of this example are shown in Table 7.4, along with MCS and FORM results. To obtain the probability of failure, P_f, one million simulations were conducted to reach a converged result in MCS, and 100 samples of LHS were used to find the P_f for the third-order PCE model. In this case, the nonlinear terms of PCE have significant effects on the regression model. The FORM result converged to $P_f = 0.0051$. From the table, it is obvious that PCE is more accurate than FORM in this example. Due to the nonlinearity of the response, it is expected that the use of PCE along with LHS instead of FORM is more applicable to this case.

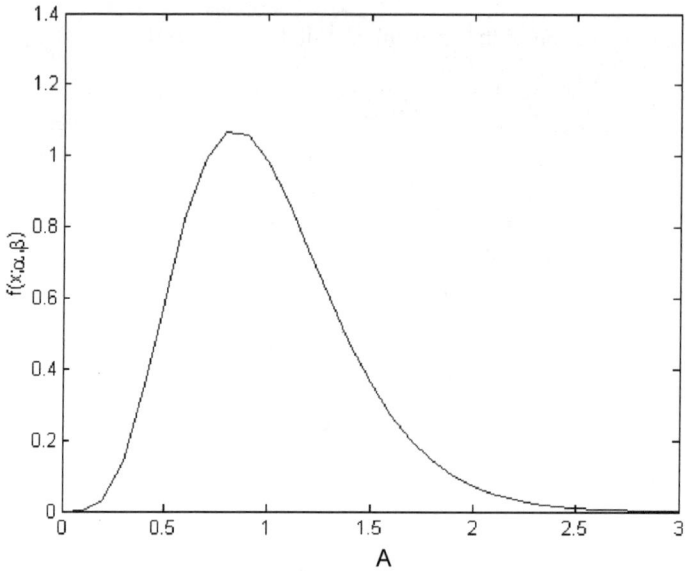

Figure 7.9. Gamma Density Function of Random Variable A

Many practical engineering situations require a wide variety of skewed distributions, which are not bell-shaped and symmetric like the normal distribution. The gamma distribution is a useful tool for representing skewed cases. For the preceeding example, the random variable A of the pin-connnected three-bar structure can be treated as a gamma distribution that has scale parameters $\alpha = 10$ and $\beta = 0.1$ (Figure 7.9), with the other variables remaining unchanged. For the description of the gamma distribution, the mean and variance of the random variable A are $\mu = \alpha\beta = 1.0$ and $\sigma^2 = \alpha\beta^2 = 0.1$. This analysis was carried out using LHS with 200 samples for two non-Gaussian techniques. In the generalized PCE method, Laguerre polynomials were used to represent the gamma distribution of the random variable A, and in the transformation technique, the transformation function shown in Table 6.2 was used. In order to estimate the quality of the non-Gaussian simulation techniques, MCS was also conducted using one million samples.

Figure 7.10 shows the results of this case for two techniques, the transformation technique (TRANS) and the generalized PCE algorithm (GPCE). By comparison, the probability of failure using MCS was estimated as $P_f = 0.0655$. This value is quite close to the results of higher order (e.g., fourth through sixth) models for the two other techniques: in the higher-order models of PCE, a maximum difference of 4.27% and a minimum difference of 1.83% were detected. To show the convergence of the results, we illustrated the results of the second through sixth-order models in Figure 7.10. The effects of fifth- and sixth-order PCE were not significant on this model, according to the F-statistics ($\alpha = 0.10$). The P_f of the fifth-order model in Figure 7.10 is slightly higher (1.0% to 2.1%) than the fourth and sixth-order model results, since the coefficients of fourth- and sixth-order PCE exhibit negative values, while the coefficient of the fifth-order PCE has a positive value. The bounded result is not significant in this case because the F-statistics indicate small effects of the fifth and sixth-order terms, and the statistics obtained in each random set of the pseudo-random number generator can be bounded with small intervals. The result of MCS also has a 1.2% bound. Figure 7.10 shows that the lower-order models of the generalized PCE algorithm have a more accurate result than the transformation technique; however, we can see that the two techniques have almost identical results for higher orders (e.g., fourth through sixth), which have sufficient accuracy compared to MCS.

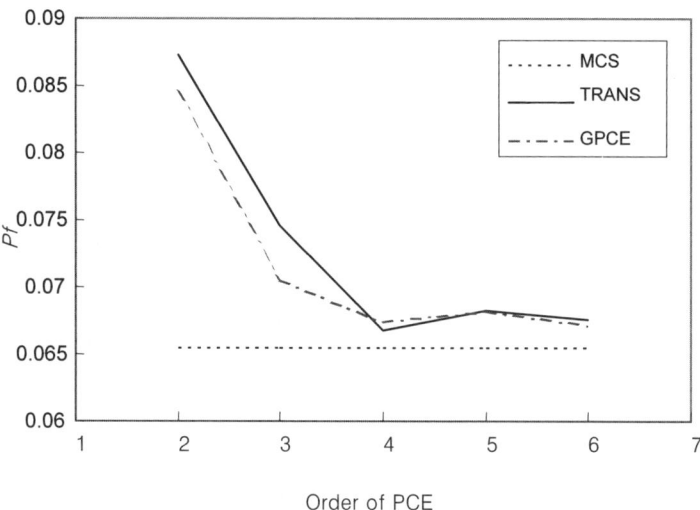

Figure 7.10. Probability of Failure

Joined-wing Example

In the joined-wing problem of the previous section, all random variables were assumed to be Gaussian distributions. Now, we consider the non-Gaussian distribution case for the same configuration of the joined-wing (Figure 7.5). Young's moduli of five locations are modeled as uncorrelated random variables using Gamma distributions, with $\alpha = 100$ and $\beta = 6.9 \times 10^8$ ($\mu = \alpha\beta = 6.9 \times 10^{10}$ Pa, $\sigma = \sqrt{\alpha\beta^2} = 6.9 \times 10^9$, COV = 0.1). The first and second buckling eigenvalues are $\lambda_1 = 1.345$ and $\lambda_2 = 1.807$ at the mean design point. The eigenvalue of 1.345

indicates a 34.5% margin of safety. To demonstrate the reliability analysis of the joined-wing, the condition $\lambda_1 < 1.34$ is chosen to represent failure.

Table 7.5. Comparison of Methods for Joined-wing Reliability Analysis

Approach	P_f	Difference (%)
MCS (10,000)	0.1327	-
LHS+PCE (200)	0.1343	1.25

Table 7.5 shows the probability of failure comparison between MCS and PCE. The MCS result was obtained by conducting 10,000 NASTRAN simulations. The non-intrusive analysis was carried out using LHS with 200 samples. In the non-intrusive formulation, fourth-order PCE and the transformation technique (Section 6.1.3) were used without considering the interaction effects of each random variable. After obtaining the undetermined coefficients of PCE, residual analysis and ANOVA analysis were conducted to determine the model adequacy. The probability of failure of the given structural system was calculated using 10,000 MCS simulations on the surrogate model of PCE, which has sufficient adequacy. The difference between the results of direct MCS and the current method was about 1.25%. This result shows that the non-intrusive formulation method, combined with the transformation technique and LHS of 200 full simulations, gives sufficient accuracy as compared to the 10,000-simulation result of MCS. A 95% confidence interval was computed by PCE for the mean response at the mean point using Equation 6.19; the resulting interval is $1.341 \leq \mu \leq 1.349$. Once the approximation is constructed, many statistical properties of the response, such as probability of failure, can be estimated without incurring significant computational costs. The 95% confidence interval for MCS (1 of 10,000) is $1.336 \leq \mu \leq 1.352$. We can interpret the interval calculated at the 95% level as 95% certainty that the true value of μ is in this interval.

7.2 Random Field

This section presents a stochastic analysis procedure for Gaussian random fields in structural reliability estimation using an orthogonal transform and a stochastic expansion with Latin Hypercube Sampling (LHS). The efficiency of the current simulation procedure is achieved by the combination of the KL transform with the stochastic analysis of PCE. In order to demonstrate the applicability of the method to large-scale practical problems, the material properties of a cantilever plate and a supercavitating torpedo are treated as random fields.

7.2.1 Simulation Procedure of Random Field

This section presents a stochastic analysis procedure for random fields using KL transform and PCE. The procedure is summarized in Figure 7.11. In this approach, the KL transform is combined with LHS to represent the spatial variability of input parameters: Equation 6.40 yields a correlated sampling set after generating a typical sampling set of LHS. Then, system analysis is conducted, such as finite element methods or analytical and empirical formulas based on the generated sampling set. To conduct the stochastic analysis using PCE, the orthogonal transform Equation 6.41 is used again to provide the dimensional reduction of the random variables. If the number of random variables is not too large, the additional step of dimension reduction can be omitted. To guarantee sufficient accuracy of the PCE model, model-adequacy-checking procedures, including ANOVA and residual analysis, are conducted. There are no significant computational costs to obtain statistical properties of the responses after constructing the PCE representation of stochastic responses.

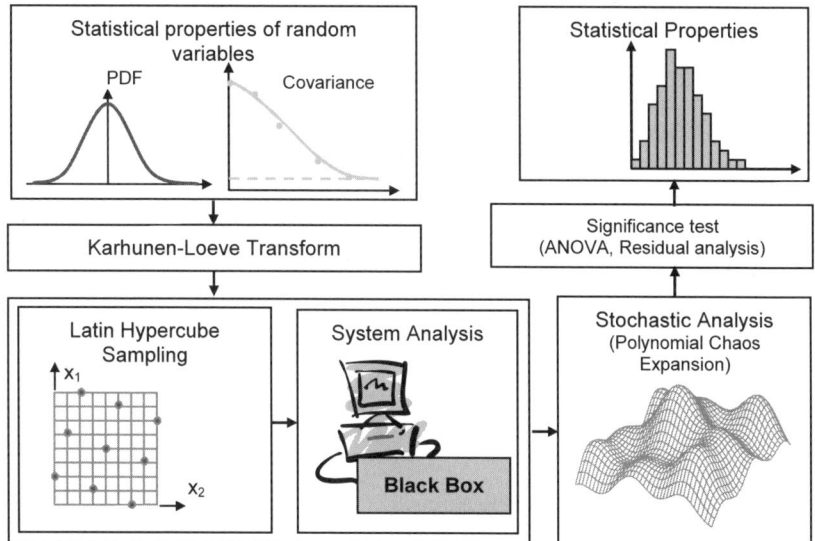

Figure 7.11. Framework of Stochastic Analysis with Random Field

During the simulation procedure, accurate representation of random fields is influenced by the number of discretization points associated with finite elements, the number of realizations, and the statistical characteristics of input parameters. Critical aspects of the simulation procedure related to the KL transform are described and demonstrated in Section 6.2.

7.2.2 Cantilever Plate Example

A cantilever plate problem (Figure 7.12) is considered to depict the aspects of the dimensional reduction using KL transform. Two point loads are applied to the tip of the plate in the orthogonal direction. This plate consists of 16 QUAD4 elements with a total length of 40 in and a total width of 10 *in*, as shown in Figure 7.12. The Young's modulus, E, of each element is considered as a random variable ($\mu_E = 7.24 \times 10^{10}$ Pa, COV = 0.1). These uncertain parameters are assumed to be Gaussian random fields. The design parameters and their statistical properties are summarized in Table 7.6. The random field was discretized by evaluating the covariance function at the element centroid and setting the properties of each QUAD4 element to a constant value. Thus, the 16 elements have different levels of Young's modulus based on the selected Gaussian covariance model. Figure 7.12 clearly shows different levels of Young's modulus using color scales for one realization.

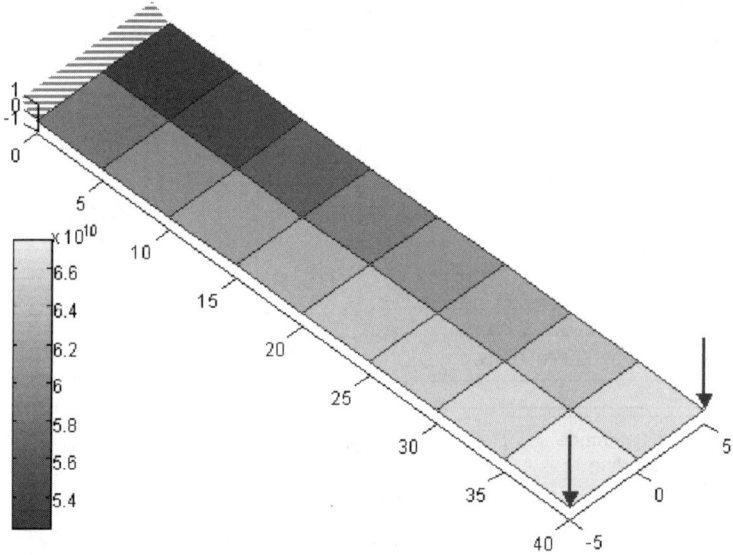

Figure 7.12. Random Field Realization of Cantilever Plate Elastic Modulus

The scree plot [13] can be an effective way to determine the appropriate dimension of reduced random variables. It shows relative eigenvalue sizes as percentages of the total sum of squares of the eigenvalues. Figure 7.13 is the scree plot of the plate's generated random field using 700 realizations in LHS with KL transform. It is clear that although the scree plot shows one large eigenvalue (78.08%), and one medium eigenvalue (18.03%), all others are small eigenvalues (lower than 2.10%). Therefore, the first two or three eigenvectors are significant and can be selected to reduce the random variables' dimension. After generating the reduced random variables using the significant eigenvectors, undetermined coefficients of PCE are obtained to provide statistical estimation of the response. To show the accuracy of the current procedure, the probability density estimations of the 4th order PCE models with one, two, and three reduced vectors of 700 realizations are compared to 10,000 simulation results of LHS. The PCE models of two (PCE2) and three (PCE3) reduced vectors yield almost identical PDF, and there are only small deviations compared to the LHS result, as shown in Figure 7.14. Although the PCE model of one reduced vector (PCE1) is only 78.08% of the sum of squares, it provides acceptable accuracy in this case.

Table 7.6. Cantilever Plate Design Data

Material Properties	Young's Modulus: $\mu_E = 7.24 \times 10^{10}$ psi, $COV_E = 0.1$ Poisson ratio = 0.33 Weight density = 0.1 lbs/in^3
Load Condition	Two 1,000 lb loads in $-z$ direction
Random Field (Young's Modulus)	Gaussian covariance model: $$C_{ij} = \sigma^2 \exp\left[-\left(\frac{x_{ij}}{l}\right)^2\right]$$ Assumed correlation length of the random field, $l = 30$ in

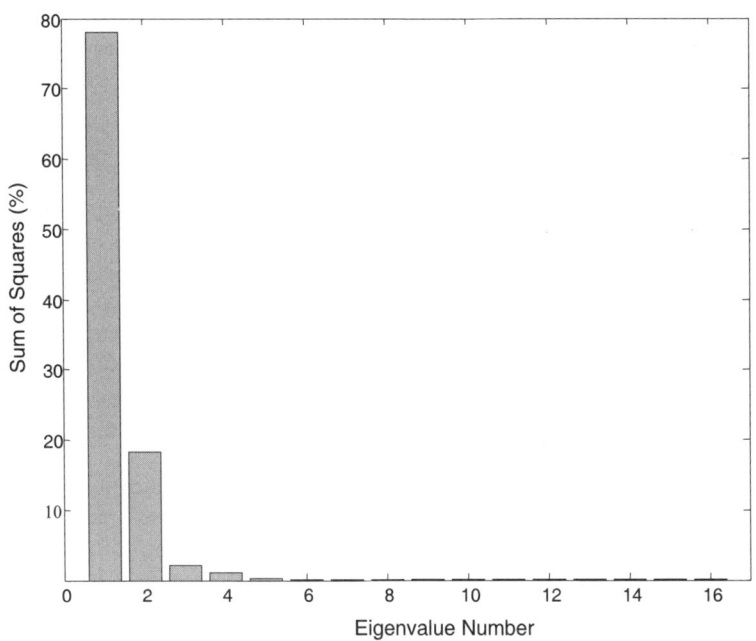

Figure 7.13. Scree Plot of Cantilever Plate

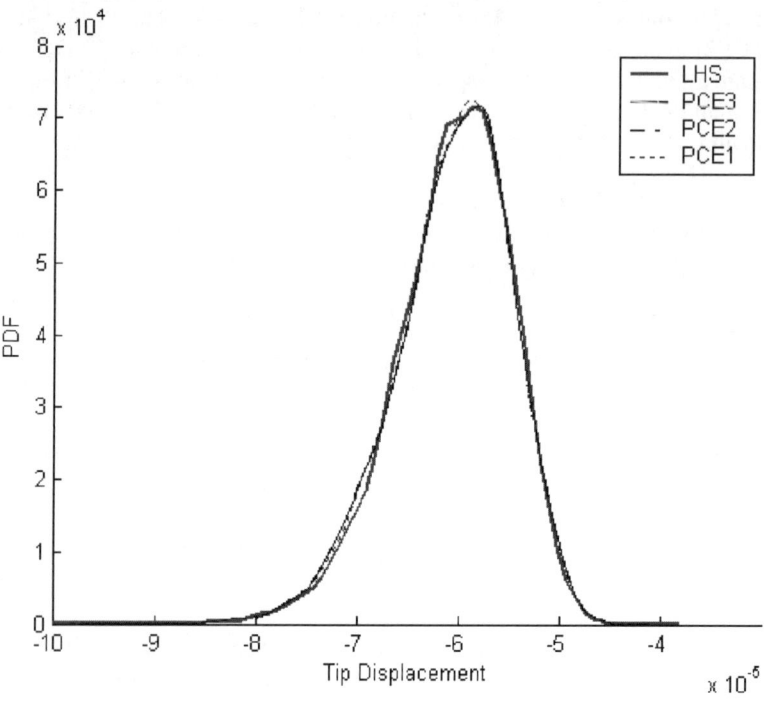

Figure 7.14. Probability Density Estimations of Tip Displacement

7.2.3 Supercavitating Torpedo Example

Undersea weapons, such as torpedoes, are designed to satisfy several competing performance criteria, including static strength, acoustic constraints, and hydrodynamic performance. Accordingly, the design of underwater structures is quite intricate, and there is little relevant research on incorporating uncertainties into their design and analysis procedures. The unexplored complexities of this problem pose challenges to designers who must quantify the confidence of performance predictions. This section illustrates that critical stochastic aspects of a complex undersea structure, such as a supercavitating torpedo (Figure 7.15), can be examined through the KL transform and PCE.

One of the U.S. Navy's priorities is to design modern undersea weapons with high reliability using multiphysics simulations and targeted testing. The supercavitating torpedo operates in a cavity of water vapor to provide the capability of traveling at speeds greater than 200 mph (Figure 7.15). A high-fidelity finite element model of the supercavitating torpedo is considered to quantify potential effects of material uncertainties. The baseline model has a cone length of 29 in, diameter of 7.99 in, and total length of 145 in, as illustrated in Figure 7.16. Loads on the torpedo model where estimated for straight operation at 173 mph. These loads were estimated utilizing boundary element analysis and potential flow theory to handle the multi-phase fluid flow analysis problem. The loads are applied mainly at the nose of the torpedo model. Additional forces and moments are applied at the locations where control fins would be mounted at the rear of the torpedo. Finally, depth pressure loads were added to the structure.

The skin and ends of the model are assembled from 640 quadrilateral plate elements and 32 triangular plate elements. In addition, 160 beam elements comprise 4 longitudinal stiffeners in the finite element model of the torpedo. The finite element model can be solved in any structural analysis package or personal research code.

Gaussian random fields of Young's modulus are introduced to examine the potential robustness of the supercavitating torpedo design. The Young's modulus of each skin element is considered as the random field varying longitudinally with Gaussian covariance ($\mu_E = 10.4 \times 10^3$ ksi, COV = 0.1, correlation length = 50 in).

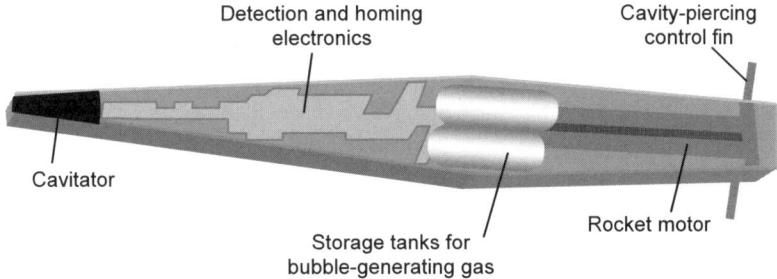

Figure 7.15. Supercavitating Torpedo Prototype

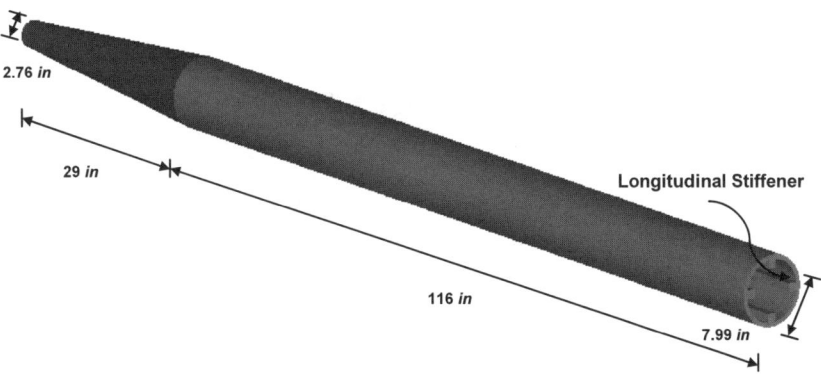

Figure 7.16. Model Dimensions of Supercavitating Torpedo

258   Reliability-based Structural Design

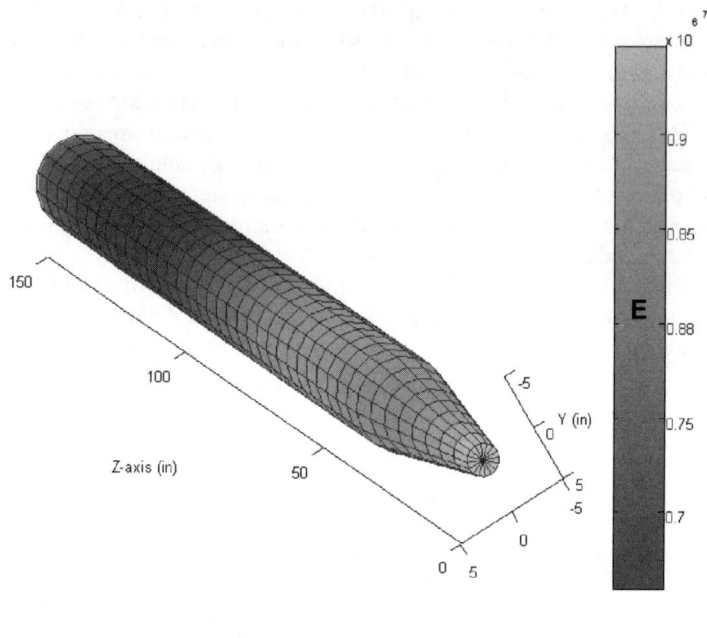

(a) Correlation Length = 50 in

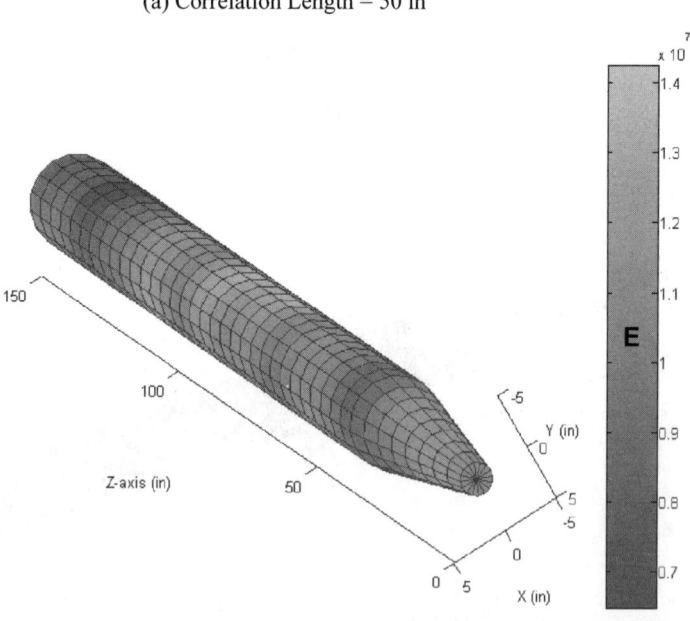

(b) Correlation Length = 20 in

Figure 7.17. Random Field Realization of Supercavitating Torpedo's Elastic Modulus

The FEM dictates that 640 random variables be used to represent the random field. An actual realization of the current random field using color scales is shown in Figure 7.17a. Although the covariance model of Young's modulus is assumed to be a Gaussian model, the nugget-effect model, or extremely short correlation length model, might apply if there is a requirement for estimating white noise phenomena. Figure 7.17b illustrates a different random field with a correlation length of 20 in for the same Gaussian covariance function. Since the covariance function is a function of the distance parameter, the realizations in Figure 7.17 appear to have similar values of the covariance at some elements that have similar distances from the origin.

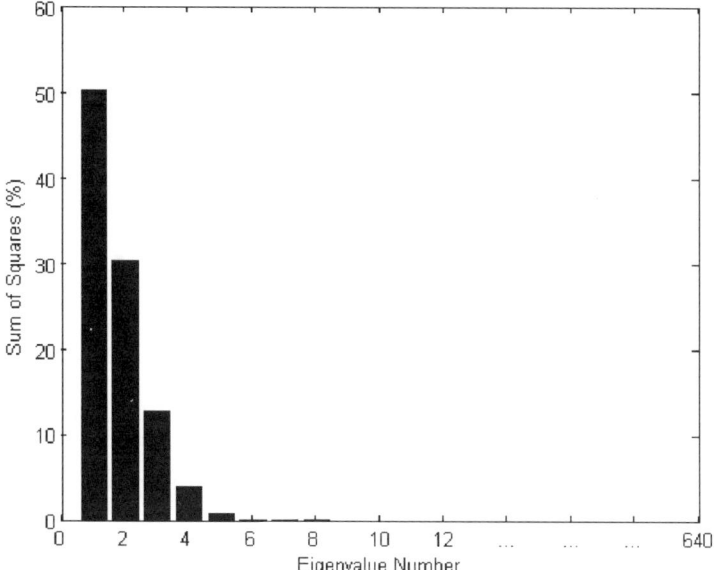

Figure 7.18. Scree Plot of Supercavitating Torpedo

The analysis procedure was carried out with LHS (Equation 6.40) using 500 realizations of the random field generated from the KL transform. To determine the appropriate dimension of the reduced random variables, the scree plot is used again. Figure 7.18 shows that, except for the first four or five eigenvectors, the other eigenvectors are insignifcant. Therefore, the first five eigenvectors are selected to reduce the dimension of 640 random variables; so, the five reduced random variables, which contain 98.67 % sum of squares of the original information, can replace 640 random variables.

After generating the reduced random variables using the significant eigenvectors, the non-intrusive formulation of fourth-order PCE is applied to obtain the statistical estimation of the maximum von Mises stress for all structural elements. After obtaining undetermined coefficients of the PCE, the model adequacy checking procedure is also conducted. Once stochastic responses of the given system's maximum stress are projected onto the PCE, many statistical properties of the response can be estimated without any additional significant computational costs. The probability density and cumulative density estimation of the maximum von Mises stress for the supercavitating torpedo are calculated using 10,000 MCS simulations with the obtained PCE model. Figure 7.19 depicts nonparametric estimates of the probability density and the cumulative density for the maximum von Mises stress. The mean and standard deviation of the maximum von Mises stress are estimated as 32,303 psi and 56 psi, respectively, and it is observed that the probability density estimation exhibits a positive skewness property. These estimated statistics provide a better picture of the design space, which allows the designer to prevent the undesired probability of failure under the given uncertain situation.

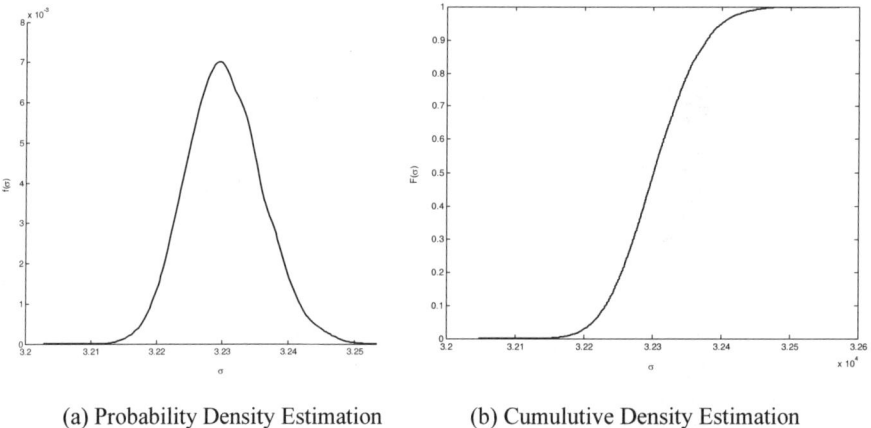

(a) Probability Density Estimation (b) Cumulutive Density Estimation

Figure 7.19. Estimated Statistics of Maximum Stress

7.3 Stochastic Optimization

With the high power of digital computers, it has become feasible to find numerical solutions to realistic problems of large-scale, complex systems involving uncertainties in their behavior. This feasibility has sparked an interest in combining traditional optimization methods with uncertainty quantification measures. These new optimization techniques, which can consider randomness or uncertainty in the data, are known as *stochastic programming* or *stochastic optimization* methods. These methods ensure robust designs, insensitive to given uncertainties, that provide the designer with a guarantee of satisfaction with respect to the uncertainties in the objective function, performance constraints, and design variables. Choi et al., [5] suggested a stochastic optimization procedure using PCE with Latin Hypercube Sampling (LHS). The non-intrusive formulation of PCE is used to construct surrogates of stochastic responses for optimization procedures. A standard optimization algorithm is then used. The following sections describe the details of this approach, and the implementation of the method is demonstrated for a three-bar truss structure and a complex engineering structure of a joined-wing aircraft.

7.3.1 Overview of Stochastic Optimization

In stochastic optimization, three common methods are available: the *chance-constrained method*, the *recourse method*, and the reliability-based method [4],[7],[11], and [19]. The chance-constrained method focuses on system reliability, which can indicate the probability of satisfying constraints. This method converts probabilistic constraints into deterministic equivalents, assuming that the distribution of the random variable is of a class of infinitely divisible distributions known as stable distributions. The stable distribution property possessed by Gaussian, Uniform, and Chi-square distributions allows the conversion of probabilistic constraints into deterministic equivalents. This assumption of a stable distribution is the major restriction of the chance-constrained method, since the solvability of the method depends on the distribution of the data [7].

In the recourse method, decision variables are partitioned into two sets, namely, first-stage and second-stage variables. The latter are also called recourse variables and involve uncertainties. The primary objective of this method is to minimize the sum of the first-stage costs and the expected value of the second-stage costs. Before the sum is minimized, the recourse variable must be found from the second-stage minimization. In this procedure, the decision maker needs to allocate a cost to the recourses in order to ensure feasibility of the second-stage problem. This feasibility requirement for the second-stage problem is burdensome and detracts from the desirability of this approach [19].

Another class of optimization under uncertainty is reliability-based design optimization. This field focuses on the probability of structural failure. The most common methods in this field are the FORM and the SORM. As discussed in Chapters 3 and 4, a potential drawback of reliability-based optimization is that FORM and SORM sometimes oscillate and converge on unreasonable values for probability of failure when the approximations of the limit state at the most probable failure point are not accurate.

A variety of algorithms has been proposed for stochastic optimization. However, there are few practical applications in industry. Stochastic optimization methods are often criticized for requiring a deep understanding of probability theory and for the need to make many simplifying assumptions in their implementation. For a more thorough discussion of stochastic programming, readers can refer to standard textbooks on the topic [2], [9]. The following section presents a new computational algorithm that uses stochastic expansions for uncertain parameters with LHS.

7.3.2 Implementation of Stochastic Optimization

Optimization under uncertainty, by its nature, is more expensive than solving deterministic problems, which alone may be computationally intensive. The computational costs of stochastic optimization problems turn out to be extremely high in many cases. This limitation encouraged researchers to introduce and adapt efficient schemes to represent uncertainty in the optimization procedure. Let us consider the following nonlinear problem of stochastic optimization:

$$\min_{x} f(x,\theta) \qquad (7.11)$$

subject to: $h_i(x,\theta) = 0$, $i = 1, ..., m$

$g_j(x,\theta) \leq 0$, $j = 1, ..., n$

where f is the objective function, x is the vector of design variables, m is the number of equality constraints h, n is the number of inequality constraints g, and θ denotes the random character of the quantities.

According to the stochastic nature of the above formulation, the representation scheme of the random quantity, θ, has a key role in the efficiency and accuracy of the stochastic optimization procedure. Many methods of stochastic optimization were tried in an effort to fit reliability, pattern recognition, filtering, and stochastic approximation methods into the framework of the optimization procedure. Tatang [18] introduced the Deterministic Equivalent Modeling Method (DEMM), along with the probabilistic collocation method, to formulate process design and optimization models. In DEMM, PCE is employed to decompose the stochastic inequality constraints into deterministic equivalents. This procedure requires an assumption regarding the appropriate order of PCE and symbolic manipulations to get the moments of the constraints. Also, the number of constraints is raised in proportion to the order of PCE through decomposition. In the probabilistic collocation method, the selected collocation points correspond to the roots of PCE and are used to obtain the coefficients of PCE. This procedure yields results equivalent to Galerkin's method for one-dimensional problems. Since the selected design points of the probabilistic collocation method are concentrated in the high probability region, the sampling procedure may not be appropriate when the tail region of the probability density function is also important.

A common approach for treating the computationally-expensive objective function and the constraints of the optimization problem is to build relatively inexpensive surrogate models using approximation methods. The choice of a surrogate-based optimization for uncertain systems can be reasonable in typical engineering applications. The non-intrusive formulation, which combines PCE and ANOVA within the framework of LHS, is an efficient tool for representing the responses of large-scale uncertain structural problems. Choi et al. [5] introduced the non-intrusive formulation procedure to build a surrogate model of uncertain systems for optimization problems. Specifically, to represent variability in stochastic constraints or objective functions, fluctuating components are introduced and approximated using the non-intrusive formulation method. The stochastic modeling process repeats and re-calibrates PCE models until sufficient model adequacies are obtained. A detailed framework of this approach is presented in Figure 7.20.

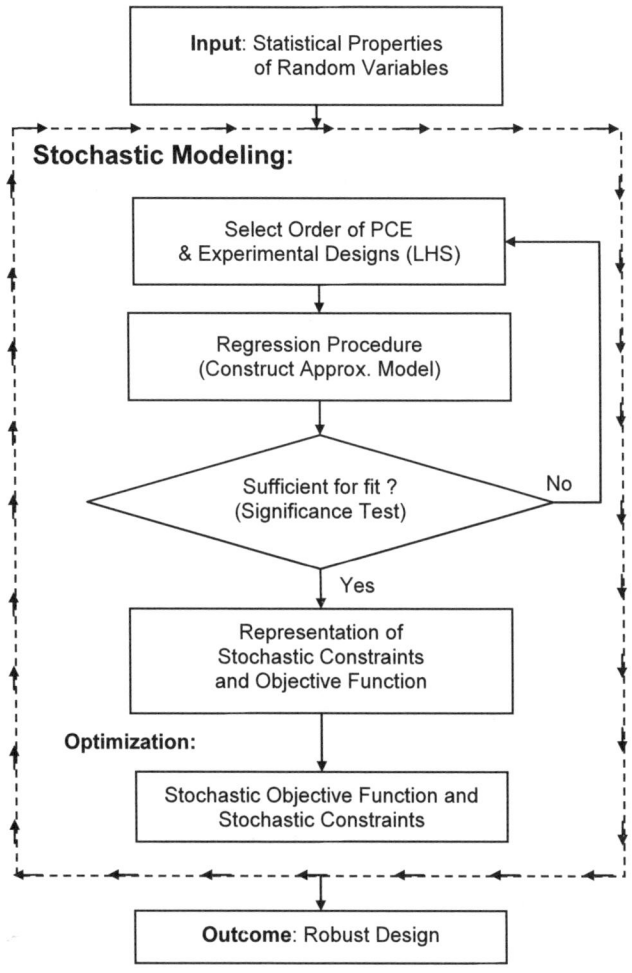

Figure 7.20. Stochastic Optimization Framework

The framework consists of two parts: stochastic modeling and an optimization procedure. In the first step, a desired number of sampling points corresponding to statistical properties is generated by LHS. The generated samples are evaluated and the results contribute toward a surrogate model of the response. If the selected model of PCE has a sufficient model adequacy, the obtained coefficients and the order of PCE can be used to represent the fluctuating components. Otherwise, a different model of PCE and additional simulations are required. Visual inspection of the residual analysis and significance testing using ANOVA are used to check model adequacy. After selecting the appropriate polynomial model of the fluctuating components used for the constraints or objective functions, the common optimization procedure can be applied.

In this procedure, the fluctuating components of the system performance constraints or objective function can be represented by a series expansion as

$$h_i(x,\theta) = \overline{h}_i(x) + <\varepsilon_i(x,\theta)> \qquad (7.12)$$

where $\overline{h}_i(x)$ are the deterministic constraints and $<\varepsilon_i(x,\theta)>$ represent the fluctuation of the random quantity around its mean value. The fluctuating components $\varepsilon_i(x,\theta)$ of the constraints or objective function can be projected onto PCE using the non-intrusive formulation described in the previous section.

Once the approximation of the fluctuation components is constructed, additional statistical information about the given system can easily be obtained over the random variable domain. The associated statistical moments of the fluctuating components can be obtained by the following expressions:

$$<\varepsilon> = b_0, \quad Var(\varepsilon) = \sum_{i=1}^{P} <\Psi_i^2> b_i^2 \qquad (7.13)$$

where b_i are the regression coefficients of polynomial chaos expansions Ψ_i, and $<\cdot>$ indicates the expected value operation as described in Section 2.1.

In the traditional design process, simplified rules, such as a safety factor that is normally based on engineers' experiences, are applied to the constraints to guarantee a desired safety of the stochastic structural systems. The size of the safety factor should depend on the level of uncertainty and the consequence to structural systems. However, this safety factor does not directly account for the random nature of most input parameters, since the simulation of a single design point fails to account for any variability. Estimated statistics of the fluctuating components (Equation 7.12, and Equation 7.13) can be an efficient substitute for or supplement to the safety factor in the design process. Thus, a modest increase in simulation cost can improve accuracy when accounting for significant fluctuating components in the overall design process. The following sections clearly demonstrate the applicability of this procedure in the design process of a three-bar truss and a joined-wing SensorCraft structure. A summary of the results obtained with this approach is also given.

7.3.3 Three-bar Truss Structure

Before experimenting with practical engineering problems, the three-bar truss structure shown in Figure 7.21 is considered to demonstrate the applicability of the procedure. Minimum material volume is the objective, subject to displacement and stress constraints in each member. For the three-bar truss, the cross-sectional areas, A_1 and A_2 are considered as design variables, as shown in Figure 7.21. The magnitude, P, and direction, θ, of the load are assumed to be random variables. Young's modulus, E, of each member is also considered to be random. These uncertain parameters are assumed to have independent Gaussian distributions, and detailed design data for optimization are given in Table 7.7. Thus, the standard optimization problem is:

Minimize: $\quad V(A_1, A_2, l) = l(2\sqrt{2}A_1 + A_2)$ \hfill (7.14)

Constraints: $\quad \sigma_1 = \dfrac{1}{\sqrt{2}}\left[\dfrac{P_u}{A_1} + \dfrac{P_v}{(A_1 + \sqrt{2}A_2)}\right] \leq \sigma_a$ \hfill (7.15)

$$\sigma_2 = \dfrac{\sqrt{2}P_v}{(A_1 + \sqrt{2}A_2)} \leq \sigma_a \qquad (7.16)$$

$$\sigma_3 = \dfrac{1}{\sqrt{2}}\left[\dfrac{P_v}{(A_1 + \sqrt{2}A_2)} - \dfrac{P_u}{A_1}\right] \leq \sigma_a \qquad (7.17)$$

$$u = \dfrac{\sqrt{2}lP_u}{A_1 E} \leq u_a \qquad (7.18)$$

$$v = \dfrac{\sqrt{2}lP_v}{(A_1 + \sqrt{2}A_2)E} \leq v_a \qquad (7.19)$$

$$0.01 \leq A_i \leq 100, \quad i = 1,2$$

where $P_u = P\cos\theta$ and $P_v = P\sin\theta$ are horizontal and vertical components of the load P; $\sigma_{1,2,3}$ are stresses in corresponding members; u and v are horizontal and vertical displacements; and subscript a indicates an allowable value of stresses and displacements.

Each response function is approximated by a third-order PCE with 50 experimental designs of LHS. After obtaining a PCE approximation of the uncertain response functions ($\sigma_{1,2,3}$, u, and v), the mean of any fluctuating components is constrained, and a well-known optimization procedure, sequential quadratic programming [7], is applied. The optimal design for the deterministic case was $X^* = (5.0000, 1.7678)$ and the objective function at the optimum was $f = 159.10$. The constraints of Equation 7.15 and Equation 7.18 are active (strictly satisfied) at the optimum. The optimal design for the uncertain system was obtained as $X^* = (5.0858, 1.6587)$ and $f = 160.43$. This result is consistent with the deterministic case in that the same constraints are active: only σ_1 and u are quite close to the limit of constraints. The optimum in the case where uncertainty is taken into account yields a slightly higher volume than the corresponding deterministic case. However, we demonstrate next that the optimal design of the uncertainty case will provide a robust design with respect to the uncertain load and material properties.

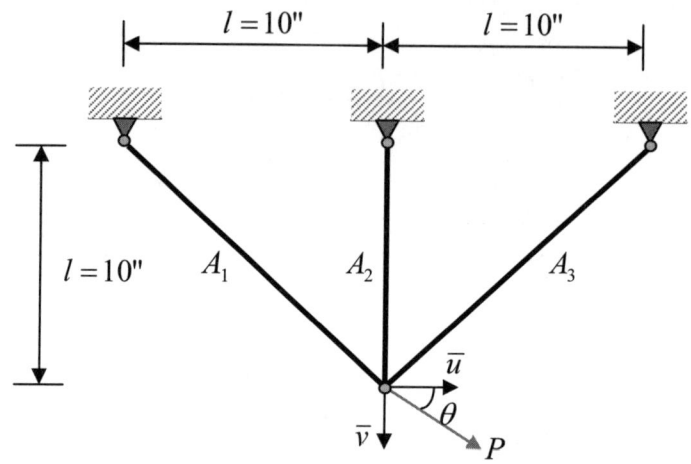

Figure 7.21. Three-bar Truss ($A_1 = A_3$)

Table 7.7. Three-bar Truss Design Data

Uncertain Variables	Magnitude of Load, P: ~ N(30000, 4500) lb
	Direction of Load, θ : ~ N(45, 7.75)°
	Modulus of Elasticity, E: ~ N(3×10^7, 4.5×10^6) psi
Objective Function	Minimize Material Volume
Limits on Design Variables: Constraints: Allowable Stress	$0.01 \leq A_{1,2} \leq 100$ in²
	$|\sigma_{1a}, \sigma_{3a}| \leq 5000$ psi, $|\sigma_{2a}| \leq 20000$ psi
Allowable Displacements	$|u_a| \leq 0.002$ in, $|v_a| \leq 0.002$ in

To show the robustness of the stochastic design for the given uncertainty, we obtained the probability of failure for the active constraints from the MCS with 100,000 realizations. The results of the probability of failure (P_f) are listed in Table 7.8. Increasing the objective function by 0.83% due to small variations of design variables yields about 4-6% additional safety in the performance constraints (Equations 7.15, and 7.18), as compared with the deterministic case. In the case of the vertical displacement constraint, v, the probability of failure is slightly increased because of the reduction of the design variable A_2. The increment of 9.57% in a much smaller P_f value can be trivial compared to the benefit of the other constraint cases, where P_f is an order of magnitude higher.

Table 7.8. Probability of Failure for Active Constraints

Constraints	P_f (Deterministic)	P_f (Stochastic)	Difference (%)
σ_1	0.4830	0.4625	-4.24
u	0.4834	0.4540	-6.08
v	0.0439	0.0481	9.57

7.3.4 Joined-wing SensorCraft Structure

The approach is also applied to the optimization of a joined-wing model under the uncertainty of material properties. Roberts et. al. [17] conducted linear and nonlinear structural optimizations of an aerodynamically trimmed joined-wing aircraft model. To consider all expected flight and ground conditions, multiple load cases, including gust, taxi crater, and landing impact load, were applied to optimize the model. The results revealed that the current joined-wing configuration is subject to critical buckling failure at maneuver-speed gust conditions. Hence, this example expands upon this model to achieve an optimized design with stress constraints in consideration of the effects of uncertain material properties.

The current joined-wing model uses 2024-T3 aluminum (Table 7.9) with a minimum skin thickness of 0.04 *in*. Young's moduli of the skin, spar, and rib elements in the vicinity of the wing joint and the two wing roots are assumed to be Gaussian and uncorrelated random variables ($\mu_E = 7.24 \times 10^{10}$ *Pa*, $\sigma_E = 7.24 \times 10^9$, COV = 0.1).

Table 7.9. 2024-T3 Aluminum Material Properties

	MPa
σ_{ty}	324
σ_{cy}	269
σ_{shear}	269
E	72,395
G	27,580

The joined-wing model is optimized to minimize weight under the maneuver gust load, which is a significantly higher load than a steady-maneuver flight condition. The design variables are the thicknesses of skin, rib, and spar elements within the structural wing box, excluding elements outside of the leading and trailing edge spars and outside of the most outboard rib. The von Mises allowable stress (σ_a) of 253×10^6 Pa was used for the stress constraints. Detailed descriptions of the deterministic results of the joined-wing optimization and further discussions of their advantages and limitations can be found in [17]. The stochastic optimization problem for the joined-wing can be written as

Minimize: $W(X)$ (7.20)

Subject to: $\sigma_i(X,\xi) \leq \sigma_a$, $i = 1, 2, ..., n$

$$\underline{x}_j \leq x_j \leq \overline{x}_j, j = 1, 2, ..., m$$

where $W(\cdot)$ denotes the weight of the wing structure; $X=\{x_1, x_2, ..., x_m\}^T$ is the vector of the design variables; $\underline{x}_j$ and $\overline{x}_j$ indicate the lower and upper bounds of design variable x_j; and $\sigma_i(\cdot)$ denotes the stress function for i^{th} element of the FEM model. The stress function $\sigma(\cdot)$ can be expressed as a function of ξ, because Young's modulus is computed as $E = \mu_E + \sigma_E \xi$. Here, μ_E and σ_E are the mean and standard deviation of Young's modulus E.

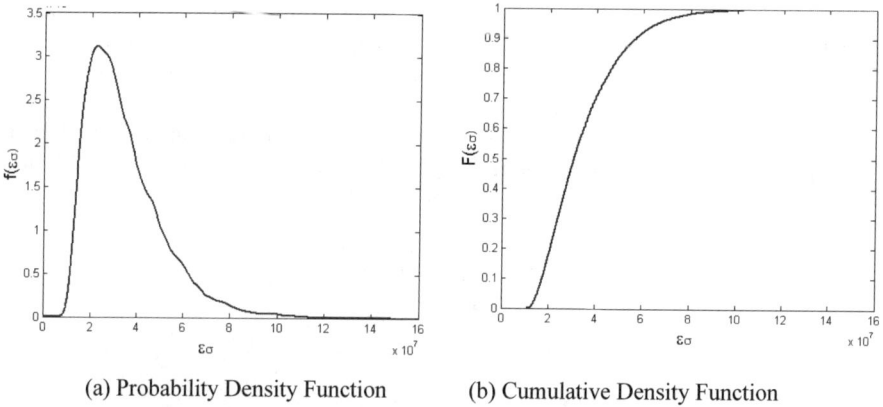

(a) Probability Density Function   (b) Cumulative Density Function

Figure 7.22. Estimated Statistics of Stress Fluctuation Components

As described earlier, the fluctuating component ε_σ of stress constraints around their mean value can be projected onto PCE, $\varepsilon = \sum_{l=0}^{P} \beta_l \Psi_l$. Then, the stochastic approximation procedure (Section 6.1.2) is used to obtain the undetermined coefficients β_l. In the non-intrusive formulation, fourth-order PCE was used with 50 simulations based on LHS, and the model adequacy was checked using residual analysis and ANOVA analysis. Figure 7.22 depicts nonparametric estimates of the

PDF and the CDF resulting from the non-intrusive analysis of the fluctuating component ε_σ of the stress constraint. The mean of the fluctuation ε_σ is estimated as 34.22×10^6 Pa in the first iteration, and the PDF clearly exhibits a non-Gaussian distribution with a positive skewness property. These estimated statistics can be included in the constraints. The effect is a reduction of the undesired probability of failure (P_f) under the given uncertain situation.

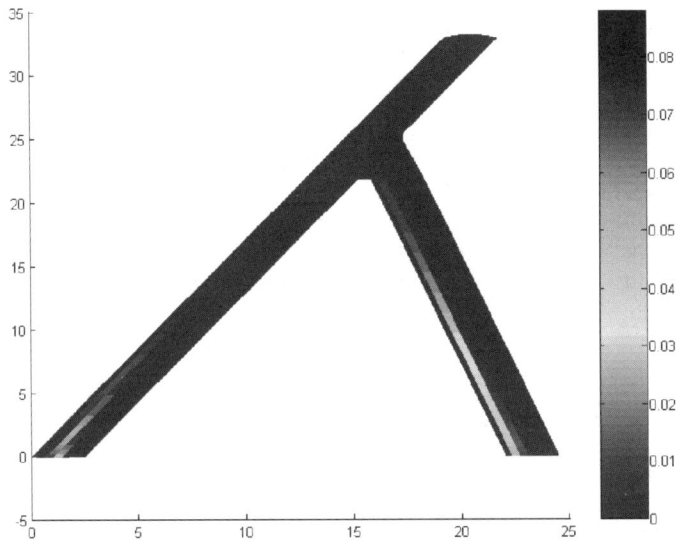

Figure 7.23. Wing Skin Thickness Distribution, Stochastic Optimization

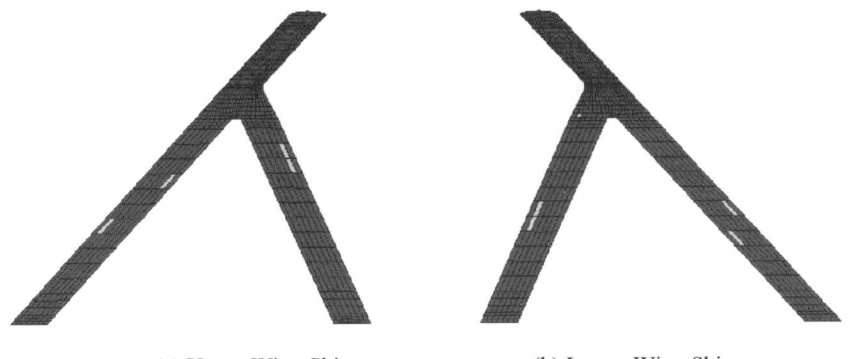

(a) Upper Wing Skin   (b) Lower Wing Skin

Figure 7.24. Identified Risk Regions for Stochastic Material Properties

Figure 7.23 depicts the thickness distributions of the upper wing skin in stochastic optimization. These results show similar distributions as compared with the results of the deterministic case [17]. The stochastic optimization increased the upper leading edge and lower trailing edge thicknesses of the wing box, just as the deterministic optimization did. During the stochastic simulations for material variability, 22 elements of the wing skin and substructure (Figure 7.24) are identified as the critical components that have maximum stresses in each and every realization used in the LHS. The optimum weights of the joined-wing for the deterministic case and the stochastic case are 2534.0kg and 2746.3kg, respectively. To obtain the P_f of the joined wing, the limit states are considered as $\sigma_a = 253 \times 10^6$ Pa for von Mises stress. The corresponding P_f of the deterministic design and stochastic design are 0.4829 and 0.0306, respectively. The P_f of the joined-wing was calculated using 100,000 MCS simulations on the surrogate model of PCE, without significant additional computational costs. In the stochastic optimization, a significant reduction of the P_f was obtained as compared to the deterministic case. This occurred because the constraint in the stochastic optimization was imposed on the mean of the random stress, which included the variability.

7.4 References

[1] Alyanak, E., Venkayya, V.B., Grandhi, R.V., and Penmetsa, R.C., "Structural Response and Optimization of a Supercavitating Torpedo," *Finite Elements in Analysis and Design*, Vol. 41, (6), March 2005, pp. 563-582.

[2] Birge, J. and Louveaux, F., *Introduction to Stochastic Programming*, Springer, New York, 1997.

[3] Blair, M. and Canfield, R.A., "A Joined-wing Structural Weight Modeling Study," AIAA-2002-1337, presented at the *43rd AIAA/ASME/ASCE/AHS/ASC Structures, Structural Dynamics, and Materials Conference*, Denver, CO, April, 2002.

[4] Charnes A. and Cooper, W.W., "Chance-Constrained Programming," *Management Science*, vol. 6, pp. 73-79, 1959.

[5] Choi, S., Grandhi, R.V., and Canfield, R.A., "Robust Design of Mechanical Systems via Stochastic Expansion," *International Journal of Materials and Product Technology*, Vol. 25, 2006, pp. 127-143.

[6] Choi, S., Grandhi, R.V., Canfield, R.A., and Pettit, C.L., "Polynomial Chaos Expansion with Latin Hypercube Sampling for Estimating Response Variability," *AIAA Journal*, Vol. 42, (6), 2004, pp. 1191-1198.

[7] Diwekar U.M., *Introduction to Applied Optimization*, Kluwer Academic Publishers, Norwell, MA, 2003.

[8] Draper N.R., and Smith H., *Applied Regression Analysis*, Wiley, New York, 1981.

[9] Ermoliev, Y., and Wets, R. J-B., *Numerical Techniques for Stochastic Optimization*, Springer-Verlag, Berlin, 1988.

[10] *GENESIS User Manual*, Volume 1, Version 7.0, Vanderplaats Reasearch & Development, Inc., Colorado Springs, CO 80906, August 2001.

[11] Haldar, A., and Mahadevan, S., *Reliability Assessment Using Stochastic Finite Element Analysis*, John Wiley & Sons, NY, 2000.

[12] Isukapalli, S.S., *Uncertainty Analysis of Transport-Transformation Models*, Ph.D. Dissertation, Rutgers, the State University of New Jersey, New Brunswick, NJ., 1999.

[13] Jackson, J., *A User's Guide to Principal Components*, Wiley, NY, 1991.
[14] Montgomery, D.C, and Peck E.A., *Introduction to Linear Regression Analysis*, Wiley, New York, 1992.
[15] Montgomery, D.C., *Design and Analysis of Experiments*, Wiley, New York, 1997.
[16] Pettit, C.L., Canfield, R.A. and Ghanem, R., "Stochastic Analysis of an Aeroelastic System," presented at *15th ASCE Engineering Mechanics Conference*, Columbia University, New York, June 2-5, 2002.
[17] Roberts, R.W., Canfield, R.A. and Blair, M., "Sensor-Craft Structural Optimization and Analytical Certification," *44th AIAA SDM Conference,* Norfolk, VA, April 2003, AIAA 2003-1458.
[18] Tatang, M.A., Direct Incorporation of Uncertainty in Chemical and Environmental Engineering Systems, Ph.D. Dissertation, Massachusetts Institute of Technology, Cambridge, MA, 1995.
[19] Wets, R.J.-B., "Stochastic Programs with Fixed Recourse: The Equivalent Deterministic Program," *SIAM Review*, Vol. 16, 1974, pp. 309-9339.
[20] Xiu, D., and Karniadakis, G., "The Wiener-Askey Polynomial Chaos for Stochastic Differential Equations," *SIAM Journal on Scientific Computing*, Vol. 24, (2), 2002, pp. 619-644.

8
Summary

The current book has focused on engineering design and optimization under risk and uncertainty. The intent was to provide a systematic background in the area of structural reliability and statistical approaches to design of mechanical systems that call for computationally expensive simula-tions. Hopefully, it will promote the development and adoption of new probabilistic decision support tools to assist risk management in the presence of uncertainty. Chapter 2 presented several probabilistic methods to represent the random nature of input parameters for structural models. Specifically, the concept of the random process/field was discussed with graphical interpretations of the discretization scheme of the covariance function. Turning to regression analysis, we studied common regression procedures and discussed the polynomial regression procedure which can be applied to stochastic approximation procedures in later chapters. The procedure of model adequacy checking was also given with a representative example of the regression problem.

In Chapter 3, we presented several probabilistic analysis methods including the FORM and SORM, MCS, Importance Sampling, LHS, and stochastic expansions. Along with classical approaches to reliability analysis, we emphasized strengths and weaknesses of each method. The ideas underlying MCS and LHS were demonstrated with intuitive illustrations of simple problems. The aim of these simple problems was to greatly reduce learning time to comprehend the basic idea of MCS and LHS to interested readers. Another representative method of uncertainty analysis, the SFEM, was also investigated, including the perturbation method, Neumann expansion method, and weighted integral method.

In Chapter 4, we studied basic problems and their solutions using FORM and SORM. The graphical interpretations of the Hasofer-Lind safety index, sensitivity factors, and the SORM were also discussed. Typical cases of each method were illustrated with basic problems that were examined in detail. We worked out the basic problems step by step in order to facilitate appreciation of the iterative nature of reliability methods. Adaptive approximations in the reliability method were developed, and results were compared to other methods.

In Chapter 6, we studied two stochastic expansions, the PCE and the KL expansion, for representing a random process or random field. The roles of both expansions in the spectral stochastic finite element method were also given in the latter part of this chapter. Basic properties of the PCE and the KL transform were illustrated with simple problems. To assist in the comprehension of the PCE, we discussed the Hermite polynomials and Gram-Charlier series. We generated correlated random variables from the KL transform. The solution of eigenproblems by using KL expansion was also given. As pointed out at that time, the direct use of stochastic expansions for representing uncertain parameters is an effective choice, because they provide analytically appealing convergence properties. We developed stochastic approximation procedures, which can represent the random field of the structural inputs and outputs by using the PCE and KL transform.

The next step was to apply these techniques to some basic problems and to practical engineering problems. In Chapter 7, we demonstrated the simula-tion procedures for Gaussian and non-Gaussian problems with examples involving the input of random fields. To show the efficiency and applicability of the developed methods, it is necessary to consider large-scale problems. Thus, we applied the procedure to a geometrically nonlinear structural model of a joined-wing aircraft and to a supercavitating torpedo with NASTRAN. Finally, we extended the developed procedure to sto-chastic optimization problems. At this point, we showed how the method can be used to 1) develop initial guidance for a safe design, 2) identify where significant contributors of uncertainty occur in structural systems, and 3) increase the safety and efficiency of the structure.

A

Function Approximation Tools

Function approximations play a major role in iterative solutions and optimization of large-scale structures. For many structural optimization problems, evaluation of the objective function and constraints requires the execution of costly finite element analyses for displacements, stresses, or other structural responses. The optimization process (Figure A.1) may require that the objective function and constraints hundreds or thousands of times be evaluated. The cost of repeating the finite element analysis so many times is usually prohibitive (Figure A.1: Step 2). However, this computational cost problem can be addressed by the use of approximations during portions of the optimization process (Figure A.1: Step 3). First, an initial design is obtained by using an exact analysis, and the information needed to construct the approximations is generated. The original design problem is changed into a sequential approximate optimization problem with an approximate representation of constraints. Then, the approximate problem is solved by an optimization algorithm (Figure A.1: Step 4). The objective function value is obtained at the optimum solution and compared with the initial value (Figure A.1: Step 5). If the convergence requirement is not satisfied, the process is repeated until convergence (Figure A.1: Step 6).

Since the approximation has replaced the expensive exact constraint calculations, significant computational savings can be realized, particularly for large-scale structural problems requiring time-consuming analyses. However, the use of fixed intervening variables for function expansion is difficult to adopt for different problems, and selection of the intervening variables is also quite difficult, requiring tremendous experience and knowledge. Therefore, the use of adaptive intervening variables for different types of problems is necessary.

Function approximations are particularly useful when the computational cost of a single evaluation of the objective functions, performance constraints, and their derivatives is very large. A typical situation is when a finite element model with millions of degrees of freedom is used to analyze a structure that is defined in terms of a handful of design variables. Reducing the number of exact structural analyses required for the design process by building surrogate simulation models makes design optimization feasible for such a case.

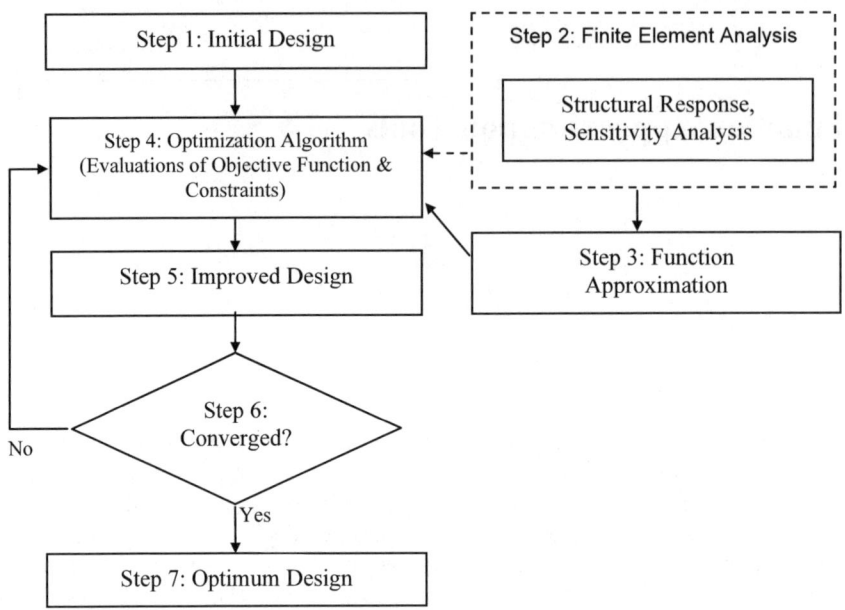

Figure A.1. Optimization Process with Function Approximation

A.1 Use of Approximations and Advantages

In general, the optimization problem is stated as

Minimize: $f(X)$ (A.1a)

Subject to: $g_j(X) \leq 0$ $(j = 1, 2, ..., J)$ (A.1b)

$$x_i^L \leq x_i \leq x_i^U \quad (i = 1, 2, ..., n)$$ (A.1c)

where X represents the vector of design variables, $f(X)$ is the objective function, $g_j(X)$ is the jth behavior constraint, and x_i^L and x_i^U are the lower and upper limits on the ith design variable, respectively; J and n denote the number of behavior constraints and design variables, respectively.

Based on the types of approximations, such as linear, reciprocal, conservative, two-point adaptive nonlinear, *etc.*, the original optimization problem of Equation A.1 is changed into a sequence of explicit approximate problems as follows:

Minimize: $f^{(k)}(X) = \widetilde{f}(X)$ (A.2a)

Subject to: $g_j^k(X) = \widetilde{g}_j(X) \leq 0 \ (j = 1,2,...,J)$ (A.2b)

$$x_i^L \leq x_i \leq x_i^U \ (i = 1,2,...,n)$$ (A.2c)

where k is the iteration number.

Because exact function evaluations are avoided in solving (A.2a) and (A.2b), computational savings are realized. Only at the convergent solution of the problem (A.2) are exact objective and constraint function calculations are needed, and the approximations are updated based on the information of the convergent point. The process is repeated until convergence. However, the approximations may result in significant errors if they do not represent the original objective and constraint functions accurately. Due to inaccurate approximations, optimization algorithms may have difficulty or may never converge without a proper choice of move limits. To avoid this problem, constructing accurate approximations is very important.

A.2 One-point Approximations

Most approximations presented in the literature were based on function and gradient information at a single point and constructed using the first-order Taylor series expansion about this point. This method is very popular because the function and its derivative values are always needed for the search direction calculation, so no additional computation is involved in developing an approximate function. There are several variations of first-order Taylor series approximations, most notably the linear, reciprocal and conservative approximations. These approximations work effectively for stress and displacement type problems; however, the truncation error of the first-order approximation might be large, even for design points closer to the expansion point. The accuracy of the first-order approximations may be increased in some disciplines by retaining higher-order terms in Taylor series expansion, such as the quadratic approximation. This requires the calculation of higher-order function derivatives that may not be available analytically or easily calculated.

Approximations are formed using the first-order and higher-order Taylor series expansion in terms of direct and reciprocal design variables. The intervening variables are fixed in these approximations. For example, the linear approximation can be considered as the first-order Taylor series expansion in terms of the intervening variables $y_i = x_i$. The reciprocal approximation is the first-order Taylor series expansion in terms of the intervening variables $y_i = 1/x_i$. As we know, for the truss structures with stress and displacement functions, using the reciprocal intervening variables can dramatically improve the approximation accuracy.

In this section, several one-point approximations (linear, reciprocal, and conservative) are introduced. One-point approximation means that the approximation is constructed based on the function value and gradient information of one point. Usually, this point is selected as the most current design point in the iteration process. The most commonly used approximations of objective and constraint functions are based on one-point information, *i.e.*, the function and its first derivative at a single design point.

A.2.1 Linear Approximation

The simplest approximation is the linear approximation, which is a first-order Taylor series expansion at a design point X_1:

$$\tilde{g}(X) = g(X_1) + \sum_{i=1}^{n} \frac{\partial g(X_1)}{\partial x_i}(x_i - x_{i,1}) \qquad (A.3)$$

where x_i is the ith component of variables X and $x_{i,1}$ is the ith component of the known point X_1. This approximation is very popular because only the function and its derivatives are needed in the search direction calculation, and no additional computation is required to develop the approximate function. However, for many applications, the linear approximation is inaccurate, even for design points X that are close to X_1. Accuracy can be increased by retaining additional terms in the Taylor series expansion, but this requires the calculation of higher-order function derivatives. A more attractive alternative is to find intervening variables that make the approximate function more accurate. One of the popular intervening variables is the reciprocal of x_i, which forms the basis of the reciprocal approximation featured next.

A.2.2 Reciprocal Approximation

The reciprocal approximation is the first-order Taylor series expansion using reciprocals of the variables $y_i = 1/x_i$ $(i = 1,2,...,n)$. It can be written in terms of the original variables x_i:

$$\tilde{g}(Y) = g(Y_1) + \sum_{i=1}^{n} \frac{\partial g(Y_1)}{\partial y_i}(y_i - y_{i,1}) \qquad (A.4a)$$

$$\tilde{g}_R(Y) = g(X_1) + \sum_{i=1}^{n} \frac{\partial g(X_1)}{\partial x_i}(x_i - x_{i,1})(\frac{x_{i,1}}{x_i}) \qquad (A.4b)$$

$$\because \frac{\partial g}{\partial y_i} = \frac{\partial g}{\partial x_i}\frac{\partial x_i}{\partial y_i}, \quad \frac{\partial y_i}{\partial x_i} = -\frac{1}{x_i^2}$$

This approximation has been proven efficient for truss structures with stress and displacement constraints because in statically determinate structures, stress and displacement constraints are linear functions of the reciprocals of the design variables X.

However, there is one problem in the reciprocal approximation given in Equation A.4: The approximation becomes unbounded when one of the variables approaches zero. A modified approximation was presented by Haftka, et al. in [2]:

$$\tilde{g}_m(X) = g(X_2) + \sum_{i=1}^{n} \frac{\partial g(X_2)}{\partial x_i}(x_i - x_{i,2})(\frac{x_{mi} + x_{i,2}}{x_{mi} + x_i}) \tag{A.5}$$

where X_2 is the current point. The values of x_{mi} are evaluated by matching them with the derivatives at the previous point X_1; that is,

$$\frac{\partial g(X_1)}{\partial x_i} = (\frac{x_{mi} + x_{i,2}}{x_{mi} + x_{i,1}})^2 \frac{\partial g(X_2)}{\partial x_i} \tag{A.6}$$

or

$$x_{mi} = \frac{x_{i,2} - \eta_i x_{i,1}}{\eta_i - 1} \tag{A.7}$$

where

$$\eta_i^2 = \frac{\partial g(X_1)}{\partial x_i} / \frac{\partial g(X_2)}{\partial x_i} \tag{A.8}$$

When the ratio of the derivatives is negative, the derivatives at the previous point X_1 are not matched. In that case, x_{mi} is set to a very large number so that the linear approximation is used for the ith variable. This modified reciprocal approximation is a two-point approximation because two-point information is used to construct the approximation given in Equation A.5.

A.2.3 Conservative Approximation

Conservative approximation, as presented by Starnes and Haftka [5], is a hybrid form of the linear and reciprocal approximations and is more conservative than both. The approximation is given as

$$\tilde{g}(X) = g(X_1) + \sum_{i=1}^{n} C_i \frac{\partial g(X_1)}{\partial x_i}(x_i - x_{i,1}) \tag{A.9}$$

where $C_i = \begin{cases} 1 & \text{if } x_{i,1}\frac{\partial g}{\partial x_i} \leq 0 \\ \frac{x_{i,1}}{x_i} & \text{otherwise} \end{cases}$ (A.10)

In the above approximation, $C_i = 1$ corresponds to the linear approximation, and $C_i = x_{i,1} / x_i$ corresponds to the reciprocal approximation.

The conservative approximation is not the only hybrid linear-reciprocal approximation possible; sometimes physical considerations may dictate the use of linear approximation for some variables and reciprocal for others. However the conservative approximation has the advantage of being convex. If the objective function and all the constraints are approximated by the conservative approximation, the approximate optimization problem is convex. Convex problems are guaranteed to have at most a single optimum.

A.3 Two-point Adaptive Nonlinear Approximations

In the previous section, the first-order Taylor series approximation is based on a single point. As the structure is being resized, new approximations are constructed at new design points. In this approach, previous analyses' information is discarded and not used to improve the later approximations. More accurate approximations [6],[7] have been developed by using more than one data point, such as two points, three points or more. These multi-point approximations use current and previous information to construct approximations. The nonlinearities in the multi-point approximations are automatically adjusted by using function values and gradients at the known points. Therefore, the multi-point approximations are adaptive and provide better accuracy than the single point approximations. Also, no higher-order gradients are needed to construct the approximations.

Adaptability refers to the capability to automatically match the nonlinearity of various functions. For one-point approximations, the nonlinearity of the approximations is fixed, since the expansion variables are fixed. In general, selecting appropriate intervening variables is extremely difficult for different engineering problems. For the stress and displacement constraints of the truss structures, the reciprocal approximation can yield better results. However, reciprocal intervening variables may not the best choice for other constraints of truss structures (e.g. natural frequency) or other structures (e.g. plate). For practical engineering problems, the use of fixed intervening variables is difficult to adopt for different constraints, and selection of the intervening variables is also quite difficult and requires experience and knowledge. Therefore, the use of adaptive approximation models or changeable intervening variables for different types of problems is desirable. The following two-point approximations are capable of adjusting their nonlinearities automatically by using two-point information.

A.3.1 Two-point Adaptive Nonlinear Approximation (TANA)

TANA is a two-point adaptive approximation, presented by Wang and Grandhi in [7] that uses adaptive intervening variables. Two-point approximation means that the approximation is constructed based on the function values and gradients information of two points. Usually, one point is selected as the most current point and the other is the previous point in the iteration process. The intervening variables are defined as

$$y_i = x_i^r, \; i = 1,2,...,n \tag{A.11}$$

where r represents the nonlinearity index, which is different at each iteration but the same for all variables. The Taylor series expansion is written at the second point, X_2 in terms of the intervening variables, y_i. Substitution of x_i gives the expansion in terms of the physical variables. The nonlinearity index is determined by matching the function value of the previous design point; that is, r is calculated numerically so that the difference of the exact and approximate $g(X)$ at the previous point X_1 becomes zero.

$$g(X_1) - \{g(X_2) + \frac{1}{r}\sum_{i=1}^{n} x_{i,2}^{1-r} \frac{\partial g(X_2)}{\partial x_i}(x_{i,1}^r - x_{i,2}^r)\} = 0 \tag{A.12}$$

where r can be any positive or negative real number (not equal to zero). The two-point adaptive nonlinear approximation (TANA) is

$$\tilde{g}(X) = g(X_2) + \frac{1}{r}\sum_{i=1}^{n} x_{i,2}^{1-r} \frac{\partial g(X_2)}{\partial x_i}(x_i^r - x_{i,2}^r) \tag{A.13}$$

This approximation has been used extensively in truss, frame, plate and turbine blade structural optimization and probabilistic design. The results presented in [6], and [7] demonstrate the accuracy and adaptive nature of this nonlinear approximation.

A.3.2 Improved Two-point Adaptive Nonlinear Approximation (TANA1)

The following intervening variables are used in the Taylor series expansion:

$$y_i = x_i^{p_i}, \; (i = 1,2,...,n) \tag{A.14}$$

where p_i is the nonlinear index, which is different for each design variable. The approximate function is represented as

$$\tilde{g}(X) = g(X_1) + \sum_{i=1}^{n} \frac{\partial g(X_1)}{\partial x_i} \frac{x_{i,1}^{1-p_i}}{p_i}(x_i^{p_i} - x_{i,1}^{p_i}) + \varepsilon_1 \tag{A.15}$$

where ε_1 is a constant, representing the residue of the first-order Taylor approximation in terms of the intervening variables, $y_i = x_i^{p_i}$. This approximation is expanded at the previous point X_1 instead of the current point X_2. This is because if the approximation is constructed at X_2, the appropriate function value would not be equal to the exact function value at the expanding point, because of the correction term ε_1. In actual optimization, X_1 is selected as the expansion point to obtain more accurate predictions closer to the current point. The approximate function and its derivative value are matched with the current point data.

By differentiating Equation A.15, the derivative of the approximate function with respect to the ith design variable x_1 is written as

$$\frac{\partial \tilde{g}(X)}{\partial x_i} = \left(\frac{x_i}{x_{i,1}}\right)^{p_i - 1} \frac{\partial g(X_1)}{\partial x_i} \quad (i = 1, 2, \ldots, n) \tag{A.16}$$

From this equation, p_i can be evaluated by letting the exact derivatives at X_2 equal the approximation derivatives at this point:

$$\frac{\partial g(X_2)}{\partial x_i} = \frac{\partial \tilde{g}(X_2)}{\partial x_i} = \left(\frac{x_{i,2}}{x_{i,1}}\right)^{p_i - 1} \frac{\partial g(X_1)}{\partial x_i} \quad (i = 1, 2, \ldots, n) \tag{A.17}$$

where p_i can be any positive real number (not equal to zero).

Equation A.17 has n equations and n unknown constants. It is easy to solve because each equation has a single unknown constant, p_i. Here, a simple adaptive search technique is used to solve them. The numerical iteration for calculating each p_i starts from $p_i = 1$. When p_i is increased or decreased by a step length (0.1), the error between the exact and approximation derivatives at X_2 is calculated. If this error is smaller than the initial error (e.g., corresponding to $p_i = 1$), the above iteration is repeated until the allowable error (0.001) or limitation of p_i is reached, and p_i is determined. Otherwise, the step length of p_i is decreased by half, and the above iteration process is repeated until the final p_i is obtained. This search is computationally inexpensive because Equation A.17 is available in a closed-form and is easy to implement. There are several other types of numerical schemes that can be used to solve for p_i and r.

Equation A.17 matches only the derivative values of the current point, so a difference between the exact and approximate function values at the current point may exist. This difference is eliminated by adding the correct term, ε_1, in the approximation. ε_1 is computed by matching the approximate and exact function values at the current point:

$$\varepsilon_1 = g(X_2) - \{g(X_1) + \sum_{i=1}^{n} \frac{\partial g(X_1)}{\partial x_i} \frac{x_{i,1}^{1-p_i}}{p_i}(x_{i,2}^{p_i} - x_{i,1}^{p_i})\} \quad (A.18)$$

ε_1 is a constant during a particular iteration. The surrogate model and derivative values are equal to the exact values at the current design point.

A.3.3 Improved Two-point Adaptive Nonlinear Approximation (TANA2)

TANA2 uses the intervening variables given in Equation A.14. The approximation is written by expanding the function at X_2:

$$\widetilde{g}(X) = g(X_2) + \sum_{i=1}^{n} \frac{\partial g(X_2)}{\partial x_i} \frac{x_{i,2}^{1-p_i}}{p_i}(x_i^{p_i} - x_{i,2}^{p_i}) + \frac{1}{2}\varepsilon_2 \sum_{i=1}^{n}(x_i^{p_i} - x_{i,2}^{p_i})^2 \quad (A.19)$$

This approximation is a second-order Taylor series expansion in terms of intervening variables, $y_i = x_i^{p_i}$, in which the Hessian matrix has only diagonal elements of the same value ε_2. Therefore, this approximation does not need the calculation of the second-order derivatives. Unlike the original second-order approximation, this approximation is expanded in terms of the intervening variables y_i, so the error from the approximate Hessian matrix is partially corrected by adjusting the nonlinearity index p_i. Equation A.19 has $n+1$ unknown constants, so $n+1$ equations are required. Differentiating Equation A.19, n equations are obtained by matching the derivative with the previous point X_1 derivative:

$$\frac{\partial g(X_1)}{\partial x_i} = (\frac{x_{i,1}}{x_{i,2}})^{p_i-1}\frac{\partial g(X_2)}{\partial x_i} + \varepsilon_2(x_{i,1}^{p_i} - x_{i,2}^{p_i})x_{i,1}^{p_i-1}p_i \quad (A.20)$$

Another equation is obtained by matching the exact and approximate function values with the previous point X_1:

$$g(X_1) = g(X_2) + \sum_{i=1}^{n} \frac{\partial g(X_2)}{\partial x_i} \frac{x_{i,2}^{1-p_i}}{p_i}(x_{i,1}^{p_i} - x_{i,2}^{p_i}) + \frac{1}{2}\varepsilon_2 \sum_{i=1}^{n}(x_{i,1}^{p_i} - x_{i,2}^{p_i})^2 \quad (A.21)$$

There are many algorithms for solving these $n+1$ equations as simultaneous equations. Again, a simple adaptive search technique is used. First, ε_2 is fixed at a small initial value (e.g., 0.5), the numerical iteration is used to solve each p_i, and the differences between the exact and appropriate function and derivative values at X_1 are calculated. Then, ε_2 is increased or decreased by a step length (0.1), p_i, and the differences between the exact and approximate function and derivative values at X_1 are recalculated. If these differences are smaller than the initial error (e.g., corresponding to $\varepsilon_2 = 0.5$), the iteration is repeated until the allowable error (0.001) or limitation of ε_2 is reached, and the optimum combination of ε_2 and p_i is determined. In the TANA2 method, the exact function and derivative values are equal to the surrogate model function and derivative values, respectively, at both points. Therefore, this approximation is more accurate than the others.

Example A.1

The three-bar truss shown in Figure A.2 is taken from [1]. The truss is subjected to stress and displacement constraints with cross-sectional areas A_A, A_B, and A_C ($A_A = A_C$) as design variables.

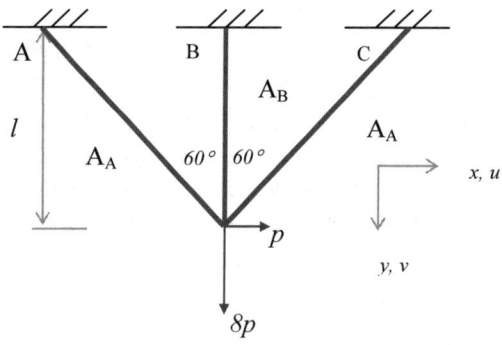

Figure A.2. Three-bar Truss

The approximation of a member C stress constraint (σ_c) is examined. The stress constraint (σ_0) using normalized variables is written as:

$$g(X) = 1 - \frac{\sigma_C}{\sigma_0} = 1 + \frac{\sqrt{3}}{3x_1} - \frac{2}{x_2 + 0.25x_1}$$

where $x_1 = A_A \sigma_0/p$ and $x_2 = A_B \sigma_0/p$.

When $g(X)$ is expanded at the point $X_1(1.0, 1.0)$, compare accuracies of the linear, reciprocal, conservative, TANA, TANA1, and TANA2 approximations. The point $X_2(1.25, 1.25)$ is the second point for TANA, TANA1 and TANA2 to calculate the nonlinearity indices r_1 and p_i.

Solution:

The first derivatives are readily computed as:

$$\frac{\partial g}{\partial x_1} = \left(-\frac{\sqrt{3}}{3x_1^2} - \frac{0.5}{(x_2 + 0.25x_1)^2}\right), \quad \frac{\partial g}{\partial x_2} = \left(\frac{2}{(x_2 + 0.25x_1)^2}\right)$$

The function value and its first derivatives at the points X_1 and X_2 are summarized in Table A.1.

Table A.1. Derivatives and Function Values at X_1 and X_2

	$X_1(1.0, 1.0)$ → $x_{1,1}=1.0$, $x_{2,1}=1.0$	$X_2(1.25, 1.25)$ → $x_{1,2}=1.25$, $x_{2,2}=1.25$
$\dfrac{\partial g}{\partial x_1}$	$\left.\dfrac{\partial g}{\partial x_1}\right|_{X_1} = -0.2574$	$\left.\dfrac{\partial g}{\partial x_1}\right|_{X_2} = -0.1647$
$\dfrac{\partial g}{\partial x_2}$	$\left.\dfrac{\partial g}{\partial x_1}\right|_{X_1} = 1.28$	$\left.\dfrac{\partial g}{\partial x_1}\right|_{X_2} = 0.8192$
$g(X)$	$g(X_1) = -0.0227$	$g(X_2) = 0.1819$

From these inputs, the approximations are constructed as:

1) Linear Approximation (Equation A.3)

$$g_L(X) = g(X_1) + (x_1 - x_{1,1})\left(\frac{\partial g}{\partial x_1}\right)_{X_1} + (x_2 - x_{2,1})\left(\frac{\partial g}{\partial x_2}\right)_{X_1}$$

$$g_L(X) = -0.0227 - 0.2574(x_1 - 1.0) + 1.28(x_2 - 1.0)$$

2) Reciprocal Approximation (Equation A.4)

$$g_R(X) = g(X_1) + \left(\frac{\partial g}{\partial x_1}\right)_{X_1}(x_1 - x_{1,1})\left(\frac{x_{1,1}}{x_1}\right) + \left(\frac{\partial g}{\partial x_2}\right)_{X_1}(x_2 - x_{2,1})\left(\frac{x_{2,1}}{x_2}\right)$$

$$g_R(X) = -0.0227 - 0.2574\left(1 - \frac{1}{x_1}\right) + 1.28\left(1 - \frac{1}{x_2}\right)$$

$$= 1 + 0.2574/x_1 - 1.28/x_2$$

3) Conservative Approximation (Equation A.9)

$$g_C(X) = g(X_1) + G_1(x_1 - x_{1,1})\left(\frac{\partial g}{\partial x_1}\right)_{X_1} + G_2(x_2 - x_{2,1})\left(\frac{\partial g}{\partial x_2}\right)_{X_1},$$

$$G_i = \begin{cases} 1 & x_{i,1}\left(\frac{\partial g}{\partial x_i}\right)_{X_1} \leq 0 \\ x_{i,1}/x_i & \text{otherwise} \end{cases}$$

$$g_C(X) = -0.0227 - 0.2574(x_1 - 1) + 1.28\left(1 - \frac{1}{x_2}\right)$$

4) TANA

From Equation A. 12,

$$g(X_1) - \left\{g(X_2) + \frac{1}{r}x_{1,2}^{1-r}(x_{1,1}^r - x_{1,2}^r)\left(\frac{\partial g}{\partial x_1}\right)_{X_2} + \frac{1}{r}x_{2,2}^{1-r}(x_{2,1}^r - x_{2,2}^r)\left(\frac{\partial g}{\partial x_2}\right)_{X_2}\right\} = 0$$

$$-0.227 - \{0.1819 + \frac{1}{r}(1.25^{1-r})(1.0^r - 1.25^r)(-0.1647)$$

$$+ \frac{1}{r}(1.25^{1-r})(1.0^r - 1.25^r)(0.8192)\} = 0$$

The nonlinearity index r of the above equation can be obtained by using numerical tools such as MATLAB® (*fzero*(.)) or Mathematica® (*FindRoot*[.]):

$$r = -1.0$$

Finally, TANA (Equation A.13) can be constructed as:

$$g_T(X) = g(X_2) + \frac{1}{r}\left\{x_{1,2}^{1-r}(x_1^r - x_{1,2}^r)\left(\frac{\partial g}{\partial x_1}\right)_{X_2} + x_{2,2}^{1-r}(x_2^r - x_{2,2}^r)\left(\frac{\partial g}{\partial x_2}\right)_{X_2}\right\}$$

$$g_T(X) = 0.1819 +$$

$$\frac{1}{-1}\{(1.25^{1+1})(x_1^{-1} - 1.25^{-1})(-0.1647) + (1.25^{1+1})(x_2^{-1} - 1.25^{-1})(0.8192)\}$$

$$= 1.0 + 0.2574 x_1^{-1} - 1.28 x_2^{-1}$$

5) TANA1

From Equation A.17 and Equation A.18

$$\left(\frac{\partial g}{\partial x_1}\right)_{X_2} = \left(\frac{x_{1,2}}{x_{1,1}}\right)^{p_1 - 1}\left(\frac{\partial g}{\partial x_1}\right)_{X_1} \rightarrow -0.1647 = \left(\frac{1.25}{1.0}\right)^{p_1 - 1}(-0.2574)$$

$$\left(\frac{\partial g}{\partial x_2}\right)_{X_2} = \left(\frac{x_{2,2}}{x_{1,1}}\right)^{p_2-1} \left(\frac{\partial g}{\partial x_2}\right)_{X_1} \rightarrow 0.8192 = \left(\frac{1.25}{1.0}\right)^{p_1-1} \quad (1.28)$$

$$\varepsilon_1 = g(X_2) - \left\{ g(X_1) + \left(\frac{\partial g}{\partial x_1}\right)_{X_1} \frac{x_{1,1}^{1-p_1}}{p_1}(x_{1,2}^{p_1} - x_{1,1}^{p_1}) + \left(\frac{\partial g}{\partial x_2}\right)_{X_1} \frac{x_{2,1}^{1-p_2}}{p_2}(x_{2,2}^{p_2} - x_{2,1}^{p_2}) \right\}$$

$$\varepsilon_1 = 0.1819 - \left\{ -0.0227 - 0.2574 \frac{1.0^{1-p_1}}{p_1}(1.25^{p_1} - 1.0^{p_1}) + 1.28 \frac{1.0^{1-p_2}}{p_2}(1.25^{p_2} - 1.0^{p_2}) \right\}$$

The nonlinearity indices p_1, p_2, and ε_1 of the above equation can be obtained by using numerical tools of MATLAB® (*fzero*(.)):

$$p_1 = -1.0, \quad p_2 = -1.0 \quad \text{and} \quad \varepsilon_1 = 0.0$$

Finally, TANA1 (Equation A.15) can be constructed as:

$$g_{T1}(X) = g(X_1) + \left(\frac{\partial g}{\partial x_1}\right)_{X_1} \frac{x_{1,1}^{1-p_1}}{p_1}(x_1^{p_1} - x_{1,1}^{p_1}) + \left(\frac{\partial g}{\partial x_2}\right)_{X_1} \frac{x_{2,1}^{1-p_2}}{p_2}(x_2^{p_2} - x_{2,1}^{p_2}) + \varepsilon_1$$

$$g_{T1}(X) = -0.0227 - 0.2574 \frac{1.0^{1+1}}{-1}(x_1^{-1} - 1.0^{-1}) + 1.28 \frac{1.0^{1+1}}{-1}(x_2^{-1} - 1.0^{-1})$$

$$= 1.0 + 0.2574 x_1^{-1} - 1.28 x_2^{-1}$$

6) TANA2

From Equation A.20 and Equation A.21

$$\left(\frac{\partial g}{\partial x_1}\right)_{X_1} = \left(\frac{x_{1,1}}{x_{1,2}}\right)^{p_1-1} \left(\frac{\partial g}{\partial x_1}\right)_{X_2} + \varepsilon_2 (x_{1,1}^{p_1} - x_{1,2}^{p_1}) x_{1,1}^{p_1-1} p_1$$

$$\rightarrow -0.1647 = \left(\frac{1.0}{1.25}\right)^{p_1-1}(-0.1647) + \varepsilon_2 (1.0^{p_1} - 1.25^{p_1}) 1.0^{p_1-1} p_1$$

$$\left(\frac{\partial g}{\partial x_2}\right)_{X_1} = \left(\frac{x_{2,1}}{x_{2,2}}\right)^{p_2-1} \left(\frac{\partial g}{\partial x_2}\right)_{X_2} + \varepsilon_2 (x_{2,1}^{p_2} - x_{2,2}^{p_2}) x_{2,1}^{p_2-1} p_2$$

$$\rightarrow 1.28 = \left(\frac{1.0}{1.25}\right)^{p_2-1} 0.1892 + \varepsilon_2 (1.25^{p_2} - 1.25^{p_2}) 1.0^{p_2-1} p_2$$

$$g(X_1) = g(X_2) + \left(\frac{\partial g}{\partial x_1}\right)_{X_2} \frac{x_{1,2}^{1-p_1}}{p_1}(x_{1,1}^{p_1} - x_{1,2}^{p_1}) + \left(\frac{\partial g}{\partial x_2}\right)_{X_2} \frac{x_{2,2}^{1-p_2}}{p_2}(x_{2,1}^{p_2} - x_{2,2}^{p_2})$$

$$+ \frac{1}{2}\varepsilon_2 (x_{1,1}^{p_1} - x_{1,2}^{p_1})^2 + \frac{1}{2}\varepsilon_2 (x_{2,1}^{p_2} - x_{2,2}^{p_2})^2$$

$$-0.0227 = 0.1819 + (-0.1647)\frac{1.25^{1-p_1}}{p_1}(1.0^{P_1} - 1.25^{P_1})$$

$$+0.8192\frac{1.25^{1-p_2}}{p_2}(1.0^{P_2} - 1.25^{P_2})$$

$$+\frac{1}{2}\varepsilon_2(1.0^{P_1} - 1.25^{P_1})^2 + \frac{1}{2}\varepsilon_2(1.0^{P_2} - 1.25^{P_2})^2$$

The nonlinearity indices p_1, p_2, and ε_2 of the above equation can be obtained by using numerical tools of MATLAB® (*fsolve*(.)):

$$p_1 = -1.0, \quad p_2 = -1.0 \text{ and } \varepsilon_2 = 0.0$$

Finally, TANA2 (Equation A.19) can be constructed as

$$g_{T2}(X) = g(X_2) + \left(\frac{\partial g}{\partial x_1}\right)_{X_2} \frac{x_{1,2}^{1-p_1}}{p_1}(x_1^{P_1} - x_{1,2}^{P_1}) + \left(\frac{\partial g}{\partial x_2}\right)_{X_2} \frac{x_{2,2}^{1-p_2}}{p_2}(x_2^{P_2} - x_{2,2}^{P_2})$$

$$+\frac{1}{2}\varepsilon_2(x_1^{P_1} - x_{1,2}^{P_1})^2 + \frac{1}{2}\varepsilon_2(x_2^{P_2} - x_{2,2}^{P_2})^2$$

$$g_{T2}(X) = 0.1819 + (-0.1647)\frac{1.25^{1+1}}{-1.0}(x_1^{-1} - 1.25^{-1}) + 0.8192\frac{1.25^{1+1}}{-1.0}(x_2^{-1} - 1.25^{-1})$$

$$= -1.0 + 0.2574 x_1^{-1} - 1.28 x_2^{-1}$$

Table A.2. Comparison of Various Approximations

x_1	x_2	$g(X)$	$g_L(X)$	$g_R(X)$	$g_C(X)$	$g_{T,T1,T2}(X)$
0.75	0.75	-0.3635	-0.2783	-0.3635	-0.3850	-0.3635
1.0	0.75	-0.4226	-0.3426	-0.4493	-0.4493	-0.4493
1.25	0.75	-0.4205	-0.4070	-0.5008	-0.5137	-0.5008
0.75	1.0	0.0856	0.0417	0.0631	0.0417	0.0631
1.0	1.0	-0.0226	-0.0226	-0.0226	-0.0226	-0.0226
1.25	1.0	-0.0619	-0.0870	-0.0741	-0.0870	-0.0741
0.75	1.25	0.3785	0.3617	0.3191	0.2977	0.3191
1.0	1.25	0.2440	0.2974	0.2334	0.2334	0.2334
1.25	1.25	0.1819	0.2330	0.1819	0.1690	0.1819

The result comparisons are shown in Tables A.2 and A.3. The reciprocal approximation yields better accuracy compared to the linear and conservative approximations. TANA, TANA1 and TANA2 have the same accuracy as the reciprocal approximation, because the nonlinearity index is exactly equal to −1.

But, TANA, TANA1, and TANA2, which use adaptive nonlinearity, can provide better accuracy than the linear, conservative, and reciprocal approximations for highly nonlinear problems where the functional dependency on design variables is difficult to predict [7].

Table A.3 Comparison of Various Approximations (Relative Errors, %)

x_1	x_2	$g(X)$	$g_L(X)$	$g_R(X)$	$g_C(X)$	$g_{T,T1,T2}(X)$
0.75	0.75	0.0	23.4424	0.0000	-5.8993	0.0000
1.0	0.75	0.0	18.9282	-6.3094	-6.3094	-6.3094
1.25	0.75	0.0	3.2072	-19.1008	-22.1611	-19.1008
0.75	1.0	0.0	51.2935	26.2369	51.2935	26.2369
1.0	1.0	0.0	0.0000	0.0000	0.0000	0.0000
1.25	1.0	0.0	-40.4622	-19.6845	-40.4622	-19.6845
0.75	1.25	0.0	4.4408	15.6837	21.3498	15.6837
1.0	1.25	0.0	-21.8564	4.3713	4.3713	4.3713
1.25	1.25	0.0	-28.1133	0.0000	7.0747	0.0000

A.4 References

[1] Haftka, R.T. and Gurdal Z., *Elements of Structural Optimization*, Kluwer Academic Publishers, 3rd edition, Boston, MA, 1993.
[2] Haftka, R. T., Nachlas, J. A., Watson, L.T., Rizzo, T., and Desai, R., "Two-point Constraint Approximation in Structural Optimization," *Computer Methods in Applied Mechanics and Engineering*, Vol. 60, No. 3, 1987, pp. 289-301.
[3] *MATLAB: User's Guide*, The MathWorks Inc., 24 Prime Park Way, Natick, MA 01760, 1999.
[4] Montgomery, D.C., *Design and Analysis of Experiments*, Wiley, New York, 1997.
[5] Starnes, J. H. and Haftka, R. T., "Preliminary design of composite wings for buckling, stress and displacement constraints," *AIAA Journal of Aircraft*, Vol. 16, No. 8, 1979, pp. 564-570.
[6] Wang, L.P., Grandhi, R.V., and Hopkins, D.A., "Structural Reliability Optimization Using An Efficient Safety Index Calculation Procedure," *International Journal for Numerical Methods in Engineering*, Vol. 38, 1995, pp. 1721-1738.
[7] Wang, L. P. and Grandhi, R.V., "Improved Two-point Function Approximation for Design Optimization," *AIAA Journal*, Vol. 32, No. 9, 1995, pp. 1720-1727.

B

Asymptotic of Multinormal Integrals

In Section 4.2.2, the asymptotic approximation of multidimensional integrals was introduced. Asymptotic expansions [1] are given for integrals of the form

$$\hat{I}(\lambda) = \int_D \exp(\lambda) f(U) f_0(U) dU \quad (\lambda \to \infty) \tag{B.1}$$

where D is a fixed domain in the n-dimensional space, and $f(U)$ and $f_0(U)$ are at least two times continuously differentiable functions.

Further, it is assumed that the boundary of D is given by the point U, with $h(U) = 0$, where $h(U)$ is also at least twice continuously differentiable. It is shown that, if $f(U)$ has no global maximum with respect to D at an interior point of D, the asymptotic behavior of $\hat{I}(\lambda)$ depends on the points on the boundary where $f(U)$ attains its global maximum on the boundary. Due to the Lagrange multiplier theorem, a necessary condition for these points is that $\nabla f(U) = k \cdot \nabla h(U)$ where k is a constant. The contribution of one of these points to the asymptotic expansion of $\hat{I}(\lambda)$ is given in Ref [1]

Defining $D = [U; g(U;1) < 0]$, $\lambda = \beta^2$, $f(U) = -|U|^2/2$, $h(U) = g(U;1)$, the formula can be applied to obtain an asymptotic expansion for $I(\beta)$. Due to the assumption made at the beginning, there is only one point U on the surface $g(U;1) = 0$ with minimal distance to the origin, i.e. only one point U_0 in which $-|U|^2/2$ achieves its maximum. Then, the asymptotic expansion is given by

$$I(\beta) \sim (2\pi)^{(n-1)/2} \exp(\frac{-\beta^2 |U_0|^2}{2}) \beta^{-(n+1)} |J|^{-1/2}, \ (\beta \to \infty) \tag{B.2}$$

with

$$J = \sum_{i=1}^{n}\sum_{j=1}^{n} U_0^i U_0^j \, cof(-\delta_{ij} - Kg_{ij}) \tag{B.3}$$

where $g_{ij} = \dfrac{\partial^2 g(U;1)}{\partial u_i \partial u_j}\bigg|_{U=U_0}$, $K = |\Delta g(U_0;1)|^{-1}$ and $cof(-\delta_{ij} - Kg_{ij})$ denotes the cofactor of the element $(-\delta_{ij} - Kg_{ij})$ in the matrix $(-\delta_{ij} - Kg_{ij})_{i,j=1,\ldots,n}$.

Since $|U_0| = 1$, the formula simplifies as

$$I(\beta) \sim (2\pi)^{(n-1)/2} \exp(\frac{-\beta^2}{2}) \beta^{-(n+1)} |J|^{-1/2} \tag{B.4}$$

For further considerations, it can be assumed due to the rotational symmetry that $U_0 = (0,\ldots, 0, 1)$ (i.e. the unit vector in the direction of the u_n-axis). Then, since U_0 is parallel to the gradient of $g(U;1)$ at U_0 according to the Lagrange multiplier theorem, the tangential space of the hyper surface at U_0 is spanned by the unit vectors in the direction of the first $n-1$ axes. Then, using the definition of the cofactor, J is given as:

$$|J| = |cof[-\delta_{nn} - Kg_{nn}]| = |\det(B)| \tag{B.5}$$

where $B = (\delta_{lm} + Kg_{lm})_{l,m=1,\ldots,n-1}$

Defining $\overline{D} = (-Kg_{lm})_{l,m=1,\ldots,n-1}$ and denoting the unity matrix by I,

$$\det(B) = \det(I - \overline{D}) \tag{B.6}$$

$\det(B)$ is given by the product of the eigenvalues of B, which are the roots of $\det(B - kI) = \det(1-k)(I - \overline{D})$. But these roots are given by $1 - k_i$ ($i = 1, 2, \ldots, n-1$), in which the k_i's are the eigenvalues of the matrix B. This gives

$$|J| = \left| \sum_{j=1}^{n-1}(1+k_j)\beta \right| \tag{B.7}$$

These eigenvalues are the main curvatures of the surface at U_0. The curvature is defined as positive. Equation B.7 shows the cases where the approximation is not applicable. Since U_0 is a point on the surface with minimal distance to the origin, the main curvatures at U_0 must not be larger than unity. As proof, consider a point U on the surface near U_0 in the direction of a principal axis of curvature at U_0, with curvature k_i larger than unity. Due to the definition of the curvature, the curve on the surface connecting U_0 and U is approximated by a part of a circle in the same direction through U_0 with radius $1/k_i$ and center $(0,\ldots, 0, 1-1/k_i)$. Using elementary trigonometric relations, for small distances $|U-U_0|$ the squared distance of U to the origin is approximately

$$|U|^2 \approx 1 + \frac{(1-k_i)|U-U_0|^2}{2} < 1 \qquad (B.8)$$

This contradicts the assumption that U_0 is a point on the surface with minimal distance to the origin with respect to the surface, and therefore, $k_i \leq 1$. Due to this,

$$|J| = \prod_{j=1}^{n-1}(1+k_j\beta) \qquad (B.9)$$

In this case where one curvature is exactly equal to unity, the approximation cannot be used. It then becomes necessary to study higher derivatives of $g(U)$ and the global behavior of the function.

B.1 References

[1] Breitung, K., "Asymptotic Approximations for Multinormal Integrals," *Journal of the Engineering Mechanics Division*, ASCE, Vol. 110, No. 3, Mar., 1984, pp 357-366.

C
Cumulative Standard Normal Distribution Table

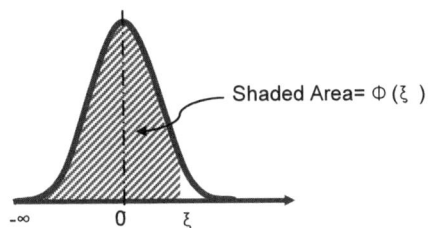

Table C.1. Statistical Table of Cumulative Standard Normal Distribution (from $-\infty$ to ξ)

$< normcdf(\xi)$ in MATLAB® $>$

ξ	0.00	0.01	0.02	0.03	0.04	0.05	0.06	0.07	0.08	0.09
0.0	0.50000	0.50399	0.50798	0.51197	0.51595	0.51994	0.52392	0.52790	0.53188	0.53586
0.1	0.53983	0.54380	0.54776	0.55172	0.55567	0.55962	0.56356	0.56749	0.57142	0.57535
0.2	0.57926	0.58317	0.58706	0.59095	0.59483	0.59871	0.60257	0.60642	0.61026	0.61409
0.3	0.61791	0.62172	0.62552	0.62930	0.63307	0.63683	0.64058	0.64431	0.64803	0.65173
0.4	0.65542	0.65910	0.66276	0.66640	0.67003	0.67364	0.67724	0.68082	0.68439	0.68793
0.5	0.69146	0.69497	0.69847	0.70194	0.70540	0.70884	0.71226	0.71566	0.71904	0.72240
0.6	0.72575	0.72907	0.73237	0.73565	0.73891	0.74215	0.74537	0.74857	0.75175	0.75490
0.7	0.75804	0.76115	0.76424	0.76730	0.77035	0.77337	0.77637	0.77935	0.78230	0.78524
0.8	0.78814	0.79103	0.79389	0.79673	0.79955	0.80234	0.80511	0.80785	0.81057	0.81327
0.9	0.81594	0.81859	0.82121	0.82381	0.82639	0.82894	0.83147	0.83398	0.83646	0.83891
1.0	0.84134	0.84375	0.84614	0.84849	0.85083	0.85314	0.85543	0.85769	0.85993	0.86214
1.1	0.86433	0.86650	0.86864	0.87076	0.87286	0.87493	0.87698	0.87900	0.88100	0.88298
1.2	0.88493	0.88686	0.88877	0.89065	0.89251	0.89435	0.89617	0.89796	0.89973	0.90147
1.3	0.90320	0.90490	0.90658	0.90824	0.90988	0.91149	0.91308	0.91466	0.91621	0.91774
1.4	0.91924	0.92073	0.92220	0.92364	0.92507	0.92647	0.92785	0.92922	0.93056	0.93189
1.5	0.93319	0.93448	0.93574	0.93699	0.93822	0.93943	0.94062	0.94179	0.94295	0.94408
1.6	0.94520	0.94630	0.94738	0.94845	0.94950	0.95053	0.95154	0.95254	0.95352	0.95449
1.7	0.95543	0.95637	0.95728	0.95818	0.95907	0.95994	0.96080	0.96164	0.96246	0.96327
1.8	0.96407	0.96485	0.96562	0.96638	0.96712	0.96784	0.96856	0.96926	0.96995	0.97062
1.9	0.97128	0.97193	0.97257	0.97320	0.97381	0.97441	0.97500	0.97558	0.97615	0.97670
2.0	0.97725	0.97778	0.97831	0.97882	0.97932	0.97982	0.98030	0.98077	0.98124	0.98169
2.1	0.98214	0.98257	0.98300	0.98341	0.98382	0.98422	0.98461	0.98500	0.98537	0.98574
2.2	0.98610	0.98645	0.98679	0.98713	0.98745	0.98778	0.98809	0.98840	0.98870	0.98899
2.3	0.98928	0.98956	0.98983	0.99010	0.99036	0.99061	0.99086	0.99111	0.99134	0.99158
2.4	0.99180	0.99202	0.99224	0.99245	0.99266	0.99286	0.99305	0.99324	0.99343	0.99361
2.5	0.99379	0.99396	0.99413	0.99430	0.99446	0.99461	0.99477	0.99492	0.99506	0.99520
2.6	0.99534	0.99547	0.99560	0.99573	0.99585	0.99598	0.99609	0.99621	0.99632	0.99643
2.7	0.99653	0.99664	0.99674	0.99683	0.99693	0.99702	0.99711	0.99720	0.99728	0.99736
2.8	0.99744	0.99752	0.99760	0.99767	0.99774	0.99781	0.99788	0.99795	0.99801	0.99807
2.9	0.99813	0.99819	0.99825	0.99831	0.99836	0.99841	0.99846	0.99851	0.99856	0.99861
3.0	0.99865	0.99869	0.99874	0.99878	0.99882	0.99886	0.99889	0.99893	0.99896	0.99900
3.1	0.99903	0.99906	0.99910	0.99913	0.99916	0.99918	0.99921	0.99924	0.99926	0.99929
3.2	0.99931	0.99934	0.99936	0.99938	0.99940	0.99942	0.99944	0.99946	0.99948	0.99950
3.3	0.99952	0.99953	0.99955	0.99957	0.99958	0.99960	0.99961	0.99962	0.99964	0.99965
3.4	0.99966	0.99968	0.99969	0.99970	0.99971	0.99972	0.99973	0.99974	0.99975	0.99976
3.5	0.99977	0.99978	0.99978	0.99979	0.99980	0.99981	0.99981	0.99982	0.99983	0.99983

D

F Distribution Table

Table D.1. Statistical Table of F distribution, $\alpha = 0.05$

v_1 \ v_2	1	2	3	4	5	6	7	8	9	10
1	161.4476	199.5000	215.7073	224.5832	230.1619	233.9860	236.7684	238.8827	240.5433	241.8817
2	18.5128	19.0000	19.1643	19.2468	19.2964	19.3300	19.3532	19.3710	19.3848	19.3959
3	10.1280	9.5521	9.2766	9.1172	9.0135	8.9406	8.8867	8.8452	8.8123	8.7855
4	7.7086	6.9443	6.5914	6.3882	6.2561	6.1631	6.0942	6.0410	5.9988	5.9644
5	6.6079	5.7861	5.4095	5.1922	5.0503	4.9503	4.8759	4.8183	4.7725	4.7351
6	5.9874	5.1433	4.7571	4.5337	4.3874	4.2839	4.2067	4.1468	4.0990	4.0600
7	5.5914	4.7374	4.3468	4.1203	3.9715	3.8660	3.7870	3.7257	3.6767	3.6365
8	5.3177	4.4590	4.0662	3.8379	3.6875	3.5806	3.5005	3.4381	3.3881	3.3472
9	5.1174	4.2565	3.8625	3.6331	3.4817	3.3738	3.2927	3.2296	3.1789	3.1373
10	4.9646	4.1028	3.7083	3.4780	3.3258	3.2172	3.1355	3.0717	3.0204	2.9782
11	4.8443	3.9823	3.5874	3.3567	3.2039	3.0946	3.0123	2.9480	2.8962	2.8536
12	4.7472	3.8853	3.4903	3.2592	3.1059	2.9961	2.9134	2.8486	2.7964	2.7534
13	4.6672	3.8056	3.4105	3.1791	3.0254	2.9153	2.8321	2.7669	2.7144	2.6710
14	4.6001	3.7389	3.3439	3.1122	2.9582	2.8477	2.7642	2.6987	2.6458	2.6022
15	4.5431	3.6823	3.2874	3.0556	2.9013	2.7905	2.7066	2.6408	2.5876	2.5437
16	4.4940	3.6337	3.2389	3.0069	2.8524	2.7413	2.6572	2.5911	2.5377	2.4935
17	4.4513	3.5915	3.1968	2.9647	2.8100	2.6987	2.6143	2.5480	2.4943	2.4499
18	4.4139	3.5546	3.1599	2.9277	2.7729	2.6613	2.5767	2.5102	2.4563	2.4117
19	4.3807	3.5219	3.1274	2.8951	2.7401	2.6283	2.5435	2.4768	2.4227	2.3779
20	4.3512	3.4928	3.0984	2.8661	2.7109	2.5990	2.5140	2.4471	2.3928	2.3479
21	4.3248	3.4668	3.0725	2.8401	2.6848	2.5727	2.4876	2.4205	2.3660	2.3210
22	4.3009	3.4434	3.0491	2.8167	2.6613	2.5491	2.4638	2.3965	2.3419	2.2967
23	4.2793	3.4221	3.0280	2.7955	2.6400	2.5277	2.4422	2.3748	2.3201	2.2747
24	4.2597	3.4028	3.0088	2.7763	2.6207	2.5082	2.4226	2.3551	2.3002	2.2547
25	4.2417	3.3852	2.9912	2.7587	2.6030	2.4904	2.4047	2.3371	2.2821	2.2365
26	4.2252	3.3690	2.9752	2.7426	2.5868	2.4741	2.3883	2.3205	2.2655	2.2197
27	4.2100	3.3541	2.9604	2.7278	2.5719	2.4591	2.3732	2.3053	2.2501	2.2043
28	4.1960	3.3404	2.9467	2.7141	2.5581	2.4453	2.3593	2.2913	2.2360	2.1900
29	4.1830	3.3277	2.9340	2.7014	2.5454	2.4324	2.3463	2.2783	2.2229	2.1768
30	4.1709	3.3158	2.9223	2.6896	2.5336	2.4205	2.3343	2.2662	2.2107	2.1646
40	4.0847	3.2317	2.8387	2.6060	2.4495	2.3359	2.2490	2.1802	2.1240	2.0772
60	4.0012	3.1504	2.7581	2.5252	2.3683	2.2541	2.1665	2.0970	2.0401	1.9926
120	3.9201	3.0718	2.6802	2.4472	2.2899	2.1750	2.0868	2.0164	1.9588	1.9105
∞	3.8415	2.9957	2.6049	2.3719	2.2141	2.0986	2.0096	1.9384	1.8799	1.8307

D.1. Statistical Table of F distribution, $\alpha = 0.05$ (contd.)

< $finv(0.95, v_1, v_2)$ in MATLAB® >

v_1 \ v_2	12	15	20	24	30	40	60	120	∞
1	243.9060	245.9499	248.0131	249.0518	250.0951	251.1432	252.1957	253.2529	254.3100
2	19.4125	19.4291	19.4458	19.4541	19.4624	19.4707	19.4791	19.4874	19.4960
3	8.7446	8.7029	8.6602	8.6385	8.6166	8.5944	8.5720	8.5494	8.5264
4	5.9117	5.8578	5.8025	5.7744	5.7459	5.7170	5.6877	5.6581	5.6281
5	4.6777	4.6188	4.5581	4.5272	4.4957	4.4638	4.4314	4.3985	4.3650
6	3.9999	3.9381	3.8742	3.8415	3.8082	3.7743	3.7398	3.7047	3.6689
7	3.5747	3.5107	3.4445	3.4105	3.3758	3.3404	3.3043	3.2674	3.2298
8	3.2839	3.2184	3.1503	3.1152	3.0794	3.0428	3.0053	2.9669	2.9276
9	3.0729	3.0061	2.9365	2.9005	2.8637	2.8259	2.7872	2.7475	2.7067
10	2.9130	2.8450	2.7740	2.7372	2.6996	2.6609	2.6211	2.5801	2.5379
11	2.7876	2.7186	2.6464	2.6090	2.5705	2.5309	2.4901	2.4480	2.4045
12	2.6866	2.6169	2.5436	2.5055	2.4663	2.4259	2.3842	2.3410	2.2962
13	2.6037	2.5331	2.4589	2.4202	2.3803	2.3392	2.2966	2.2524	2.2064
14	2.5342	2.4630	2.3879	2.3487	2.3082	2.2664	2.2229	2.1778	2.1307
15	2.4753	2.4034	2.3275	2.2878	2.2468	2.2043	2.1601	2.1141	2.0658
16	2.4247	2.3522	2.2756	2.2354	2.1938	2.1507	2.1058	2.0589	2.0096
17	2.3807	2.3077	2.2304	2.1898	2.1477	2.1040	2.0584	2.0107	1.9604
18	2.3421	2.2686	2.1906	2.1497	2.1071	2.0629	2.0166	1.9681	1.9168
19	2.3080	2.2341	2.1555	2.1141	2.0712	2.0264	1.9795	1.9302	1.8780
20	2.2776	2.2033	2.1242	2.0825	2.0391	1.9938	1.9464	1.8963	1.8432
21	2.2504	2.1757	2.0960	2.0540	2.0102	1.9645	1.9165	1.8657	1.8117
22	2.2258	2.1508	2.0707	2.0283	1.9842	1.9380	1.8894	1.8380	1.7831
23	2.2036	2.1282	2.0476	2.0050	1.9605	1.9139	1.8648	1.8128	1.7570
24	2.1834	2.1077	2.0267	1.9838	1.9390	1.8920	1.8424	1.7896	1.7330
25	2.1649	2.0889	2.0075	1.9643	1.9192	1.8718	1.8217	1.7684	1.7110
26	2.1479	2.0716	1.9898	1.9464	1.9010	1.8533	1.8027	1.7488	1.6906
27	2.1323	2.0558	1.9736	1.9299	1.8842	1.8361	1.7851	1.7306	1.6717
28	2.1179	2.0411	1.9586	1.9147	1.8687	1.8203	1.7689	1.7138	1.6541
29	2.1045	2.0275	1.9446	1.9005	1.8543	1.8055	1.7537	1.6981	1.6376
30	2.0921	2.0148	1.9317	1.8874	1.8409	1.7918	1.7396	1.6835	1.6223
40	2.0035	1.9245	1.8389	1.7929	1.7444	1.6928	1.6373	1.5766	1.5089
60	1.9174	1.8364	1.7480	1.7001	1.6491	1.5943	1.5343	1.4673	1.3893
120	1.8337	1.7505	1.6587	1.6084	1.5543	1.4952	1.4290	1.3519	1.2539
∞	1.7522	1.6664	1.5705	1.5173	1.4591	1.3940	1.3180	1.2214	1.0000

Note: $F(\alpha, v_1, v_2)$, v_1 is the numerator, and v_2 is the denominatory

D.2. Statistical Table of F distribution, $\alpha = 0.10$

$< finv(0.90, v_1, v_2)$ in MATLAB$^\circledR >$

v_1 \ v_2	1	2	3	4	5	6	7	8	9	10
1	39.8635	49.5000	53.5932	55.8330	57.2401	58.2044	58.9060	59.4390	59.8576	60.1950
2	8.5263	9.0000	9.1618	9.2434	9.2926	9.3255	9.3491	9.3668	9.3805	9.3916
3	5.5383	5.4624	5.3908	5.3426	5.3092	5.2847	5.2662	5.2517	5.2400	5.2304
4	4.5448	4.3246	4.1909	4.1073	4.0506	4.0098	3.9790	3.9549	3.9357	3.9199
5	4.0604	3.7797	3.6195	3.5202	3.4530	3.4045	3.3679	3.3393	3.3163	3.2974
6	3.7760	3.4633	3.2888	3.1808	3.1075	3.0546	3.0145	2.9830	2.9577	2.9369
7	3.5894	3.2574	3.0741	2.9605	2.8833	2.8274	2.7849	2.7516	2.7247	2.7025
8	3.4579	3.1131	2.9238	2.8064	2.7265	2.6683	2.6241	2.5894	2.5612	2.5380
9	3.3603	3.0065	2.8129	2.6927	2.6106	2.5509	2.5053	2.4694	2.4403	2.4163
10	3.2850	2.9245	2.7277	2.6053	2.5216	2.4606	2.4140	2.3772	2.3473	2.3226
11	3.2252	2.8595	2.6602	2.5362	2.4512	2.3891	2.3416	2.3040	2.2735	2.2482
12	3.1766	2.8068	2.6055	2.4801	2.3940	2.3310	2.2828	2.2446	2.2135	2.1878
13	3.1362	2.7632	2.5603	2.4337	2.3467	2.2830	2.2341	2.1954	2.1638	2.1376
14	3.1022	2.7265	2.5222	2.3947	2.3069	2.2426	2.1931	2.1539	2.1220	2.0954
15	3.0732	2.6952	2.4898	2.3614	2.2730	2.2081	2.1582	2.1185	2.0862	2.0593
16	3.0481	2.6682	2.4618	2.3327	2.2438	2.1783	2.1280	2.0880	2.0553	2.0282
17	3.0262	2.6446	2.4374	2.3078	2.2183	2.1524	2.1017	2.0613	2.0284	2.0009
18	3.0070	2.6240	2.4160	2.2858	2.1958	2.1296	2.0785	2.0379	2.0047	1.9770
19	2.9899	2.6056	2.3970	2.2663	2.1760	2.1094	2.0580	2.0171	1.9836	1.9557
20	2.9747	2.5893	2.3801	2.2489	2.1582	2.0913	2.0397	1.9985	1.9649	1.9367
21	2.9610	2.5746	2.3649	2.2333	2.1423	2.0751	2.0233	1.9819	1.9480	1.9197
22	2.9486	2.5613	2.3512	2.2193	2.1279	2.0605	2.0084	1.9668	1.9327	1.9043
23	2.9374	2.5493	2.3387	2.2065	2.1149	2.0472	1.9949	1.9531	1.9189	1.8903
24	2.9271	2.5383	2.3274	2.1949	2.1030	2.0351	1.9826	1.9407	1.9063	1.8775
25	2.9177	2.5283	2.3170	2.1842	2.0922	2.0241	1.9714	1.9293	1.8947	1.8658
26	2.9091	2.5191	2.3075	2.1745	2.0822	2.0139	1.9610	1.9188	1.8841	1.8550
27	2.9012	2.5106	2.2987	2.1655	2.0730	2.0045	1.9515	1.9091	1.8743	1.8451
28	2.8939	2.5028	2.2906	2.1571	2.0645	1.9959	1.9427	1.9001	1.8652	1.8359
29	2.8870	2.4955	2.2831	2.1494	2.0566	1.9878	1.9345	1.8918	1.8568	1.8274
30	2.8807	2.4887	2.2761	2.1422	2.0493	1.9803	1.9269	1.8841	1.8490	1.8195
40	2.8354	2.4404	2.2261	2.0910	1.9968	1.9269	1.8725	1.8289	1.7929	1.7627
60	2.7911	2.3933	2.1774	2.0410	1.9457	1.8747	1.8194	1.7748	1.7380	1.7070
120	2.7478	2.3473	2.1300	1.9923	1.8959	1.8238	1.7675	1.7220	1.6843	1.6524
∞	2.7055	2.3026	2.0838	1.9449	1.8473	1.7741	1.7167	1.6702	1.6315	1.5987

Note: $F(\alpha, v_1, v_2)$, v_1 is the numerator, and v_2 is the denominatory

D.2. Statistical Table of F distribution, $\alpha = 0.10$ (contd.)

$< finv(0.90, v_1, v_2)$ in MATLAB® $>$

v_1 \ v_2	12	15	20	24	30	40	60	120	∞
1	60.7052	61.2203	61.7403	62.0020	62.2650	62.5291	62.7943	63.0606	63.3280
2	9.4081	9.4247	9.4413	9.4496	9.4579	9.4662	9.4746	9.4829	9.4912
3	5.2156	5.2003	5.1845	5.1764	5.1681	5.1597	5.1512	5.1425	5.1337
4	3.8955	3.8704	3.8443	3.8310	3.8174	3.8036	3.7896	3.7753	3.7607
5	3.2682	3.2380	3.2067	3.1905	3.1741	3.1573	3.1402	3.1228	3.1050
6	2.9047	2.8712	2.8363	2.8183	2.8000	2.7812	2.7620	2.7423	2.7222
7	2.6681	2.6322	2.5947	2.5753	2.5555	2.5351	2.5142	2.4928	2.4708
8	2.5020	2.4642	2.4246	2.4041	2.3830	2.3614	2.3391	2.3162	2.2926
9	2.3789	2.3396	2.2983	2.2768	2.2547	2.2320	2.2085	2.1843	2.1592
10	2.2841	2.2435	2.2007	2.1784	2.1554	2.1317	2.1072	2.0818	2.0554
11	2.2087	2.1671	2.1231	2.1000	2.0762	2.0516	2.0261	1.9997	1.9721
12	2.1474	2.1049	2.0597	2.0360	2.0115	1.9861	1.9597	1.9323	1.9036
13	2.0966	2.0532	2.0070	1.9827	1.9576	1.9315	1.9043	1.8759	1.8462
14	2.0537	2.0095	1.9625	1.9377	1.9119	1.8852	1.8572	1.8280	1.7973
15	2.0171	1.9722	1.9243	1.8990	1.8728	1.8454	1.8168	1.7867	1.7551
16	1.9854	1.9399	1.8913	1.8656	1.8388	1.8108	1.7816	1.7508	1.7182
17	1.9577	1.9117	1.8624	1.8362	1.8090	1.7805	1.7506	1.7191	1.6856
18	1.9333	1.8868	1.8369	1.8104	1.7827	1.7537	1.7232	1.6910	1.6567
19	1.9117	1.8647	1.8142	1.7873	1.7592	1.7298	1.6988	1.6659	1.6308
20	1.8924	1.8449	1.7938	1.7667	1.7382	1.7083	1.6768	1.6433	1.6074
21	1.8750	1.8272	1.7756	1.7481	1.7193	1.6890	1.6569	1.6228	1.5862
22	1.8593	1.8111	1.7590	1.7312	1.7021	1.6714	1.6389	1.6042	1.5668
23	1.8450	1.7964	1.7439	1.7159	1.6864	1.6554	1.6224	1.5871	1.5490
24	1.8319	1.7831	1.7302	1.7019	1.6721	1.6407	1.6073	1.5715	1.5327
25	1.8200	1.7708	1.7175	1.6890	1.6590	1.6272	1.5934	1.5570	1.5176
26	1.8090	1.7596	1.7059	1.6771	1.6468	1.6147	1.5805	1.5437	1.5036
27	1.7989	1.7492	1.6951	1.6662	1.6356	1.6032	1.5686	1.5313	1.4906
28	1.7895	1.7395	1.6852	1.6560	1.6252	1.5925	1.5575	1.5198	1.4784
29	1.7808	1.7306	1.6759	1.6466	1.6155	1.5825	1.5472	1.5090	1.4670
30	1.7727	1.7223	1.6673	1.6377	1.6065	1.5732	1.5376	1.4989	1.4564
40	1.7146	1.6624	1.6052	1.5741	1.5411	1.5056	1.4672	1.4248	1.3769
60	1.6574	1.6034	1.5435	1.5107	1.4755	1.4373	1.3952	1.3476	1.2915
120	1.6012	1.5450	1.4821	1.4472	1.4094	1.3676	1.3203	1.2646	1.1926
∞	1.5458	1.4871	1.4206	1.3832	1.3419	1.2951	1.2400	1.1686	1.0000

Note: $F(\alpha, v_1, v_2)$, v_1 is the numerator, and v_2 is the denominatory

Index

adaptive approximation, 136
adaptive safety-index algorithm, 112
advanced first-order second-moment method, 178
AFOSM, *see* advanced first-order second-moment method
aleatory uncertainty, 3
analysis of variance, 46
antithetic variates method, 68
Askey scheme, 58, 211
asymmetry, 17
asymptotic
 approximation, 130, 291
 expansion, 291
autocorrelation, 37
autocovariance, 37
average, 13
ANOVA, *see* analysis of variance, 46, 60, 239, 241, 263

backward selection procedure, 47
basic variable, 65
bathtub curve, 35
bell curve, 21
bins, 70
Breitung formulation, 130
Brownian motion, 203
buckling
 analysis, 246
 eigenvalue, 246

CDF, *see* cumulative distribution function
central
 limit theorem, 20
 measures, 13
chance-constrained method, 261
Chebyshev polynomials, 209
Chebyshev-Hermite polynomial, 213
chi-square, 28
coefficient of variation, 15, 83
collinear, 46
collinearity, 209
conditional
 expectation method, 68
 probability, 13
confidence interval, 46, 211
conservatiave approximation, 279
constrained minimization problem, 155
constraint, 6
continuous random variable, 10
correlation coefficient, 15
COV, *see* coefficient of variation
Covariance, 15
 functions, 41
cross-covariance, 37
cumulative distribution function, 10

degree of belief, 3
delta-correlated model, 42
DEMM, *see* deterministic equivalent modeling method

dependent variables, 44
derived density function, 12
design variable linking, 197
deterministic
    approach, 2
    equivalent modeling method, 262
dimensionality reduction, 219
direct approach, 155
direction cosine, 90
discrete random variable, 10
disk burst margin, 144
dispersion measures, 14

eigenfunctions, 42, 221
eigenvalue
    matrix, 221
    problems, 226
eigenvalues, 42, 221
element stiffnes matrix, 73
ensemble, 37
epistemic uncertainty, 3
equal probability intervals, 70
equivalent normal transformations, 111
ergodic, 38
error sum of squares, 45
expectation operator, 14
expected
    value operation, 206
    value, 13
exponential, 28
    covariance, 226
    distribution, 34
    model, 42
exterior penalty function, 162, 163
extreme value distribution, 29
    Type I, Type II, Type III, 29

F-statistic, 47
factor of safety, 2
failure
    probability sensitivity, 183
    probability, 65
    region, 52
    surface, 52
fatigue crack growth, 142
feasible directions algorithm, 157

FEM, *see* finite element method
finite element method, 72
first-order
    approximation, 126
    reliability method, 57
    second moment, 56, 81
first moment, 13
flatness, 17
fluctuating components, 264
FORM, *see* first-order reliability method
forward selection procedure, 47
FOSM, *see* first-order second moment
FOSM reliability index, 88
fourth central moment, 17
Frechet distribution, 30
frequency, 3
    diagram, 10
function approximation, 275

gamma
    distribution, 28, 250
    function, 28, 31
Gaussian
    distribution, 20
    model, 42, 259
    random field, 77, 230
generalized
    Fourier coefficients, 206
    PCE, 211, 212, 250
geometric stiffness matrix, 246
global stiffness matrix, 73
Gram-Charlier
    method, 215
    series, 213
Gram-Schmidt algorithm, 125
Gumbell distribution, 29

Hasofer and Lind
    iteratioin method, 88
    safety index, 56, 86
    transformation, 56
Hasofer Lind-Rackwitz Fiessler
    method, 89, 99, 101
Hermite polynomials, 213, 215
higher-order reliability method, 57

histogram, 10
HL safety index, *see* Hasofer and Lind safety index
HL transformation, *see* Hasofer and Lind transformation
HL-RF method, *see* Hasofer Lind-Rackwitz Fiessler method
homogeneous
 chaos, 204
 processes, 38,
HORM, *see* higher-order reliability method
Hotelling transform, 219
hypergeometric orthogonal polynomials, 58

ill-conditioned problem, 46, 209
importance sampling, 68
inacuracy, 3
independent, 13
 polynomial, 77
 variables, 44
indicator function, 65
indirect approach, 155
interior penalty function method, 160
interval information, 3
intrusive formulation, 59, 219
inverse
 normal distribution function, 7
 transformation method, 62
irregularities, 204

Jacobi polynomials, 58
joined-wing, 244, 251, 267
joint PDF, 12
joint probability, 12

Karhunen-Loeve
 expansion, 39, 221, 226, 230
 transform, 42, 218, 220, 237
KL expansion, *see* Karhunen-Loeve expansion
KL transform, *see* Karhunen-Loeve transform
Kronecker delta, 206, 221
Kuhn-Tucker necessary conditons, 180

kurtosis, 17

lack of fit, 46
lack of knowledge, 3
Lagrange multiplier, 69, 180, 291
Laguerre polynomials, 58, 209, 250
Laplace method, 130
Latin Hypercube sampling, 60, 68, 70, 219, 237, 263
least-squaares
 scheme, 206
 criterion, 206
LHS, *see* Latin Hypercube sampling
likelyhood of events, 3
limit-state, 6, 3, 52
linear
 approximation, 278
 regression, 44
linking matrix, 198
lognormal distribution, 25

margin of safety, 53
marginal cumulative distribution function, 100
marginal density, 12
marginal PDF, 13
MCS, *see* Monte Carlo simulation
mean value first order second moment method, 56, 81
mean-square, 204
median, 14
Mill's ratio, 131
mode, 14
model adequacy, 47
Monte Carlo simulation, 60
most probable failure point, 56, 86
move limits, 198
MPP, *see* most probable failure point
multidimensional integrals, 291
multidisciplinary optimization, 153
multiple
 Hermite polynomials, 204
 MPP, 90
multi-point approximation, 280
multivariate sampling method, 70
MVFOSM, 56, 81, 83, 91
 reliability index, 83

Neumann
  expansion method, 75
  series, 231
non-intrusive, 59, 219, 262
  formulation, 206, 209
nonlinear index, 111, 112
non-overlapping designs, 70
non-probabilistic approach, 4
non-stationary, 37
normal
  density function, 215
  probability plot, 47
normally distributed variable, 22
nugget-effect model, 42, 259

objective uncertainty, 3
one-dimensional Hermite
  polynomials, 205
one-point approximations, 277
orthogonal
  polynomials, 39
  transform method, 220
  transformation, 125
  eigenvector matrix, 221
orthogonality, 204

Paris
  constant, 142
  equation, 142
PCE, *see* polynomial chaos expansion
PDF, *see* probabilisty density
  function
penalty
  function methods, 160
  parameter, 160
perfect correlation, 15
performance measure approach, 201
perturbation method, 73
pin-connectec three-bar truss, 248
PMA, *see* performance measure
  approach
PMF, *see* probability mass function
pobabilisty theory, 2
polynomial chaos expansion, 39,58,
  203, 219, 237, 263
principal
  component analysis, 219

  eigenvector matrix, 222
probabilistic
  approach, 4
  collocation method, 58, 59
probability, 10
  density, 3
  density function, 3, 9
  mass function, 10,
  of failure, 53
product of marginals, 13
pseudo-objective function, 160
push-off, 158

quadratic Taylor series expansion,
  131
quasi-random points, 66

random
  coefficient, 39
  function, 36, 39
  process, 37
  responses, 36
  variables, 9
  vectors, 9
random field, 36, 219, 252, 253
  discretization, 39, 40
Rayleigh distribution, 21
realization, 37
reciprocal approximation, 278
recourse method, 261
regression
  analysis, 43
  coefficient, 44
  sum of squares, 45
reliability, 4
  analysis, 51
  index, 52, 82
reliability-based design optimization,
  200, 261
residual
  analysis, 47, 239, 242
  plots, 47, 243
  sum of squares, 45
residuals, 44, 47
response surface method, 58
restricted pairing, 72
RF method, 101

safe region, 52
safety factor, 54, 264
    design, 2
safety index, 52, 53, 88
sampling methods, 60
scree plot, 254
search direction vector, 157
second central moment, 14
second-order
    second moment, 56
    approximation, 128
    reliability moment, 57, 124, 136
    Taylor series, 124, 283
seed values, 63
sensitivity
    analysis, 178
    factor, 90, 97
sequential unconstrained
    minimization techniques, 160
serviceability limit-states, 52
SFEM, *see* stochatic finite element method
simple random sampling, 60
skewness coefficient, 17
skewness, 17
SORM, *see* second-order reliability moment
SOSM, *see* second-order second moment
spatial variability, 37
spatially-correlated data, 219
spectral
    decomposition, 42
    stochastic finite element method, 58, 220, 229
split sampling method, 68
squared exponential model, 42
SSFEM, *see* spectral stochastic finite element method
standard deviation, 14
standard
    normal cumulative distribution function, 22
    normal density function, 98
    normal distribution, 22
    normal random variable, 206
    normal values, 71

stationary, 37
statistical
    tolerancing, 20
    trial method, 60
statistically independent, 15
step size, 157, 158
stochastic
    analysis procedure, 237
    approach 2, 3
    approximation, 209
    expansions, 58
    Glerkin FEM, 60, 220
    processes, 204
    response surface method, 59
    finite element method, 72. 73
stratified sampling method, 68, 70
structural reliability, 6
subjective uncertainty, 3
SUMT, *see* sequential unconstrained minimization techniques
supercavitating torpedo, 256
systematic sampling method, 68

TANA, *see* two-point adaptive nonlinear approximation
TANA2, 111, 283
Taylor series, 39, 56, 73, 81, 89, 99, 158, 277, 280
Tchebycheff-Hermite polynomial, 213
ten-bar truss, 138
termination criteria, 200
third central moment, 17
three-bar truss, 264
throw away approach, 198
total sum of squares, 45
traditional design process, 2
transform matrix, 221
transformation technique, 212, 250
travel distance, 158
Tvedt's formulation, 133
two-member frame, 146
two-point adaptive nonlinear approximation , 110, 111, 280

ultimate limit-states, 51
uncertainty analysis, 3

uncorrelated, 15
uniform random variable, 71
unrestricted pairing, 72
usable
    direction, 158
    feasible dirctions, 158

variability, 3
variance, 14
    reduction techniques, 68

weakest link, 31
weakly stationary, 37
Weibull distribution, 30, 31
weighted integral method, 77
well-stratified designs, 72
white noise, 42, 259
Wiener process, 203